LEAN AND MEAN

ALSO BY BENNETT HARRISON

(With Marcus Weiss and Jon Gant) *Building Bridges:*
Community Development Corporations
and the World of Employment Training (1994)

(With Lucy Gorham) *Working Below the Poverty Line* (1990)

(With Barry Bluestone) *The Great U-Turn: Corporate Restructuring*
and the Polarizing of America (1988)

(With Barry Bluestone) *The Deindustrialization of America:*
Plant Closings, Community Abandonment, and the Dismantling
of Basic Industry (1982)

(With George M. von Furstenburg and Ann Horowitz) *Patterns of*
Racial Discrimination (1974)

Urban Economic Development (1974)

Education, Training, and the Urban Ghetto (1972)

(With Harold L. Sheppard and William J. Spring) *The Political Economy of*
Public Service Employment (1972)

(With Thomas Vietorisz) *The Economic Development of Harlem* (1970)

LEAN AND MEAN

THE CHANGING LANDSCAPE OF CORPORATE POWER IN THE AGE OF FLEXIBILITY

BENNETT HARRISON

THE GUILFORD PRESS
New York   London

An amended version of chapter 2 appeared as "The Myth of Small Firms as the Predominant Job Generators," in *Economic Development Quarterly* 8 (February 1994). An amended version of chapter 4 appeared as "The Italian Industrial Districts and the Crisis of the Industrial Form: Parts I and V," in *European Planning Review* 2, nos. 1 and 2 (1994). An amended version of chapter 5 appeard as "Concentrated Economic Power and Silicon Valley," in *Environment and Planning* A 26 (February 1994).

Published by The Guilford Press
A Division of Guilford Publications, Inc.
72 Spring Street, New York, NY 10012

Printed in the United States of America

This book is printed on acid-free paper.

Last digit is print number: 9 8 7 6 5 4 3 2 1

LIBRARY OF CONGRESS CATALOGING-IN-PUBLICATION DATA

Harrison, Bennett.
    Lean and mean: the changing landscape of corporate power in the
age of flexibility / Bennett Harrison
        p.   cm.
    Includes bibliographical references and index.
    ISBN 1-57230-252-6
    1. Industries, Size of.   2. Industrial concentration.
3. Corporate reorganizations.   I. Title.
HD69.S5H367   1997
338.6'4—dc20   93-40497
                                                              CIP

CONTENTS

Concentration without Centralization: How the Big Firms Are Reorganizing Global Capitalism. Why Small Firms Do Not Drive Economic Growth and Create the Most New Jobs. Why Are We So Ready to Accept the Small Firms Story? What's Wrong with the Small Firms Story? Trouble in Paradise: Hierarchy and Inequality in the Industrial Districts. Why Should We Care? Rethinking National and Regional Economic Development in a World of Production Networks.

Alternative Explanations for the Apparently Growing Importance of Small Firms. The Conventional Wisdom, Revisited. Reassessing the Evidence on Small Firm Shares of Job Creation and Employment Growth. Small Firm Dynamics. Wages and Working Conditions in Smaller Firms. The Making of a Myth.

3 ARE SMALL FIRMS THE TECHNOLOGY LEADERS? 53

Product Innovation and Small Firms. Do Computer-Controlled Manufacturing Process Innovations Privilege Small Firms? Biotechnology Networks. Toward the Greater Standardization of Computer Programming: Japan's Software Factories. Technological Change and the Continuing Tendencies toward Standardization, Size, and Scale.

4 THE EVOLUTION (AND DEVOLUTION?) OF THE ITALIAN INDUSTRIAL DISTRICTS 75

Small Firm–Led Growth Models and the Northern Italian Industrial Districts. Penetration of the Industrial Districts by Outside Financial Conglomerates: The Case of Sasib. The Emergence of Hierarchy within a District-Based Production Network: The United Colors of Benetton. The Negative Competitive Consequences of Excessive Fragmentation: The Case of Prato. Further Evidence of Crisis and Concentration in the "Third Italy."

5 IS SILICON VALLEY AN INDUSTRIAL DISTRICT? 106

Silicon Valley as an Industrial District: The Debate. Inequality and Dual Labor Markets in the Silicon Valley Production System. Concentrated Power and the Role of the Military in Silicon Valley's Economic Development. The Three Faces of Silicon Valley.

PART III: THE EMERGING SYSTEM OF GLOBALLY NETWORKED PRODUCTION

6 "FLEXIBILITY" AND THE EMERGENCE OF LARGE FIRM–LED PRODUCTION NETWORKS 125

Crisis and the Corporate Search for Flexibility. Production Networks as a Means of Achieving Flexibility. Types of Interfirm Production Networks. Territorially Based Production Systems as Networks. Some Basic Definitions. Types of Input-Output Systems. The Territorial Dimension. Governance Structures. Networked Production Systems. Developmental Tendencies. Putting the Context Back In.

Government Industrial/Technology Policy and Interfirm Alliances. Planning for Conversion from Military Production. Manufacturing Modernization. Regional, State, and Local Economic Development in a World of Networked Production. Coordinating Federal and State/Local Approaches to Working with Production Networks. The Dark Side, Redux.

11 POSTSCRIPT: REASSESSING *LEAN AND MEAN* ON THE EVE OF THE NEW MILLENIUM

The Stubborn Myth about Small Business as the Engine of Economic Development. Further Evidence of a Dark Side to Flexibility. As for Agglomeration and Industrial Districts, It Depends. What about This Idea of "Globalization"? (A Bit) More on Economic and Social Policy in the Age of Lean and Mean Business.

ACKNOWLEDGMENTS

The research for this book began in the spring and summer of 1989, during my residencies as a visiting scholar at Balliol College, Oxford University, United Kingdom, and at the Laboratory of Industrial Policy, NOMISMA, in Bologna, Italy. I am grateful for financial support from a Fulbright West European Research Grant, from the M.I.T.-Balliol Exchange Program, from the United Nations Centre for Regional Development in Nagoya, Japan, and, subsequently, from the National Science Foundation and the Carnegie Bosch Institute for Research on International Management.

Among the many colleagues who made my investigations in England and Italy possible and pleasurable, I must first of all acknowledge my hosts: Dr. Joseph Stoy, the senior tutor of Balliol College, and Dr. Patrizio Bianchi, the director of the *Laboratorio* at NOMISMA. Others who engaged me in close discussion of these issues before, during, or following the fieldwork include Aris Accornero, Ash Amin, Fiorenza Belussi, Annaflavia Bianchi, Michael Blim, Susan Christopherson, Gordon Clark, Phil Cooke, Ron Dore, Michael Enright, Patrizia Faraselli, Rich Florida, Luigi Forlai, Duncan Gallie, Francesco Garibaldo, Gary Gereffi, Amy Glasmeier, Francis Green, Giuseppina Gualtieri, Gary Herrigel, Stuart Holland, Bryn Jones, Horst Kern, Bernd Kiel, Ricky Locke, Ned Lorenz, Martha MacDonald, Ann Markusen, Flavia Martinelli, Doreen Massey, Jonathan Morris, Fabio Nuti, Paolo Perulli, Mario Pianta, Andy Sayer, Anno Saxenian, Erica Schoenberger, Allen Scott, David Soskice, Guy Standing, Nigel Thrift, Todd Watkins, Frank Wilkinson, Stephen Wood, and, last but not least, my "mates" in the spring 1989 Oxford University seminar on flexibility and restructuring: Andrew Glyn, David Harvey, and Erik Swyngedouw.

The usual gang of labor and regional economists, geographers, and planners—especially Barry Bluestone and Mike Storper—played their usual helpful roles in working with me to piece together (and to find the weak points in) the argument. It is a special treat to have dear friends with whom one may strongly disagree and still stay in the relationship.

A set of new acquaintances from the business schools and sociology departments headed me in the direction of studying organizations and interorganizational networks, and thinking about the many different aspects of "flexibility." Thanks to Mark Lazerson, Jim Lincoln, Gary Loveman, Chick Perrow, Alex Portes, and especially Woody Powell for their influential writing, for being kind enough to read sections of my manuscript, and for engaging me—usually, critically—in defending my ideas.

In my world, new ideas are honed in the context of peer-group academic seminars. I am grateful for having had an opportunity to try out these thoughts in seminars at the United Nations Centre for Regional Development in Nagoya, Japan; at Bristol, Cambridge, Oxford, Sussex, and Wales universities in the United Kingdom; at the International Labour Office in Geneva; at NOMISMA in Bologna; and at the University of Venice. In the United States, earlier versions of the work in progress were presented to seminars at the Institute of Industrial Relations and the Department of City and Regional Planning at the University of California–Berkeley; the Industrial Relations Seminar and the Industrial Performance Center at M.I.T.; a joint geography-sociology seminar at the Johns Hopkins University; the Lewis Center for Regional Studies at UCLA; the Program on Regional and Industrial Economics at Rutgers University; the Russell Sage Foundation; the Sloan Foundation; the Woodrow Wilson School at Princeton University; and at my home base seminar on Technology and Social Change at Carnegie Mellon University.

Over the last four years, through an inevitably disruptive relocation from M.I.T. to Carnegie Mellon University, when it was often difficult to get my hands on just the right document or file folder at just the right moment, secretaries at both universities kept me together. I especially thank Donna Dowd, Meg Gross, and Sheryl Huweart. Documents in Italian were translated for me at Carnegie Mellon by Lisa Stanziale. Jane Judge at Basic Books was an exceptionally astute project editor. And Constance Reiter at CMU helped to fill in the holes in my bibliography.

After a decade of collaborative writing, I had forgotten how difficult and lonely it is to go *solo*. It would have been literally impossible without the pure friendship (quite apart from the skilled professionalism) of Bob Kuttner, and the substantive engagement, expertise, faith, and personal courage of Maryellen Kelley.

Finally, for his (nearly) infinite wisdom and patience on this and former projects, I am eternally grateful to the former publisher and editor-in-chief of Basic Books, Martin Kessler.

Thanks, all.

FOREWORD

When Bennett Harrison's *Lean and Mean* was first published in 1994, it literally changed the conventional wisdom about scale and efficiency. As the economy has become simultaneously globalized and decentralized, a great many commentators considered the large enterprise an industrial dinosaur. It was too big, too bureaucratic, too slow to respond to changing markets, changing technologies, and local variations. The travails of many of the world's largest corporations—IBM, General Motors, U.S. Steel, the state-owned enterprises of Europe and Latin America—seemed to confirm this conventional view.

Conversely, small businesses enjoyed special favor. Supposedly, these were the true repositories of entrepreneurship and innovation. They generated a disproportionate share of new jobs. And with the age of the internet, a small local enterprise with a new product could reach a global marketplace more nimbly than its behemoth competitors. Small was beautiful, the very incarnation of the resurgent free market. Ideologically, these small businesses, the hero of the piece, took on a kind of populist virtue. If big was often bad, small was therefore good.

Harrison revealed this story to be less than a half truth, in its multiple dimensions. First, the very large enterprise was far from moribund. Large corporations had in fact emerged from the trauma of the 1970s and 1980s as lean and mean competitors. And they had learned how to maximize the benefits of both large scale and decentralization. As Harrison noted, globalization did reward flexibility and decentralization, but these could occur—indeed, often occurred best—within the networks of large, global enterprises. As Harrison found, "Indeed, the more the economy is globalized, the more it is accessible only to companies with a global reach."

Globalization had led large corporations to adopt new forms of management. They had to change in order to survive. But survive they did. The new global company combined a capacity for very flexible relationships with local units, with renewed commitment to global marketing, investment in research, development and worker training, and state-of-the art management. Indeed, it was only large global enterprises that could command the financial resources to stay ahead of the competition, by ploughing money back into research and technology, as well as best-practice management. It was the big companies that had the staying power to ride out business cycles.

What was new, however, was a different blend of big and small. Big companies had ceased to be the private bureaucracies of the gray flannel suit era. They had downsized; they had de-layered; they now relied more on networks of suppliers. Thus they precisely combined the benefits of scale and size with the advantages of localism that cheerleaders for small business had mistakenly attributed uniquely to small enterprises.

In the new economy, large firms were at the center of inter-firm production networks. It was the large businesses, as buyers of intermediate goods and global sellers of finished ones, that set the standards and diffused the technology. It was the large firms that had the deep pockets and did the marketing. Harrison found variations on this theme in Japan's keiretsu, Italy's dynamic fashion industry, and American semiconductor makers.

To be sure, all businesses start out small. And some small businesses indeed grow rapidly into large ones. In some sectors, where the product is something close to pure knowledge, that growth can come very fast. Of course, once a small business becomes a large one—the contrast between Microsoft and Apple comes to mind—it had better maximize all its advantages or size alone will be no guarantee of success.

At the same time, the small business sector is characterized by a very high failure rate. In *Lean and Mean*, Harrison not only researched the evolution and revival of the large transnational firm, he persuasively debunked the job-generation claims about small business. Champions of small business, most famously David Birch, relying mostly on Dun and Bradstreet data, wove a powerful myth—that small business generated as much as 88 percent of the jobs. As Harrison demonstrated, such claims often confused gross job creation with net job creation and tended to conflate enterprises with sites. A local outlet affiliated with a large conglomerate may have fewer than twenty employees, but it is not a small business. Harrison demonstrated that the share of jobs in large and in small business has remained remarkably constant over the years.

The lionization of small business, as Harrison suggested, had a political function. It comported nicely with the dominant ideology of the age: Get government out of the way, and economic miracles will sprout. Larger businesses were more likely to have links with a mixed economy. They were more likely to have research connections or contracts with government; more likely to rely on government trade policies; and more likely to have decent social benefits and even unions. Indeed, some of the most technologically dynamic industries—aerospace, biotechnology, information technology, telecommunications—had and continue to have beneficial entanglements with government.

However, if large enterprises have made a surprising comeback by blending the advantages of the global with the local, all is not entirely well in Harrison's account. It is true, certainly, that the industrial burial of American industry was premature. However, in learning to be leaner and meaner, big business has become especially meaner to its employees. The downsizing has occurred with little thought to the effect on communities, career paths, and human lives.

Once, labor economists differentiated between primary and secondary labor markets. Primary labor markets, usually in large companies, tended to offer professionalization, decent pay, and job security. Secondary markets, typically in small and marginal businesses, offered casual labor—no long-term relationships, no fringe benefits, no assurance of employment at all. In the era of stable oligopolies that characterized the postwar boom, large industrial firms, often regulated ones, were the habitat of primary labor markets. Some of the benefits were anchored in regulation and in a welfare state; others in collective bargaining. The institutional stability of regulation and oligopoly gave unions additional bargaining power.

In the new economy, some jobs in large firms still follow this model, but they are a dwindling share. And the large, networked firm that Harrison describes has learned to pursue total quality without a true social compact with its workforce. Harrison concluded by suggesting that large companies have substantial latitude to pursue either a high road or a low road with their workers. Either road can lead to profit and technical excellence.

But flexible production with suppliers and strategic partners that leaves all but the most high-level workers out of the bargain is simply too tempting a choice for many firms. A variation on the theme is the "virtual corporation," with a handful of core employees—and a lot of casual labor.

There is, however, an alternative path that some corporations do choose to pursue. The challenge, Harrison concluded, is for society to find ways to "make lean corporations less mean."

In a postscript written specially for this new edition, Harrison adds new data and insight on the role of small business, which is now enjoying yet another vogue as the putative savior of the inner city. If small businesses, Harrison writes, "are the most vulnerable to cyclical instability, have the shortest expected survival rates, and tend to pay the lowest wages," then the solution is not tax breaks but policies that influence the institutional context in which small businesses succeed or fail. These include public/private "investment in infrastructure, technical assistance, targeted venture capital . . . and closer relations with large-firm networks."

These, of course, are policy questions. And, absent policy changes to spread prosperity, the dark side of lean and mean production will likely get darker. The conventional prescriptions of better education, training, and diffusion of "best practice," Harrison suggests, are necessary but not sufficient. Macroeconomic policy needs to spur faster growth and tighter labor markets; the terms of engagement for global trade need to be fair, to workers in both the first and the third worlds; unions need a new lease on life. All of this has a profound impact on local conditions, but these policies must be fashioned nationally and even supranationally. Harrison's fine book is an important spur to this evolving debate.

ROBERT KUTTNER

PART I

OVERVIEW

1

BIG FIRMS,
SMALL FIRMS, NETWORK FIRMS

There are more than 1,200 booths arrayed across the football field–
length floor of the David P. Lawrence Convention Center in Pittsburgh, Pennsylvania. The smells and tastes of cigarette smoke, coffee,
and Coca-Cola fill the air. Everywhere, people (mostly men, but a surprising number of professional women, as well) are giving lectures,
inspecting one another's wares, exchanging telephone numbers, and
making deals.

We are attending a trade show of companies in the steel business.
Companies from around the world are advertising their competence in a
wide variety of activities. Some actually make steel bars, sheets, and
related products. Others manufacture the machinery, parts, or computerized control systems. Still others offer the mill owners services ranging
from design and plant maintenance to personnel management. And
some specialize in disposing of the hazardous waste materials thrown
off in the process of making steel.

As my friends and I pick up brochures and stop to chat with company
representatives, we look for the presence of small, high-tech, independent entities. They are hard to find. Either the firms represented have
themselves been created by consortia of companies from different countries, or they are branches, subsidiaries, or divisions of foreign multinationals whose parentage appears on the brochures and posters only in
the small print (if at all). Sandwiched in between the row on row of
cross-national companies, we occasionally encounter a certifiably inde-

pendent, local small firm bearing the placard "Benton Harbor, Michigan," "Oakland, California," "Portland, Maine," or "Birmingham, Alabama."

But there is little doubt about who dominates these proceedings. It is the GEs, the IBMs, the Digitals, the Westinghouses, the 3Ms, the Hitachis, the Sumitomos, the Rockwell Internationals, the SKFs, the Bachmanns, the Ebners, the Herkules, the Siemens—alone, and together with their worldwide networks of large and small "partners." A similar convention of purveyors of construction or financial services—and of computer and semiconductor manufacturers, as well—would have an equally multinational character, dominated by the big firms.

To read the daily newspapers, this judgment must seem awfully surprising, to say the least. Headlines report on the crises of such giant corporations (and household names) as IBM, General Motors, and Sears Roebuck. We are bombarded with expert opinion about how these and other big firms have lost their competitive edge because of organizational rigidities and obsolete technological capabilities. The big firms, we are told, have become too inflexible, too rigid, and unable to adjust to the brave new world of heightened global competition, where only the fleet of foot—rather than the strong—survive.

At best, these observations tell only part of the story of how business is evolving in the closing years of the twentieth century. In the fields of computer hardware and software, IBM may again be in trouble—it has happened before—but those other standard-bearers in the industry, Intel (whose microprocessors drive most personal computers) and Microsoft (whose operating systems direct those Intel and other chips) go from victory to victory, and both are members in good standing of the Fortune 500 (to the extent that Intel's long-run command of the industrywide microprocessor standard *is* being challenged by other chip makers, the challenge is coming from such consortia as the Somerset group, created recently by IBM, Apple, and Motorola—all very big companies, indeed). The declining significance of the catalogue business of retailer Sears Roebuck has been succeeded not by the emergence of a thousand small niche distributors but by even more successful mass retailers and distributors such as Lands End and Wal-Mart.

Other giant American companies have found ways to flourish in the new, more uncertain, more competitive environment. AT&T and Xerox are regularly cited by business analysts and executives as successful multinational corporations. And except for its problems with the same mammoth pension liabilities that are haunting companies in all of the mature industries, the Ford Motor Company has substantially transformed itself for the better—in only a decade.

We are constantly being told that technological change now systematically favors (or is mainly the product of) small companies. The idea is pervasive, but it simply is not correct. Take that quintessential high-tech activity: the design and manufacture of computers. It is no secret that in Japan, the computer industry has from the beginning been dominated by the NECs, the Toshibas, and the Fujitsus. But dominance by major firms is also true in America. In 1987 (the most recent year for which the appropriate data were published by the U.S. Bureau of the Census), 85 percent of all the individual enterprises in the computer industry in the United States did indeed employ fewer than 100 workers. Only about 5 percent of all computer makers had as many as 500 employees. Yet that comparative handful of firms—that 5 percent at the top— accounted for fully 91 percent of all employment and of all sales in the computer industry in that year.[1]

Meanwhile, in eastern Asia, the giant *keiretsu* of Japan and the *chaebol* of South Korea—huge industrial, service, and financial conglomerates— enter new domains of economic activity, from entertainment and health care to aerospace and medical technology, by adding more divisions to their already enormous holdings. If the Japanese economy is in some difficulty these days, the source lies mainly in the bursting of the speculative financial and real estate "bubble" of the 1980s; the rise of the exchange value of the yen, which has seriously dampened the exports on which that country's overall economic development strategy has long been based; and the global recessions which are just ending in the United States, if not yet in Europe. Recent Japanese successes may indeed have been "miraculous," but no economy can grow without customers. Nevertheless, few knowledgeable students of Japan doubt the long-run technological and financial viability of Mitsubishi, Sumitomo, Fujitsu, or Toyota.

And even as Europe rides the next wave of consolidation of its Economic Community—now sure to extend some day to the Ural Mountains of Russia, albeit at a slower pace than was popularly expected when the Berlin Wall first fell—that continent is experiencing a veritable blizzard of mergers and acquisitions, and all manner of cross-border strategic alliances, involving both the public and the private sectors.

To see just how much the economic development action remains where it has been throughout the twentieth century—under the control of big corporations and their partners—one need only look at two commodities that are central to the daily lives of every North and South American (and Asian, and European) household: television sets and cars. The cost of developing the next generation of high-definition televisions is astronomical, and once the U.S. Federal Communications Commission (FCC) selects a standard, the winning design will immediately have a guaranteed mass market for TV sets oriented to that protocol. That is why some of the world's biggest high-tech corporations decided to form teams to develop the new standard system. Initially, one team included the French giant Thomson, the Dutch electronics conglomerate Philips, and NBC, probably the most famous American pioneer in recording technology. Other teams were led by General Instrument, working with M.I.T., and by Zenith, which joined forces with AT&T. But given the huge stakes, and with explicit shepherding by the FCC, the three teams announced in May 1993 a "grand alliance," under which they would share technical know-how and divide the eventual winnings.[2] This is not a story about the local Chamber of Commerce or the Elks Club. Rather, it is a story about big corporations and government industrial policy.

And what about cars? By the spring of 1993, it had become apparent that Toyota, Ford, and Honda were making great strides in developing truly global production systems. Parts manufactured in one location were being delivered to final assemblers based in another. Assembly lines located on every continent were turning out automobiles that were being shipped not only to local markets but across continents—even (in the case of Honda and Toyota) back to Japan, itself! The German car makers are moving in the same direction. Why? The answer: to hedge against unexpected currency fluctuations and to take even greater advantage of economies of large-scale production.[3] Again, this is hardly a story about industry growth driven by small business. If such direct foreign investments into *this* country have slowed down in recent years, blame it on the recession at home—*not* on the plans and deep pockets of the foreign giants.

Yet despite such examples, a multitude of writers continue to preach the virtues of small firms as the engines of contemporary economic growth. We are told that, as discretionary incomes increase and living standards reach historically unprecedented levels around the world, consumers increasingly seek more customized, fashion-oriented goods and services. Mass markets become saturated, the demand for such commodities as clothing and furniture becomes increasingly fragmented, and mass education and mass communications both facilitate and promote a growing heterogeneity in customers' tastes. In a fashion-conscious world, agility in identifying new wants and in getting new products to market becomes the key to winning the competitive wars.

These developments are said to conjoin to favor technically adroit, well-informed small enterprises—or at least give them a new fighting chance. Why? The answers we are offered are partly behavioral and partly technical. The bureaucratic organization of the big firms militates against agility. And the fragmentation of markets deprives the big firms of the opportunity to exploit various technical advantages that, over the course of the last century, were made possible by the drive toward standardization and mass production.

That's the theory. The facts show otherwise. With the usual few headline-capturing exceptions, small firms turn out to be systematically *backward* when it comes to technology. For example, on every continent, the big companies and establishments are far more likely than are the small ones to invest in, and to deepen their use of, computer-controlled factory automation.

And the argument that the proliferation of niche markets is inexorably driving a small firm renaissance reflects a misunderstanding of the nature of contemporary markets. As the Johns Hopkins University geographer David Harvey so elegantly observes, our postmodern world is awash in the layering of styles from different eras, in everything from music to architecture to movies, and those with the money (and the time) to spend can mix and match a seemingly infinite variety of goods and services.[4] But what does this tell us about the underlying organization of *production*? As it turns out, very little.

For one thing, as documented by Professor Theodore Levitt of the Harvard Business School, the biggest companies can and do market their wares in many countries simultaneously, portraying what are for the most part standardized commodities in the most colorful, idiosyncratic local languages.[5] The technology expert Yves Doz offers many examples of such "world products": British raincoats, Italian sweaters, Swiss watches (Rolex or Swatch), French wines, and Japanese consumer electronics—all produced and distributed by large corporations and their business partners.[6] Moreover, these products all come in several flavors, in both high-style *and* less expensive versions. What the architects of the romance of small business are ignoring is that the big firms can produce for both mass *and* niche markets—a neat trick that few small firms can pull off. Thus, Toyota can deliver both its big-selling, inexpensive Corolla *and* the high-priced, world-class Lexus.

Moreover, the existence of niches favoring certain high-cost, customized products in no way threatens the profitability of the lower-priced market segments (that is why they are called "niches"). Thus, one may choose to focus one's attention on the highly style-conscious, small-batch furniture turned out by the best of the small Italian enterprises, in workshops where craftsmen (and they are still nearly always men) enjoy a substantial degree of control over their work. But do not forget IKEA, the Swedish mass producer of attractive low-end, low-cost furniture. IKEA is a global corporation that sells its goods through an international network of warehouselike department stores. And lest we make the mistake of thinking that mass production and marketing are relevant only at the low-value end of the spectrum, consider Roche-Bobois, which displays its relatively expensive but mass-produced furniture in attractive showrooms in more than a dozen countries.

To be sure, small firms and individual business establishments do have a role to play in the evolving industrial structure of world capitalism. And managers most certainly care about "flexibility." But as I show later, the role that small firms are playing is typically that of follower, not leader. And while it may be enhancing the agility and profitability of individual firms, the search for flexibility—by the managers of both big and small companies—is also leading to practices that are undermining the employment security and incomes of a growing fraction of the population, exacerbating inequality and contributing to the underlying sense of futility that now characterizes politics worldwide.

Concentration without Centralization: How the Big Firms Are Reorganizing Global Capitalism

Announcements of the demise of concentrated economic power in the form of the large, resourceful, multidivisional, multiproduct, multiregional, often multinational corporation are premature. Yet the difficulties facing traditional big business are formidable. How then, *have* the survivors managed to cope? How do newly emergent large firms make it in a world that was thought to belong to the smallest of the small?

Rather than dwindling away, concentrated economic power is changing its shape, as the big firms create all manner of networks, alliances, short- and long-term financial and technology deals—with one another, with governments at all levels, and with legions of generally (although not invariably) smaller firms who act as their suppliers and subcontractors. True, production is increasingly being decentralized, as managers try to enhance their flexibility (that is, hedge their bets) in the face of mounting barriers to market entry and of chronic uncertainty about political conditions and customer demands in distant places. But decentralization of production does not imply the end of unequal economic *power* among firms—let alone among the different classes of workers who are employed in the different segments of these networks. In fact, the locus of ultimate power and control in what Robert B. Reich, the U.S. Secretary of Labor and a Harvard University lecturer, calls "global webs"[7] remains concentrated within the largest institutions: multinational corporations, key government agencies, big banks and fiduciaries, research hospitals, and the major universities with close ties to business. That is why I characterize the emerging paradigm of networked production as one of *concentration without centralization*.[8]

In this period of great ferment and experimentation by business, the big companies are of course doing many other things than just forming external alliances and networks. They are, for example, radically decentralizing R&D, moving their laboratories from under the noses of central management out into the operating divisions of the firms.[9] I am deliberately limiting my inquiry to an examination of how companies are reorganizing their *external* relationships—with one another and with other key entities such as universities, governments, hospitals, banks, and of course the unions. But it is worth remarking in passing that the same underlying principles that apply to these external relationships—especially the consistency of decentralized *activity* with concentrated *control* over resources—apply equally to the internal restructuring now under way.

The revitalization of the big firm sector, in the wake of the worldwide economic turmoil of the 1970s and early 1980s, has been erected on four basic building blocks. First, managers are vigorously paring down the mix of activities (and the number of employees) they deem central to the firm's existence, while relegating the rest to positions at a greater "distance" from corporate headquarters. The much-discussed "lean production" strategy of the car companies exemplifies this approach.[10] In the long, slow, disappointingly flat recovery from the recession of 1991, company after company has followed the lead of the car makers in trimming its in-house operations to just its "core competencies,"[11] farming out other work to rings of outside suppliers. Specialists describe this as the transformation of the firm into core-ring and other network forms of business organization. The most successful competitors in Japan, Europe, and the United States are those that are adopting some version of this network form of industrial organization.[12]

Second, more and more companies are finding new ways to use computerized manufacturing and management information systems to coordinate these far-flung activities and increase the flexibility with which they enter and exit different markets, alter production designs, and monitor their employees' performance. Computer programmable automation (PA) on the factory floor or in the warehouse permits firms to enhance product quality while reducing machine set-up and other costs of production. Personal computer and workstation information technology facilitates much greater coordination and control of production across organizational boundaries (and national borders), as well as within firms and agencies. Indeed, the modern inventory management and work reorganization practice known as "just in time" delivery of parts and supplies to the big firms that assemble everything from automobiles to toaster ovens would be literally impossible without the enhanced coordination afforded by the new technical capabilities.[13] Ever more distant organizations are thus linked to one another via computer, along with the accompanying standardized software: packages of computer programs that allow nonspecialists to interact electronically to an extent that would have been unimaginable only a few years ago.

Third, the most successful of the big firms have been busily constructing so-called strategic alliances among one another, both within and, especially, across national borders.[14] Particularly in Japan[15] and Western Europe,[16] but also within the production network that makes up the U.S. Department of Defense's domestic military procurement system (as Professor Maryellen Kelley of Carnegie Mellon University and Professor Todd Watkins of Lehigh University are exploring),[17] these alliances typically incorporate at least the first tier of the networks of generally smaller firms that supply parts, design services, and manufacture components for the big firms at the center of the "partnership."

Fourth, within more and more of the big firms and their principal subcontractors, we are witnessing systematic attempts by managers to elicit the more active collaboration of their most expensive-to-replace workers in the "mission" of the corporation. In Europe, these arrangements are grounded in institutions (works councils, unions, and social democratic political parties) that reinforce the independence of workers, increasing their willingness to accept more "flexible" task assignments and to reveal their "tacit knowledge" of the production process to their managers and supervisors. In Asia, at least in the biggest corporations, management purchases worker collaboration by awarding core workers substantial (if not always literally lifetime) employment security.[18] In the United States, quality circles, joint labor-management problem-solving committees, and teams have had a much rockier road, with considerably less willing participation by labor and often fickle commitment from management.[19]

But the competitive success of the large corporations is not without its own contradictions. In particular, the restructuring experiments pursued by the big companies and their strategic partners since the 1970s are polarizing the population, contributing to the growing inequality among white-collar workers as well as between blue-collars and white-collars. The polarization is now evident and palpable. It manifests itself in terms of income, status, and economic security. Economists and sociologists used to refer to this tendency toward polarization as labor market "dualism," and most believed that it had substantially disappeared as heightened international competition steadily eroded the stability of traditional monopolistically or oligopolistically organized industries. But as the big firms have turned increasingly to decentralized production and to the use of alliances among existing entities as a (partial) substitute for expanding their own facilities, the institution of segmented labor markets has actually been elaborated.[20]

It works this way: According to a central tenet of best-practice flexible production, managers first divide permanent ("core") from contingent ("peripheral") jobs. The size of the core is then cut to the bone—which, along with the minimization of inventory holding, is why "flexible" firms are often described as practicing "lean" production. These activities, and the employees who perform them, are then located as much as possible in different parts of the company or network, even in different geographic locations. A good example is the siting of the "back offices" of the big insurance companies, banks, and corporate headquarters. These facilities house masses of typically poorly paid, overwhelmingly female clerical workers, tucked away in suburban "office parks," far from the downtown corporate headquarters to which they are linked, where their companies' higher-level functions are performed.[21]

Although represented as state-of-the-art management, the practice of lean production (the principle applies as much to the service sector as to manufacturing) involves the explicit reinforcement or creation *de novo* of sectors of low-wage, "contingent" workers, frequently housed within small business suppliers and subcontractors.[22] The advent of these generally big firm–led core-ring production networks is almost surely adding to the national (and increasingly international) problem of "working poverty," in which people work for a living but do not earn a living wage. As a result, both within the big firms and their most trusted partners and suppliers, and ultimately over the economy as a whole, core employees become increasingly segregated from outside peripheral employees—a gap that is measurably reflected in the by now widely acknowledged phenomenon of growing earnings inequality among American (and, as we shall see, some overseas) workers.[23] I call this the dark side of flexible production.

To sum up the argument: I am suggesting that the emerging global economy remains dominated by concentrated, powerful business enterprises. Indeed, the more the economy is globalized, the more it is accessible only to companies with a global reach. The spread of networking behavior signifies that the methods for *managing* that reach have changed dramatically, *not* that there has been a reemergence of localism, as others have argued. Dressed in new costumes, and armed with new techniques for combining control over capital allocation, technology, government relations, and the deployment of labor with a dramatic decentralization of the location of actual production, the world's largest companies, their allies, and their suppliers have found a way to remain at the center of the world stage.

WHY SMALL FIRMS DO NOT DRIVE ECONOMIC GROWTH AND CREATE THE MOST NEW JOBS

My argument about the revitalization and transformation of the big firms and their production networks must sound even more surprising to a public that, for more than a decade, has been told repeatedly that *small* companies are now the engines of economic growth and development. According to the new conventional wisdom, the large corporation was in many respects becoming something of a dinosaur, increasingly unable to compete in a "postindustrial" world characterized by continually fluctuating consumer demands, heightened international competition, and the need for more flexible forms of work and interfirm interaction.

As the big firms collapsed under their own weight, we were told, a panoply of small, flexible enterprises were rushing in to fill the ecological void. Small enterprises were said to be creating most of the new jobs in all of the world's highly industrialized countries. The world described by an earlier generation of scholars—Joseph Schumpeter, Raymond Vernon, John Kenneth Galbraith, and Alfred Chandler—was thought to be collapsing before our very eyes. Now it was the turn of the small, agile companies to drive technological progress.

But hard evidence shows that the importance of small businesses as job generators and as engines of technological dynamism has been greatly exaggerated. In the United States and Germany, after we factor out the ups and downs of the business cycle, the share of all jobs accounted for either by small companies or by individual workplaces with fewer than 100 employees (the official criterion for "small" that is used by the Paris-based Organization for Economic Cooperation and Development—the OECD—when making international comparisons) has hardly changed since the 1960s. Moreover, many de jure independent small companies turn out in varying degrees to be de facto dependent on the decisions made by managers in the big firms on which the smaller ones rely for markets, for financial aid, and for access to political circles. As we shall see later, there are also sound technical reasons why precisely the kinds of short period changes in the size distribution of firms that so appeal to the "small is beautiful" ideologies systematically exaggerate the relative importance of the tiniest companies, and overstate the fragility of the biggest corporations.

Still, on every continent, stories on the front pages and in the business sections of the leading newspapers and magazines feature seemingly endless anecdotes about an explosion in the number of small businesses. Thus, for the American economy as a whole, for mature as well as for high-tech industry, the consultant David Birch reckons that "very small firms [with fewer than 20 employees] have created about 88 percent of all net jobs in [1981–85]."[24] And *Business Week*, always an opinion setter on economic matters, announced in a lead story that "Small Is Beautiful Now in Manufacturing,"[25] while across the Atlantic, the London-based *Economist* editorialized:

> The biggest change coming over the world of business is that firms are getting smaller. The trend of a century is being reversed. . . . Now it is the big firms that are shrinking and small ones that are on the rise. The trend is unmistakable—and businessmen and policy makers will ignore it at their peril.[26]

Economists use the concept of economies of scale to describe the potential savings in unit production costs as facilities are operated at higher volumes. Scope economies are said to exist when the joint cost of making more than one product on the same basic equipment, or "plat-form," in the same facility is less than the cost of turning out the same set of products in separate facilities. Historically, these economies of scale and scope joined financial and supervising economies in reinforc-ing the tendency toward larger units of production and distribution.

Now, thanks to the advent of new, more flexible computer-based tech-nologies, from electronic bar-code readers at the supermarket checkout counter to numerically controlled machine tools and flexible manufactur-ing systems for the factory floor or the laboratory workbench, these inter-nal economies of scale and scope are said to be disappearing. In the words of the management consultant Tom Peters (of *In Search of Excellence* fame), "old ideas about economies of scale are being challenged. . . . Scale itself is being redefined. Smaller firms are gaining in almost every market."[27] The commentator George Gilder is only the most prominent popularizer of the even more extravagant claim that the smallest compa-nies are now even *more* technologically sophisticated than the old giants.[28]

This faith in the innovative or job-creating power of small business extends beyond the borders of the United States. Thus, some observers see the cornucopia of small machine and electronics manufacturing shops that dot the alleyways of Japan's industrial cities, supplying its giant industrial corporations, as the true source of that nation's inter-national competitive advantage.[29] In the United Kingdom, analysts attribute the continuing spread of the dreaded "British disease" of incompetence in world-class manufacturing to the *absence* of networks of technically skillful, small subcontracting firms, and both the Conser-vative government and the Labour opposition seek explicitly to encour-age the creation of a business climate conducive to the proliferation of such companies.[30]

A rather more interesting variation on the small firm theme calls our attention to the survival from an earlier era (or in some places, the recent emergence) of networks of mostly small, loosely linked but spa-tially clustered firms. The businesses that make up these so-called industrial districts are described as typically utilizing a craft form of work organization. The alleged widespread adoption of small-scale com-puterized automation helps to make these networks of what the M.I.T. economist Michael J. Piore and the M.I.T. sociologist and creator of the field of "industrial politics" Charles Sabel call "flexibly specialized" firms capable of rapidly reconfiguring themselves to meet the continu-ally fluctuating demands of the world market.[31] In the modern era, the industrial districts were first discovered in north-central Italy in the 1970s, then elsewhere in Europe, and they have now become the object of both study and policy prescription in many different regions of Europe, North America, and eastern Asia.[32]

At a time when many Western and Third World political leaders continue to entertain the philosophy that further government involvement in the economy only erodes economic efficiency—probably the most long-lasting and pernicious legacy of the Reagan-Thatcher years—in Japan, North America, and Europe, local and regional governments have been actively supporting their industrial districts with a variety of infrastructural and business services. Tying it all together are (we are told) a sense among the locally oriented small firm owners and managers of shared long-run interest; of mutual *trust* deriving from repetitive mutual business contracting said to be *embedded* within deeply rooted local social relationships associated with political, familial, and (in some places) religious life; and the practice of *reciprocity* among all the actors in the community.[33] Giacomo Becattini, the prominent, elegant Florentine economist, calls this the "industrial atmosphere," borrowing an evocative language first coined by the British economist Alfred Marshall, who depicted the late-nineteenth-century steelmaking district around the town of Sheffield, and, more or less at the same time, by Alfred Weber, the German father of industrial location theory.[34]

More than anything else, it is the embedding that is thought to confer on these new growth poles of generally small enterprises the ability to capture simultaneously economies of scale and scope, but at the level of the district as a whole rather than within individual firms.[35] Therein lies their alleged competitive advantage over the large, vertically integrated, centralized, and concentrated monopolies that dominated economic life in the industrialized world for most of the twentieth century. In the elegantly argued and widely influential view of Piore and Sabel, the world has come to a "second industrial divide," at which a resurgence of the nineteenth-century districts has been made possible by the growing complexities of a global economy, which make it ever more difficult for large, concentrated economic organizations to compete.[36]

WHY ARE WE SO READY TO ACCEPT THE SMALL FIRMS STORY?
Throughout the 1980s, and up to the present, local and regional policy makers in one country after another, and in such international organizations as the OECD and the German Marshall Fund, have taken their

cues from one or another of these well-publicized theories of small firm–led economic growth and development. This has prompted many elected and appointed officials to shift their attention from the "old-fashioned" efforts to attract new branch plants of multinational corporations, to promoting the "incubation" of small businesses and, in some cases and places, to trying to actually *create* districts of interdependent, technologically up-to-date small firms.

It is easy to see why both the entrepreneurial and the industrial district versions of the theory of small firm–led growth would become so popular in the 1980s, especially among policy makers. The 1970s constituted a historical moment when the managers of large corporations in many places seemed to have lost their strategic bearings. Moreover, by the end of that decade, and well into the 1980s, the very legitimacy of government was being challenged by newly reenergized conservative political movements that had succeeded in capturing political power and decisive influence on public opinion, especially in the United States and the United Kingdom but to some extent also in the Federal Republic of Germany, Austria, and eventually even Sweden, the country with the world's most highly developed welfare state. Both tendencies strengthened interest in and celebration of entrepreneurship, small business, free enterprise, deregulation, and decentralized ("free") markets.

This is the environment that has proved so hospitable to the questionable statistics of Birch and to the laissez-faire ideological tracts of Gilder, both of whom are outspoken advocates of economic development policies fashioned around the allegedly driving force of the dynamic small firm. It is the same environment that has given rise to such public policy conceptions as the enterprise zone, the industrial incubator, government deregulation, tax preferences for venture capital funds, and science parks—all in the interest of promoting and nurturing the growth of small businesses. This approach has acquired a special champion in the American media, in the influential magazine *Inc.*, and in Europe, in the London *Economist*.

Interestingly, the Left in many places has also become enchanted with many of the elements of such a program—although for different reasons. To many, the big firms seemed hopelessly inaccessible. Moreover, as socialist and social democratic parties and groups attained some degree of control over municipal, state, or provincial governments (and even the national government, as in France and Spain) during the depths of the recession of the late 1970s through early 1980s, it was both ideologically attractive and seemingly feasible politically to articulate a "progressive localism" that encouraged the development of cooperatives and other kinds of small firms. Nowhere was this inclination more apparent than in the United Kingdom. There, in the early 1980s, the small firm "renaissance"—and the Italian model in particular—caught the interest of several leaders of the Greater London Council and the Greater London Enterprise Board.[37]

In the United States, efforts to self-consciously construct (or preserve) industrial districts got under way in a number of states, including Massachusetts, Pennsylvania, and Michigan. The Washington-based Corporation for Enterprise Development became a prominent voice for the planned emulation of the Italian model. And in New York City, the architect C. Richard Hatch, who knows Italy intimately, advocated the transferability of the Italian model with the same fervor that, in the late 1960s, he brought to the argument that Third World import substitution strategies also made sense for the economic development of black urban ghettos in the United States.[38]

These initial efforts to create small firm–led American production networks on the Italian model have met with mixed results, at best. But the advocates have been working harder than ever to improve their performance and to gain political support from the White House and from a recently revitalized U.S. Small Business Administration. American interest in these small firm–oriented economic development models has been heightened recently by the pressing political problem of furthering the conversion of military contractors to civilian activities, and by the critical need to rescue communities that had become highly dependent on military bases and contractors.

WHAT'S WRONG WITH THE SMALL FIRMS STORY?

For all their obvious popularity, the new verities about vigorous small entrepreneurs and networks of loosely coupled, flexibly specialized small firms offer at best a partial picture of the truly fundamental developments in the restructuring of business across the industrialized world. In other respects, these descriptions of who is driving economic development in post-1970s global capitalism are plain wrong.

Take the job generation question. As a general proposition, across the industrial world, the biggest companies and plants unquestionably are downsizing, especially in manufacturing (on the other hand, at least in the United States the average individual facility in the *service* sector has actually been getting *bigger*).[39] *Why* are manufacturers getting smaller? We know that managers are outsourcing work they used to perform in-house.[40] They are also partnering with other existing firms, as a way of accessing new technical know-how, markets, territories, and capital without having to make new capacity-expanding investments themselves. In this regard, Bo Carlsson, a Case Western Reserve University industrial economist, reports that

> the share of multi-unit companies in U.S. manufacturing employment increased throughout the postwar period until the late 1970s. But after 1977, the share of multi-unit companies declined for the first time. . . . This suggests that subcontracting and outsourcing have become more important forms of disintegration in recent years.[41]

But then the increasing number of small firms turns out to be in part a function of the core-ring, lean production strategies of the *big* companies. It is the strategic downsizing of the big firms that is responsible for driving down the average size of business organizations in the current era, *not* some spectacular growth of the small firms sector, per se.[42] What we have witnessed over the last decade constitutes the lopping off of the tip of an iceberg more than it does a meltdown of the old prevailing structure.

In fact, in the United States, Germany, and Japan, once we account properly for the usual ups and downs of the business cycle, the shares of national employment in both small *establishments* (that is, in individual plants, stores, and offices) and small *enterprises* (entire companies) have hardly changed at all for several decades. The Japanese data for the most recent years actually record a slight *decline* in the small firm (and plant) shares of jobs. Only in the United Kingdom did the small firm (and establishment) shares grow steadily between the mid-1970s and the mid-1980s. But even there, this looks to be mainly the result of the sharp decline in the fortunes of the biggest corporations during the disastrous economic years of 1973 through 1983, these corporations' subsequent laying off of middle managers as well as shop floor workers, and the permanent shuttering of their older, most inefficient large factories—not some explosive growth of small business, per se.

Nor do countries with a high proportion of their overall manufacturing employment in small firms display systematically superior economic performance. Across the member nations of the OECD there is no correlation whatsoever between the relative importance to each country of small manufacturing firms and either the national unemployment rate or the rate of growth of overall national manufacturing employment.[43]

New attention to the *dynamics* of job creation and destruction over time does even more damage to a naive small firms story. For example, recent research from the United States and Germany points to a consistent tendency of the *largest* firms in any cohort to experience the fastest rates of growth over time, and the smallest chances of going out of business during any given interval of time.[44]

Still another kind of evidence has emerged that casts doubt on Birch's thesis that very small start-up businesses are the principal source of economic vitality in modern industrial economies. In 1989, after years of providing Birch with his data on company and establishment births and deaths in American industry, the Dun and Bradstreet Corporation (D&B) decided that it had had enough of being quoted so often as the source of the claims that small firms were creating most of the jobs in the United States.[45] So the company set its in-house economists to reassessing what their own numbers seemed to be saying.

What they found was startling. Among the 245,000 new companies that were started up in the United States in 1985—in the middle of the Reagan-era military- and real estate–driven economic boom—75 percent of the employment gains by 1988 occurred in those firms that, at birth, had *already* employed more than 100 workers when they were first launched. Moreover, this group of businesses constituted only *three-tenths of one percent* of the 1985 cohort.[46]

A similar calculation for the United Kingdom, conducted by David Storey of the University of Warwick, showed that, of the more than 560,000 firms estimated to have employed fewer than 20 persons in 1982, 10 percent had gone out of business by 1984, and 88 percent still had fewer than 20 employees at that time. Only 2 percent of the 1982 cohort grew beyond 20 employees during the two years following their start-up.[47] When Storey and his colleagues looked across the entire European Economic Community, they concluded that

> even over a period as long as twelve years, only a very small minority (less than ten percent) of [new] firms grow out of the smallest size category [fewer than 20 employees], and less than one percent of [new] firms grow sufficiently to become large enterprises (with more than 100 employees).[48]

Finally, there is the matter of just how independent the small firms really are, especially in relation to the big firms for whom they act as suppliers and subcontractors. In their writing for the International Institute for Labour Studies (IILS) in Geneva, the Harvard Business School's Gary Loveman and IILS Director Werner Sengenberger conclude that "large enterprises often have very many legally independent subsidiaries. While the subsidiaries are *de jure* independent, they are *de facto* part of the large enterprise and should be accounted for, accordingly." For example, one German study found that "the 32 largest German manufacturing enterprises had in excess of 1,000 legally independent subsidiaries, and the number grew by almost 50 per cent from 1971 to 1983."[49] Once again, what we are seeing is evidence of how *production* may be decentralized, while power, finance, distribution, and control remain concentrated among the big firms.

For all the widespread interest in small firms as job generators, Birch and Gilder have failed to address a rather obvious companion question: How well do small companies do, vis-à-vis the largest firms, in providing their workers with a respectable standard of living? That is, how do wages, benefits, and such working conditions as occupational health and safety differ (if indeed they do) by the size of the organization? Certainly, for the purposes of evaluating the public policy implications of government subsidization of the small business sector, whether through grants, loans, tax incentives, or relaxation of environmental and other regulatory controls, these would seem to be important concerns.

In a pamphlet written in 1979 for the Council of State Planning Agencies, David M. Gordon, professor of economics at the New School for Social Research, decisively demonstrated that the part of the U.S. economy offering the generally lowest paid, least stable, and least well protected job opportunities—what sociologists and economists (following Piore) call the secondary labor market—consisted mainly (although certainly not entirely) of small businesses, especially in the service sector.[50] This fine bit of applied policy analysis went almost totally unnoticed, however, and a decade passed before the well-being of employees of small enterprises was explicitly considered in the context of the debates over job generation.

For the United States, the definitive study of whether the big or the small firms offer better working conditions is to be found in a book published in 1990 by the economists Charles Brown, James Hamilton, and James Medoff.[51] In contrast to Gordon's earlier pamphlet, *this* little volume has had a big impact on academic thinking about the small firm question. The authors found, first, that across American industry, "workers in large firms earn higher wages, and this fact cannot be explained completely by differences in labor quality, industry, working conditions, or union status." Second, employees of the big firms also enjoy "better benefits and greater job security than their counterparts in small firms." Third, in American political life, small firms are more likely than large ones to be granted exemptions from environmental or health and safety regulations, with the inevitable negative implications for their employees. Finally, workers in small firms are more likely to quit their jobs, and the fact that they express a greater desire to join unions (if they do not already belong to one) than do those working for big companies reinforces the strong impression that the workers themselves perceive conditions as being better in the larger organizations.

I have been presenting evidence on how the distributions of jobs, wages, and benefits differ between big and small businesses. There are still other flaws in the small firms story. Earlier, I alluded to the belief that smaller companies and establishments are now actually *more* technologically innovative than the supposedly rigid and inflexible big firms. Some writers assert that computer-programmable machinery systematically favors smaller units of production.[52] Or, as Gilder argues, the dramatically shrinking scale (in other words, the miniaturization) of microelectronic components leads inexorably to a commensurate shrinking of the "optimal" scale of the firms that make them.[53]

Much new theory and empirical evidence adds up to a powerful challenge to this contention. There *is* no particular size of firm, nor any special scale of production, that any given technology invariably favors or requires—nor any one "best" design of jobs that employers everywhere will introduce in connection with some new round of automation. For example, the Carnegie Mellon University economist Wesley Cohen and his colleagues have shown that small firms do best as product rather than as process innovators, and then only under certain market conditions. In other settings, the large producers still have a measurable advantage, given their greater resources.[54] Carnegie Mellon's Kelley has conducted a number of econometric studies from which she concludes that big firms are far more likely to adopt and use both complex *and* small-scale factory automation than are smaller companies and factories. In fact, during the 1980s the technology gap between the smallest and the largest American manufacturers actually grew *wider*, according to private industry trade association data.[55]

Gilder's biggest error in reasoning is his disregard of the fact that even playing in the league where "microcosmic" technology is being created requires ever larger scale production and concentrated control over finance capital. Consider the remarks of Intel's board chairman, Gordon E. Moore, in announcing a corporate plan to turn Intel's Albuquerque, New Mexico, chip factory into the world's biggest facility of its kind: "This is our first billion-dollar factory, but it won't be our last. Chip factories are getting bigger and more expensive as our manufacturing technologies continue to become more complex. The entry fee to be a major player in the global semiconductor market of the '90s is $1 billion—payable in advance."[56]

Summing up research from three continents, Loveman and Sengenberger conclude that "the economic performance of small enterprises [across the OECD] is, on average, inferior to that of large enterprises; productivity levels as well as profit rates appear to be lower, the capacity for innovation and technological improvement smaller . . . the average social standard of the quality of jobs and the conditions of work are [also] inferior in the small firm."[57] Clearly, the conventional wisdom about the superiority of the small firms in the age of flexibility is due for a major overhaul.

TROUBLE IN PARADISE: HIERARCHY AND INEQUALITY IN THE INDUSTRIAL DISTRICTS

Small, independent firms are neither as bountiful nor as beautiful as the new conventional wisdom has led us to believe. But what about those *networks* of small firms, those locally oriented industrial districts in such regions as north-central Italy and California's Silicon Valley? Here, too, we can see signs of concentration without centralization—geographically and organizationally dispersed production, but with strategy, marketing, and finance ultimately controlled by (or, in the case of the Italian districts, coming increasingly under the control of) the big firms.[58]

As a growing number of Italian researchers and local government officials are themselves observing, as they pursue the economic development of their own areas, powerful "lead firms" from within and without the districts now threaten to alter the collaborative nature of interfirm relations inside the districts. Mergers and acquisitions are on the rise. Financial conglomerates are dictating production procedures to what used to be truly independent small firms.

I interpret neither the appearance (or reappearance) of hierarchy, nor unequal power and the remote control of key elements of a district's economy by outside corporations, as a sign of regional economic *failure*. Instead, such changes are more a sober reminder that, for all their intended local orientation, the districts are operating within much more extensive fields. In the context of a global system populated by big companies perpetually on the prowl for new profitable opportunities, the very success of a district can itself bring about changes that give rise to its opposite, and we observe the re-creation of hierarchical organization.

All of these concerns are relevant to constructing a richer, more balanced reassessment of the evolution of the Western Hemisphere's most dramatically successful high-tech region—Silicon Valley. Those who wish to characterize the Valley as an industrial district on the Italian model are not wrong. But they are offering only a partial perspective. Silicon Valley has many faces, each of which manifests a different aspect of the emerging post-1970s system of networked production that I have named concentration without centralization—including the dark side.

In the most romantic characterizations,[59] Silicon Valley's astonishing success as the home base for a myriad of companies that design, produce, and export computers, workstations, microchips, disk drives, and software is mainly a story about an adventuresome gang of creative, supremely—even belligerently—independent entrepreneurs, many of them refugees from other, less free-wheeling parts of the country and the world, practicing textbook-style free market economics.

Silicon Valley shows another face to other observers. As seen by AnnaLee Saxenian, a city planning professor at the University of California at Berkeley, Silicon Valley is a full-fledged industrial district, a dense thicket of mostly small and medium-sized (but also some quite large) firms that alternately cooperate and compete with one another.[60] These networks of producers are said to be embedded in a local political economy that provides job training, finance capital, and an incessant flow of ideas and information about the latest design and production techniques. Well connected to the rest of the world, Silicon Valley's flexibly specialized firms nevertheless have a "Marshallian" orientation, in the sense that the district may trade with the rest of the world (and quite successfully, thank you), but *production* relationships remain (according to Saxenian) highly localized.

From yet a third perspective, Silicon Valley increasingly faces outward. According to a bevy of astute observers—including the management professor David Teece, the regional economist Ann Markusen, the technologists Kenneth Flamm and Martin Kenney, the management consultant Charles Ferguson, the political scientist Richard Gordon, and the urban planner Richard Florida—Silicon Valley as a production *system* was substantially created by major multinational corporations and remains profoundly dependent on them, and on the fiscal and regulatory support of the national government—especially as represented by the Department of Defense (as recently as the mid-1980s, Santa Clara County, the heart of Silicon Valley, remained one of the three top recipients of defense contracts in the United States).[61]

This orientation toward the outside world is reflected in the growing extent to which much of the actual manufacturing content going into Silicon Valley products is being sourced from distant locations, both inside the United States and abroad. And growing direct foreign investment in the Valley, especially by such Japanese giants as Fujitsu, seems aimed more at tapping into the region's exceptionally rich knowledge base in order to improve home-country performance (or to dampen anti-Japanese political sentiments in Washington) than at furthering the foreign investors' desire to join a local club of firms interested mainly in the economic development of Silicon Valley, per se.

Quite apart from having underappreciated the complexity of the relationships between the industrial districts and the outside world, writers on the subject have unaccountably ignored the extent to which these local economies depend crucially on a highly unequal organization of work—the dark side, once again. These admirers of the industrial districts in California, Italy, and elsewhere have focused their attention on the craft-based firms with the "good" jobs, while ignoring the low-wage, superexploitative elements of the network that help to keep the whole system going by suppressing labor costs. Once these elements are incorporated into the picture—once the district is understood to be only one segment of a geographically extensive network, only part of which is "flexibly specialized"—the ethical appeal of such a system becomes more problematic.

On every continent, the great majority of the good jobs within the districts themselves are held by men of the dominant color and ethnicity. Minorities, women, and immigrants are overwhelmingly treated as outsiders, consigned to jobs situated in the back rooms of these district's shops or outside these regions altogether, in the small factories and sweatshops that occupy the periphery of the geographically extensively production systems of which the districts, per se, constitute only the core. By drawing too narrow a box around the activity taking place solely *within* the districts, advocates are understating the degree of inequality among workers and between regions.

In Italy, a good example is Benetton, the maker of colorful clothing sold in spritely little franchise shops in seventy-nine countries—more than 300 shops in Japan, alone.[62] Most of the design and the high-end production work continue to be situated in or around Treviso, near Venice, where the firm was founded in the 1960s and where it is still headquartered. By contrast, nearly all of the labor-intensive assembly, pressing, and embroidery work is contracted out. A first tier of midsized firms perform R&D, design, or high-level manufacturing functions, collaborating closely with (but working for the most part on orders from) technicians and managers in the core corporation. In turn, these subcontractors are expected to manage successively lower order tiers of suppliers, situated within the Veneto region and farther away, in southern Italy, Turkey, and other low-wage areas.

These lower-tier suppliers are typically very small, highly specialized, and almost never unionized, and they are generally owned and run by small-town or rural men who employ a workforce consisting predominantly of women under the age of twenty-five. Labor costs in the lower-tier workshops are below the national average. Whether the national labor laws with respect to health and safety, minimum wages, paid vacations, and the like are observed depends mainly on whether the local political parties, the owners' confederations, and the unions enforce them. As for skills, managers expect newly hired workers to be able to carry out assigned tasks within, at most, a year of coming on the job. The pace of production in the small contract shops can be extremely intense.

Finally, at the lowest level within this interregional and obviously hierarchical production system stands the *home worker*: lacking skills (or, at any rate, power), receiving the lowest wages, and having no legislated health and safety protection. Such home work appears to be more prevalent in the south of Italy than in the more urbanized north, but even on this question, the visitor gets contradictory stories.

The inequalities are not quite so stark in Silicon Valley, but there are important underlying similarities in the labor process and in its geography. As early as the 1970s, it was becoming apparent that the workforce employed inside the semiconductor companies at the heart of the Silicon Valley economy was highly stratified. As Saxenian documented in her earliest published research,[63] at the top of the hierarchy are the highly educated, well-paid managers, engineers, and other professionals. At the same time (and often within the same factories and laboratories), nearly half of all workers in the Valley's high-tech companies perform production and maintenance tasks, four-fifths of which are officially classified as semiskilled or unskilled. Wages in these jobs are dramatically lower, and benefits often nonexistent.[64]

During the 1960s and 1970s, immigration into Santa Clara County reflected this stratification. Well-educated engineers and scientists moved into the western foothills near Stanford University, to be closer to their offices and labs, as well as to the more expensive luxury homes and amenities. At the same time, the industry's demand for production workers stimulated an equally large in-migration of unskilled, predominantly Mexican, Chicano, and Asian workers. These workers were shunted off to new residential areas far from the heart of the Valley, especially in and around the explosively growing city of San Jose.

Now, as has been shown by the University of California political scientist Richard Gordon, the San Jose urban planner Linda Kimball, and the UCLA professors Paul Ong, Allen Scott, and Michael Storper, there are whole neighborhoods of Los Angeles—hundreds of miles away from Santa Clara County—where both documented and undocumented workers perform unskilled and semiskilled assembly tasks, often at home, for contractors to the high-tech firms of Silicon Valley. In *those* neighborhoods, the quality of housing and public services is as far below that of the northern reaches of the Valley as one could possibly imagine.[65] These urban ghettos are as much a part of the famed "Silicon Valley production system" as are the engineering laboratories at Stanford, or the military R&D facilities within Lockheed's Missiles and Space Division in Santa Clara County.

Why Should We Care?
Rethinking National and Regional Economic Development in a World of Production Networks

There are at least five good reasons why policy makers and those who advise them should reexamine their beliefs about the myth that, at least in business, small is both beautiful and bountiful and, in turn, start paying more attention to the implications for economic development and social justice of the evolution of network forms of business organization.

One reason has to do with deciding who is to be eligible for participation in national industrial and technology partnerships in the years ahead. Second, to the extent that network forms of business exacerbate inequality and polarization, both long-run economic growth and a gradual improvement in the distribution of income are compromised. A third reason for concern is the inability of production networks, however lean and mean, to resolve the macroeconomic problem of balancing sectorial supply and demand at the global level. A fourth concern involves the fashioning of local and regional economic development policies in a world of lean production and business networks. Finally, the network form is wreaking havoc with traditional approaches to the regulation of business, so that a serious rethinking is called for. Let us explore these matters, one by one.

National industrial and technology policy planners are increasingly having to confront the uncomfortable fact that a growing number of "our" corporations are joining forces with "their" corporations, through all manner of technology partnerships, strategic alliances, cross-licensing deals, and other manifestations of the network form of business organization. How to even think about network forms of production has become a necessity to a national government determined to work actively with the private sector in promoting economic growth.

At the center of this matter stands the debate over what Robert Reich first posed as the question: "Who Is Us?"[66] Briefly, Reich argues that, for all practical purposes, U.S. multinational corporations no longer exhibit any important loyalty to any particular national government—let alone to any particular site or region. Indeed, with the offshore divisions of American companies eclipsing their domestic operations in terms of profitability in recent years, it is arguable that companies no longer want or need a "home base." Instead, Reich believes, they have become (or are in the process of becoming) entities unto themselves, dependent only on being assured of a continual supply of highly skilled, technically well-trained professionals and technicians, whom he calls "symbolic analysts."

But if business no longer has particular national loyalties, then why favor American companies over (say) the Japanese, when providing U.S. government assistance for training or for R&D? Why not offer public support for whoever's companies are interested in setting up shop on our shores?

The belief that the idea of a home base no longer holds much meaning for the big corporations is not shared by Laura D'Andrea Tyson, who chairs the President's Council of Economic Advisers. She observes that the lion's share of the output of U.S.-owned companies is still produced within our borders, and that the same is true of Japanese and German industry. Moreover, according to the logic of the "new trade theory," in which one company's (or sector's or region's) initial lead in international competition may be sustained over a long period of time, helping your own country's firms to get there first can have big payoffs.[67] In short, governments can help to develop the home base of their companies, and a stronger home base can promote a lasting competitive advantage for the companies that call that base home.

For some years, the Harvard Business School's strategy guru, Michael Porter, has been developing the case for why a home base of mutually supportive institutions is, if anything, even *more* important to companies today than in the past. For Porter, firms that succeed in global competition will be those that are able to access highly trained workers and other key inputs into the production process, such as infrastructure and finance. A sizable and growing home market (peppered with what Porter calls "sophisticated and demanding buyers") confers upon companies based there the economies of scale that give them a cost (and therefore price) advantage over foreign competitors. The availability of world-class suppliers helps a firm develop and maintain a competitive edge, as does a sharp, pervasive rivalry among both local competitors and suppliers, which pressures firms to invest and innovate, and to pursue the continual upgrading of the quality of labor and other inputs. Porter believes that the assembling and managing of these complex interrelationships among customers, competitors, suppliers, and the governments that help to train labor and build the requisite infrastructure are more easily accomplished within the same geographic and cultural context.

Who is right? It is true that the offshore operations of American firms are more profitable, on average, than their domestic operations. This may well be the result of superior productivity in those foreign operations. But that begs the question: From what arrangements does that comparatively more effective overseas performance arise? William Lazonick, an economic historian at the University of Massachusetts at Lowell, thinks the answer may be that overseas divisions of U.S. companies are deliberately being situated within the more highly developed, competitively efficient home bases of their foreign competitors! If IBM of Japan is doing better than IBM in the United States, or if Ford of Europe is a more efficient producer than its American parent, it may well be because these

overseas divisions have become deeply involved in collaborative manufacturing with local suppliers, customers, competitors, and governments in
the Far East and Europe, respectively. In other words, Porter's supportive
environments *are* being constructed, but by other governments and their
major firms, not here at home—and that matters.

There is a role for nation-based industrial policy, namely, to work with
business in erecting supportive industrial environments whose sheer
complexity makes it unlikely that they will come into existence, or be
maintained, solely by market forces. Moreover, such environments may
attract additional inward foreign direct investment, as well—bringing
with it new jobs, taxes, and technical know-how.[68] Which companies are
"ours," which are "theirs," and for whom such environments would be
"home" may be less important than getting on with the business of creating and nurturing the supportive environments in the first place.

This brings us to a second policy consideration. If living in a world of
lean and mean companies and their global networks of suppliers, strategic partners, and financiers is inescapable, then policy makers are sooner
or later going to have to come to grips with the dark side of flexibility—
if only to get themselves reelected. Because flexibility depends so fundamentally on the perpetuation of contingent work (that is, part-time,
part-year, temporary, and contract work), the shift toward network forms
of industrial organization promises to *strengthen*, not arrest, the politically volatile trend toward income polarization.

The class and associated wage structures that characterized nineteenth-century industrial capitalism could be depicted as a *pyramid*
with a narrow top and a wide base. The rapid growth of a wage-earning
middle class during the twentieth century (and especially in the years
following World War II) effectively transformed that distribution into
one with the shape of a *diamond*, featuring a small number of very rich
individuals at the top, a declining fraction of very poor people at the
bottom, and a burgeoning middle group.

But economists, sociologists, and journalists now almost universally
(if reluctantly) agree that since the 1970s the distribution of income has
been changing its shape again, becoming an *hourglass* with an expanding upper end of well-paid professionals (including Reich's symbolic
analysts), a growing mass of low-paid workers at the bottom, and a
shrinking middle class made up of downwardly mobile former factory
workers and middle managers made redundant by the philosophy of
lean production. Moreover, it appears that this unsettling trend toward a

polarization of earnings is occurring worldwide, albeit at varying rates. The consequences differ, also, since countries have such very different "safety nets" in place to prop up those who cannot make it in the labor market.[69]

Finding ways to maintain civilized labor and living standards in a world economy increasingly populated by forms of industrial organization that exacerbate such polarization will be no small feat. A few brave international organizations, such as the United Nations' Geneva-based International Labour Office (and the IILS, within it) have for some time been organizing meetings of delegates from different countries, companies, and unions, and issuing useful reports with ideas for coping with life in a world of networked production.[70] The international secretariats of the various trade union confederations, and any number of shop stewards' associations, are doing the same.

Their mission is, however, greatly undermined by the weakness of the unions—and of social democratic and other Left political parties, generally—within both the industrialized and the newly industrializing countries. This weakness is reinforced by the relative ease with which network forms of business organization can enter and exit different localities. These facts of life in the age of flexibility have even prompted some American trade unionists and labor economists—and a full-fledged presidential commission—to give some thought to how employees' collective rights and needs in the workplace of the future might be protected in a world increasingly devoid of labor unions as we have known them.[71]

Yet a third policy concern has to do with those global crisis tendencies that network forms of industrial organization are *not* capable of resolving. In particular, while networks and strategic alliances may have improved the coordination of global *production*, they can do nothing by themselves to stabilize and coordinate global *demand*. Worldwide, nations' central banks remain as disarticulated as ever. Private speculation in foreign currencies continues to undermine such experiments in multinational statecraft as the effort to build a United Europe. International trade negotiators continue to find themselves caught between the Scylla of the norm of free trade and the Charybdis of economic nationalism. And the creation of more and more contingent workers who will never be able to participate fully as consumers in the emerging global marketplace creates a permanent drag on world economic growth. As has been consistently argued by the economic journalist Robert Kuttner, without *some* sort of global demand management, no amount of institutional tinkering on the supply side will be sufficient to restore long-run global economic growth.[72]

Nor are the problems of balancing supply and demand as independent of one another as most contemporary popular treatments of international economic policy make it appear. Even the orthodox contemporaries of Keynes understood the inherent instability of an economic system in which every current expression of demand by businesses, in the form of investment in new plant and equipment, expands the economy's productive capacity, thereby requiring still greater increments in final demand to contain unemployment and to maximize efficient utilization of that capacity. An economy increasingly populated by network firms is not free of this contradiction.

Add innovation to the story, and balancing acts become even more difficult to pull off. Thus, the greater the expense to a firm of installing new capital equipment embodying the latest technologies, the greater the pressure on management to operate it around the clock or to otherwise increase volume or the diversity of products, in order to amortize the investment. But this puts even further pressure on the economy as a whole to create sufficient aggregate demand to absorb the additional output.[73] Again, the point is that, while they may facilitate the coordination of problems on the supply side of the market, and arguably accelerate the rate of innovation by making knowledge (like other specialized assets) more accessible to more players, network forms of business organization do nothing to solve the underlying macroeconomic problem of reconciling supply and demand. Indeed, by weaving together more and more locations all over the world, production and research networks probably reinforce a tendency toward global excess productive capacity, thereby making the balancing of aggregate supply and demand even more difficult than it already was.

And then there is the matter of trying to fashion local and regional economic development policies in a world of lean and mean networks of businesses—a fourth reason for wanting to better understand how the network form is evolving. Like many other writers on these subjects, I have an interest in business strategy that is motivated primarily by my desire to promote local economic development with equity— jobs with justice. Economic development planners—from local officials to presidents, members of national legislatures, and their advisers— must be able to make accurate guesses about the extent to which

global economic growth in the years ahead is likely to be shaped principally by the strategic decisions of the managers who govern networks
reflecting and wielding concentrated economic power, or by the small
craft-based enterprises, acting either as atomistic competitors or in
locally embedded industrial districts. The implications for national and
regional economic development policy are very different.

For example, new social contracts between subnational governments,
unions (where they exist or can be organized), and small businesses
seem both attractive and politically feasible—especially to the extent
that the small firms are federated in one form or another, making collective political negotiation at least potentially tractable. It is not surprising that experiments along these lines are presently being launched in
many places across the industrialized world, and that the hopes of many
local planners and politicians are riding on the outcomes. At last count,
at least a dozen states in the United States, alone, were engaged in such
activities.[74]

But if it is the big firms and their strategic partners that actually
dominate the economic development process, the role for locally initiated development policy itself becomes embedded into larger and
broader fields of play. Thanks to the pioneering writing of Richard Barnet, the co-founder of the Institute on Policy Studies, of the economists
Stephan Hymer and Raymond Vernon, and the more recent essays by
the economic geographer Gordon Clark, we have long been aware of
this dilemma—this fundamental imbalance between the mobility of
multilocational businesses and essentially local publics.[75] This contradiction constituted the core conception that underlay my own research
conducted in the early 1980s with Barry Bluestone. In fact, the working
title of the manuscript that became *The Deindustrialization of America*
was "Capital Versus Community."

As the big firms come increasingly to rely on licensing, joint venturing, and strategic partnering, rather than on direct investment in new
branch plants, the already intense competition among localities for new
private investment, especially but not only in high-tech sectors, will
surely be heightened. With each passing year, even home-grown companies nurtured by local economic development agencies will face the
growing likelihood of being drawn into the orbits of distant corporations
whose managers have the power to decide the long-term fortunes of the
no longer quite so "local" firm.

Clark has suggested that, rather than emphasize the creation of linkages among producers located *within* a region, localities might try to launch projects that deliberately seek to better connect their individual companies and activities to the new global firms and their networks (of course, these approaches are not mutually exclusive). Reich, Bluestone, and I also define "productive" local economic development policy (as contrasted with a tax- and wage-cutting race to the bottom, in which localities compete with one another by *reducing* their standard of living) as the building of stronger attractors for catching multilocational or networked capital. This is done by providing high-quality infrastructure— roads, bridges, waste disposal, telecommunications, transportation— and highly skilled labor. We and Clark agree that those places that are unable to make such connections, or that fall below the "scanning" thresholds of the vision of the global corporations, are likely to experience slower long-run economic growth or even stagnation in the future.

I have been emphasizing the importance of designing policies that more closely connect regions, districts, and even neighborhoods to the international economy. And I have identified crisis tendencies and contradictions within the existing industrial districts. But let me quickly add that none of this denies the potential significance of geographic clustering, especially for promoting the diffusion of innovative production methods to and among the small and medium-sized firms that systematically lag behind the larger enterprises and establishments in the extent to which they are adopting and using such new technologies as programmable factory automation.

Many innovative approaches to the development of clusters of simultaneously locally oriented and outwardly oriented firms are already under way. These typically have in common the public provision of shared overhead services and the facilitation of cooperation among the privately owned small firms themselves. From Bologna, Lyons, and London to New York City, Milwaukee, and Tokyo, such projects now appear with increasing frequency, generating much local enthusiasm. Fine and good. Nothing I have said here should be interpreted as denying the usefulness of local efforts toward building or nurturing production networks (including clusters of small firms) that might be collectively competitive where no (or few) individual businesses could be. Such a strategy can and should be part of a comprehensive policy mix.

But then an informed perspective on the forces that have undermined, or that threaten to undermine, such projects elsewhere becomes, if anything, even *more* important. International and interregional comparative research conducted over the past decade has yielded a library full of success stories, and not a little cheerleading. We owe those planners and policy makers who are responsible for managing the newest efforts a comparable glimpse of the *problems* and outright failures, as well.

The greatest danger, I think, is that an insufficiently critical focus on the "possibilities for prosperity" (to quote the subtitle of Piore and Sabel's book on the competitive advantages of small, flexible firms) through locally based development could distract us from the need to continually monitor and find new ways to socially regulate the behavior of the multinational enterprises and their alliances—yet a fifth policy concern. Obviously, we do not want to destroy the remarkable innovative capacity of production networks. But given how profoundly the networks and alliances are undermining traditional forms of regulation, anyway—by virtue of their transforming our concepts of everything from property rights to our shared understanding of what constitutes a working "career"[76]—we might as well begin to experiment deliberately with new regulatory arrangements.

How to relate the local to the global? How to encourage innovative management behavior without simultaneously adding to long-term unemployment and working poverty? How to regulate businesses whose organizational boundaries (and, as a consequence, legal liabilities) are becoming increasingly fuzzy? It is questions such as these—not how to further celebrate the romance of small firm–led economic growth and development—that constitute the critical issues for government, business, and labor policy in the closing years of the twentieth century.

PART II

REASSESSING THE IDEA THAT SMALL FIRMS ARE THE ECONOMIC DEVELOPMENT DRIVERS

2

THE MYTH OF SMALL FIRMS
AS JOB GENERATORS

The typical American business establishment has been getting smaller over the past two decades. In 1967, the average factory, office, or store employed 1,100 people. By 1985, that number had fallen to just 665. Such declines were common across all the industries and regions of the country. Japan and Western Europe have undergone a similar decline in the average size of the workplace, especially in the manufacturing sector.[1]

Particularly in the United States, had the late 1970s and the 1980s been a period of extensive retooling of older industry, with companies investing heavily in new technology, the simplest explanation for the "downsizing" of American business would have been that managers were substituting machines for people. In other words, while the largest firms (and plants) may have been declining in importance as measured by *employment*, they might not necessarily have declined in their relative contribution to national income, output, or profits. But in fact, especially in the United States, productivity and investment were stagnating badly throughout these years.[2] Some other explanation was needed.

And sure enough, an explanation *did* appear, at just the right time. Even as Ronald Reagan was settling into the American White House and Margaret Thatcher was well into her governance of England, policy makers here and abroad were being told that the number-one reason for the decline in the size of the average business was an extraordinary boom in the small firm sector of all economies. In short, the key to "job generation" now lay firmly in the hands of the small entrepreneur. Within an amazingly few years, everyone who was anyone "knew" that "small firms create most of the new jobs." Lawmakers in Washington, D.C., London, Paris, and Brussels were scrambling to rewrite international, national, and regional programs, laws, and regulations to favor the sector thought to be the most responsible for creating work for citizens who desperately needed employment. In the words of Dale Bumpers, the influential senator from Arkansas who chairs the Senate's Small Business Committee: "As you all know, small business creates about 80 percent of the jobs in this country."[3] The story line was simple and straightforward: Small was bountiful. Small was beautiful. Small was *in*.

Only it was not—and is not—true.

ALTERNATIVE EXPLANATIONS FOR THE APPARENTLY GROWING IMPORTANCE OF SMALL FIRMS

In fact, entrepreneurship is not the only (let alone the most important) explanation for downsizing—even in theory. Thus, for example, case study research into the behavior of large corporations revealed the widespread practice after the late 1960s of so-called productive decentralization: that is, the hiving off or externalization of activities that had formerly been performed inside the big firms themselves. In Europe, in companies such as Fiat and Olivetti, and in the United States, in ones such as General Motors and General Electric, managers seemed determined to reduce their dependence on militant unionized workers (these were years of major strikes and general labor unrest throughout much of the industrialized world). Some of the laid-off skilled workers eventually started up their own small businesses and became suppliers to the very corporations that had originally discarded them.[4]

All of these studies called attention to the strategy being followed by a number of large, multiplant firms to gradually run down capacity in their older, unionized "brownfield" facilities while shifting work to their newer "greenfield"—and generally nonunion—operations, both domestic and foreign. These shifts were subsequently confirmed in detailed case studies of specific U.S. multiplant corporations operating both union and nonunion facilities.[5] "Vertical disintegration" of the kind first observed in Europe—the practice of large companies contracting to purchase goods and services, formerly produced in-house, from generally small firms whose owners used to (and, in some cases, continued to periodically come back to) work for the big firms—was documented in such disparate American economic sectors as aircraft manufacture, machine tool building, computers, and the production of motion pictures for theaters and television.[6] All of these developments added to the widespread perception that smaller units of production were coming to account for a growing share of economic activity.

Vertical disintegration represents a policy of a going concern increasing its potential profitability by changing *where* it makes its products. Another dramatically different explanation for declining average firm size was that, battered by three recessions from 1970 through 1982, the largest companies, plants, and divisions were simply shutting down in record numbers, with the result that the American economy as a whole (following the experience of the United Kingdom) was effectively deindustrializing. As a simple matter of empirical fact, there is no question that just such a "shake-out" did occur, although the extent of the job losses and the implications for U.S. competitiveness remain hotly contested. For the United States, Barry Bluestone and I estimated that perhaps 30 million jobs were lost during the 1970s to closures of business establishments across all sectors of the economy (not just in manufacturing).[7] Subsequent research based on the U.S. Bureau of Labor Statistics' dislocated worker surveys, beginning in 1979, converged on the estimate that an average of between 1.5 million and 2 million jobs were lost to closures in the United States in each year of the 1980s. As we now know, layoffs of even managers and other middle-class white-collar employees became a regular part of the American business landscape throughout the 1990–91 recession and have continued during the subsequent recovery.[8]

Of course, a long-run shift of economic activity from sectors with high average firm or plant sizes—such as durable manufacturing, with its giant steel mills and auto and aircraft assembly plants—to sectors with typically smaller companies or establishments, particularly within the service sector, could also have been responsible for the observed changes in the relative importance of smaller units of production. In other words, even if *nothing* shut down and *no* company "disintegrated," a gradual growth in the proportion of all jobs belonging to industries with smaller typical establishments would lead to the observed reduction in the national average firm and plant size.

Still another possibility is that the largest companies have indeed been deliberately reducing their size and the scope of their activities, but not for the reasons offered by the theorists of deindustrialization. In the business press, this sort of strategic gambit is called "reverting to core competencies" or pursuing "lean production."[9] Shutdowns or divestitures of segments of the firm now deemed peripheral to the company's core operations are motivated not so much by a reaction against labor unions, high wages, or the congestion and crime problems of brownfield locations (such as older big cities) as by a systematic strategic analysis that there are weak (or no) synergies among broadly diversified activities. Thus, investment capital is withdrawn from the least related pursuits. As a result, the big firms—especially those that had for many years actively pursued a conglomerate strategy—were indeed downsizing, but the explanation was not vertical disintegration, union avoidance, or the search for a "good business climate," per se.

And then there is the possibility that average establishment (and possibly enterprise, or firm) size has been falling for the reasons offered by the theorists of the entrepreneurial revival and of the competitive superiority of small firms. Perhaps there *has* been a burst of small business start-ups, in places as otherwise disparate as Japan, northern Italy, and California, fueled by private venture capitalists and public regional development authorities, and that this growth has been so substantial— and these small firms *so* very successful—as to overtake the lethargic big firms on three continents.

Thus, at least *five* plausible explanations exist for the presumed changes in the relative importance of large and small companies and establishments as sources of job creation:

- vertical disintegration of the big firms, to escape unions, high wages, and "bad business climates";
- closures concentrated among the largest companies and units;
- a secular shift from manufacturing (with its generally larger facilities) to services (with its generally smaller facilities);
- the strategic downsizing of large conglomerates as part of a retreat into core competencies; and
- a genuine disproportionate growth in the activity of small firms.

All five explanations could have contributed to the observed decline in the average size of business operations across the industrialized world, and to increases (if in fact they occurred) in the share of employment accounted for by small enterprises and plants. In reality, research conducted on recent developments worldwide has uncovered plenty of evidence of vertical disintegration, closures, the manufacturing-to-services shift, and the strategic retreat into core competencies.

But what about that fifth possibility—an explosive growth of genuinely small business? Let's take a closer look.

THE CONVENTIONAL WISDOM, REVISITED

I have shown that at least five very different developments might each have produced the observed decline in the size of the average North American, Japanese, and European business. In fact, the work that first brought the whole subject to national attention in the Anglo-American world more than a decade ago contained no such suggestion of or concern about underlying causes. This was the work of David Birch, the person most responsible for promoting the thesis that economic growth is now *led* by very small firms.

Birch, a research associate affiliated first with Harvard University and later with M.I.T., was the first policy analyst to exploit the rich fund of information contained in the credit reports on the great majority of American businesses assembled by the Dun and Bradstreet Corporation (D&B). In papers first distributed privately in the late 1970s[10] (but not published until much later),[11] Birch tabulated employment counts from D&B on individual establishment births, deaths, growth, and decline over three-year intervals, beginning in 1969. He also constructed "family trees" linking individual establishments to their parent firms.

Out of these efforts, Birch claimed to have demonstrated that, over three-year intervals, small companies, plants, stores, and offices were responsible for the vast majority of new jobs created in the United States. (Birch's definition of *small* varies from one consulting report and magazine interview to another; the cutoff for which he has been most often cited is twenty employees or fewer.) These numbers were immediately picked up, embellished, celebrated, and advertised to a much wider audience by the popular conservative writer George Gilder.[12] Across the Atlantic, the prestigious Paris-based Organization for European Cooperation and Development (OECD) both publicized Birch's findings about the United States and generated roughly similar stories for Europe, using similar methods.[13] Birch finally decided to publish a comprehensive statement of his views (although not of his methods) in a 1987 book, drawing on data from D&B records for 1981 through

1985.[14] Here, he concluded that fully 88 percent of all net job creation during this period could be attributed to the activities of companies with fewer than twenty employees.[15]

One of Birch's most dramatic and influential conclusions was that it seemed to be the prolific *birth rates* of new start-up businesses in all sectors and in all regions of the United States that were responsible for the most job generation. Since closure rates were found to be remarkably similar across regions, the secret to regional growth success lay in how many *new* businesses a place could produce. The policy implications were obvious: development planners should do everything they can to promote the start-up of as many new small businesses as possible. The closures will gernerally be beyond their control.

REASSESSING THE EVIDENCE ON SMALL FIRM
SHARES OF JOB CREATION AND EMPLOYMENT GROWTH

Birch has many valiant defenders who have an abiding interest in the celebration of entrepreneurship. For example, John Case, an excellent and fair-minded editor of *Inc.* magazine, has gone to great lengths to support Birch's basic conclusions.[16]

Nevertheless, those conclusions have been sharply called into question by a number of thoughtful critics at the Brookings Institution, in the media (especially at the *Wall Street Journal*, of all places!), and ultimately within D&B itself.[17] Nor has Birch's curiously cavalier attitude toward these criticisms shed further light on the question, or sustained confidence in his findings. For example, in an interview with the *Wall Street Journal*, he once remarked that his earlier claims were a "silly number," that he could "change that number at will by changing the starting point or the [time] interval," and that "anybody can make it come out any way they want."[18]

There are, to be sure, serious technical limitations inherent in these D&B data (which, to be fair to the company, were never intended to be used to closely monitor business births and deaths, but rather to provide credit reports to firms requesting them). Birch's own secrecy about his methods and his constantly changing announcements about what the "right" numbers are do not help. Fortunately, we have other evidence and a fresh crop of researchers on whom to draw for another look at the question.

One of the most valuable of the new studies on job generation covers the performance of small firms in nine different countries: France, Italy, the United Kingdom, West Germany, Japan, Norway, Hungary, and Switzerland, as well as the United States. These studies were conducted through the auspices of the International Institute for Labour Studies (IILS), a research group housed within the International Labour Office in Geneva. The project was coordinated by Michael Piore, an M.I.T. economist, and Werner Sengenberger, the IILS director and a German economist. Sengenberger and Gary Loveman, a Harvard Business School professor and former student of Piore, drew from the individual country reports to summarize the basic comparative statistical trends in the employment size distributions of firms and establishments.[19]

In an overview paper summarizing all of the studies, Loveman and Sengenberger took the official OECD cutoff of 100 employees as their operational definition of what constitutes being "small" (to facilitate international comparisons, this is the definition that I, too, employ throughout this book, wherever possible). They interpreted the accumulated evidence as indicating that average firm size *has* been shrinking since the 1970s, as has average plant size, especially in manufacturing, and that the average number of production sites per firm and the number of legally distinct firms are growing in most countries and most sectors.

Moving from data description to interpretation, they then claimed that the share of jobs in companies and plants employing fewer than 100 workers has undergone a V-shaped pattern over time, declining in the years after World War II (the period of consolidation of the large, vertically integrated, bureaucratically managed corporation) and then trending upward in the 1970s and 1980s. After examining several alternative plausible theoretical explanations, the authors opted for "flexible specialization," the idea (championed by Piore and his M.I.T. colleague Charles Sabel) that a growing uncertainty about global demand for new products and processes and an increasing heterogeneity in tastes are conferring particular competitive advantages on small-scale firms that can quickly adapt themselves to different market niches and flexibly shift their specialized roles within often localized production networks of like-minded enterprises.[20]

Whatever the validity, generalizability, and predictive power of this theory of flexible specialization (about which I say more in later chapters), we must first establish whether and to what extent the phenomenon it was intended to explain is in fact occurring! I think that Loveman and Sengenberger have exaggerated the magnitude of this

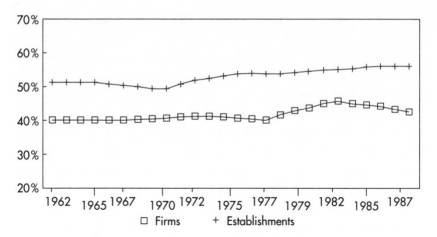

FIGURE 2.1   SHARE OF ALL JOBS IN BUSINESSES WITH
FEWER THAN 100 EMPLOYEES, UNITED STATES

Note: I have used straight-line interpolation to complete both time series.

Sources: Gary Loveman and Werner Sengenberger, "Introduction: Economic and Social Reorganization in the Small and Medium-Sized Enterprise Sector," in *The Re-Emergence of Small Enterprises: Industrial Restructuring in Industrialized Countries,* ed. Werner Sengenberger, Gary Loveman, and Michael J. Piore (Geneva: International Institute for Labour Studies, International Labour Office, 1990), table 2.8; and for 1987: U.S. Bureau of the Census, *1987 Enterprise Statistics: Company Summary,* bulletin ES87-3 (Washington, D.C.: Government Printing Office, June 1991), table 2, and U.S. Bureau of the Census, "County Business Patterns, 1987," unpublished worksheets.

turnaround in the data. Indeed, for the biggest countries, which also constitute the United States' most important foreign competitors, I find it impossible to spy any uptick at all.

Consider the United States. When we update Loveman and Sengenberger's time series for this country with new data from the most recently published (1987) enterprise statistics from the U.S. Census of Manufacturers (information that was not available to the IILS project at the time it began to publish), it appears that the long-run share of small firms has barely changed since at least the late 1950s. The trends for the United States are displayed in figures 2.1 and 2.2. Loveman and Sengenberger's data series terminated in 1982, the trough of the deepest recession since the Great Depression. As they themselves acknowledge,

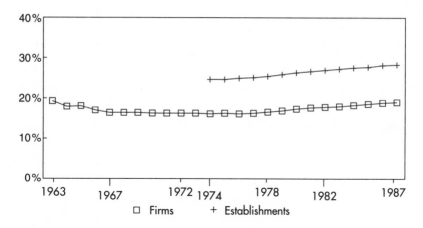

FIGURE 2.2   SHARE OF MANUFACTURING JOBS
IN BUSINESSES WITH FEWER THAN 100 EMPLOYEES,
UNITED STATES

Sources: Gary Loveman and Werner Sengenberger, "Introduction: Economic
and Social Reorganization in The Small and Medium-Sized Enterprise Sector," in *The Re-Emergence of Small Enterprises: Industrial Restructuring in
Industrialized Countries,* ed. Werner Sengenberger, Gary Loveman, and
Michael J. Piore (Geneva: International Institute for Labour Studies, International Labour Office, 1990), tables 4 and 11 (with straight-line interpolation);
U.S. Bureau of the Census, *1987 Enterprise Statistics: Company Summary,*
bulletin ES87-3 (Washington, D.C.: Government Printing Office, June
1991), table 2; and U.S. Bureau of the Census, "County Business Patterns,
1987," unpublished worksheets.

the employing behavior of large firms tends to be procyclical, meaning
that, during national downturns in the economy, it is the big firms that
lay off the largest number of people. As for outright shutdowns, one
General Electric, Prudential Insurance, or Ford Motor Company closing
can cost more jobs than the deaths of thousands of small businesses.
The only way to be sure that this does or does not mean a permanent
downsizing of the big firms or growth of the small ones is to wait for the
economy to turn around.[21]

Figure 2.1 shows that, after accounting for the U.S. recovery after 1982,
the proportion of Americans working for small companies and for individual establishments (particular factories, stores, warehouses, or offices) has
barely changed at all, since at least the early 1960s. Over the past quarter
of a century, there may have been some small upward trend in the

number of individual establishments with fewer than 100 employees. For entire firms, there is no trend at all.

Now consider only the manufacturing sector. Figure 2.2 also indicates a small recent upward trend in the share of all workers employed within small establishments (individual plants). But as with the economy as a whole, if any trend in the employment share of whole firms exists, it is barely discernible. And in any case, recall that there are a multitude of alternative explanations for what scant upward trends have occurred: from deindustrialization, strategic downsizing by the big corporations, and vertical disintegration so that managers could externalize the procurement of high-cost or highly specialized resources, to financial restructuring that has forced many companies to retire part of their past indebtedness by divesting themselves of even productive assets—the business equivalent of selling off the family jewels in order to pay the mortgage.

As for developments abroad, when it comes to European and Japanese data, we have usable information only for manufacturing, and only for selected years. I have graphically analyzed these data elsewhere in great detail.[22] Here, I will only report that, for Japan, the story for the years from the end of World War II until the late 1960s or early 1970s was one of extraordinary growth of the big firms, reflected in the long and rather steep decline in the share of Japanese employed by companies with fewer than 100 workers. Some turnaround occurred during the 1970s, but the trend reversed itself again between 1980 and 1983. When we consider the record of the United States' other major international competitor, the former West Germany, there are no trends at all. The share of small firms and plants in total manufacturing employment appears not to have changed since at least the early 1960s. Only in the United Kingdom has the period since 1970 witnessed a significant upward trend in the relative importance of small manufacturing companies and factories, as large U.K. manufacturers continue to wither or disappear. But again, recall that there are many possible explanations for such developments, of which a renaissance of small business start-ups is only one.

Loveman and Sengenberger admit that their estimates of the rising shares of small firm and plant employment (a "rise" that I challenge) may systematically exaggerate the relative importance of smaller independent units of production. The reason is that large companies typically have many legally independent subsidiaries: generally, smaller businesses that are substantially (if not entirely) dependent on them for finance, contracts, and business and political goodwill. I reported in chapter 1 on statistics from Germany, showing this sort of de facto dependence of small suppliers on their big customer firms to be a growing trend.

In other words, in many cases the legally independent small firms from which the big companies purchase parts, components, and services may not be all that independent, after all, but should rather be treated as de facto branch plants belonging to the big firms. *Production* may be decentralized into a wider and more geographically far-flung number of work sites, but *power, finance,* and *control* remain concentrated in the hands of the managers of the largest companies in the global economy. In chapter 6, we take a closer look at this emerging world system of what I call "concentration without centralization."

Finally—and here, I return to the data on the United States—statistics that emphasize how most companies employ small numbers of workers provide only half the story. The other, all-too-neglected half concerns how many people in the country are employed by the (small number of) big firms. Thus, across all sectors, it is true that only 1 percent of all companies operating within the United States in 1987 had 500 or more employees. But this tiny group of big firms employed 41 percent of all private sector workers. Within the manufacturing sector, the disproportion was even more lopsided. The 1 percent of all manufacturing companies with 500 or more workers accounted for fully 70 percent of all people working in manufacturing jobs.[23]

SMALL FIRM DYNAMICS

Let us now return to Birch's oft-cited figures on job generation. Over the last several years, researchers have subjected Birch's claims to rigorous scrutiny and have found flaws in both his methods (insofar as they can be inferred from his writing) and his interpretations. But most seriously, until his most recent media interviews, Birch missed perhaps the greatest lesson from his own data: that the lion's share of job creation over time is contributed by a tiny fraction of new firms. That is, hardly any of those small start-ups ever grow at all!

To be fair to Birch, the early D&B data he was using included many spurious "births" and "deaths," due to the recoding of records when the ownership of a firm, plant, or division changed hands. In a period such as the last two decades, when changes in ownership have proliferated enormously through mergers and acquisitions, a data set with such a flaw will systematically exaggerate the incidence of start-ups—precisely the part of the story on which most of Birch's argument depends.

But subsequent careful analyses of *cohorts* of companies have since been conducted, first by a University of California at Berkeley city planning professor, Michael Teitz, and his students;[24] later by the U.S. Small Business Administration; and eventually by D&B economists, using a file that was systematically corrected for spurious births and deaths. All of these studies revealed that the great majority of jobs created by any cohort over time are attributable to the expansion of a handful of firms.

For example, according to D&B's own inquiry, among the 245,000 businesses that were started up in the United States in 1985 (in all sectors of the economy, not just manufacturing), 75 percent of the employment gains made by 1988 occurred in just 735 fast-growing firms, a group that constituted three-tenths of one percent of the cohort. Moreover, these fast-growing firms were all *born* employing more than a hundred workers.[25]

A related argument has been contributed by another Berkeley professor, the economist Jonathan Leonard.[26] He shows how short-term gains and losses in employment shares by classes of small and large firms or plants could be nothing more than a case of "regression to the mean." The idea is that, to the extent that producers have a kind of long-run equilibrium level (or trend rate of growth) of employment from which, at any point in time, they may have departed for one reason or another, over time they may be expected to return to that equilibrium level (or trajectory). In periods of rapid structural change such as the late 1970s and early 1980s, the perturbations are likely to have been even more pronounced than usual for established firms, while the many new start-ups had nowhere to go but up (or out) and the largest mature facilities were most likely to be cut back (or closed). This is actually consistent with Birch's own emphasis in his more recent writing on the churning that takes place in the birth-death-expansion-contraction process. But its implication is that short-period estimates of such a dynamic process will *always* seem to indicate that the smallest organizations are the ones that are growing the most, while the largest seem to be shrinking the most.

There is yet another methodological problem with the way in which Birch and the Small Business Administration use these data. For each three-year period of analysis, they assign each establishment to a size category for the *base year* of the period. Suppose that an initially very small firm with fewer than 20 employees does so well that, by the end of the period, it has been able to hire five hundred additional workers (Birch likes to call such fast-movers "gazelles"; everyone's favorite historical example is Lotus, the famous spreadsheet software company). This obviously attractive development gets treated in their accounting scheme as job generation allocated entirely to the *small firm* (less than 20) sector—in other words, the small firm sector added 500 jobs—since that's where the firm was situated when it was first assigned a size. But this is confusing, since what we are observing is not growth within a stratum, but rather a transformation from small to big, *across* size strata.

Now, suppose a large company with base period employment of (say) 1,000 falls apart, sells off its inventories and machinery, and by the end of the three-year period is down to 18 employees, most of them guarding the premises against vandalism. In Birch's scheme, all of these losses are attributed to the big firm sector. That in itself is not incorrect. But when you put the two examples together, it is easily seen that this methodology systematically inflates the relative importance of the little guys.[27]

Still another cut at the problem starts by differentiating single plant firms from the more organizationally complex units belonging to multi-plant companies (a distinction that the Carnegie Mellon University professor of management and technology Maryellen Kelley and I have discovered to be an enormously significant predictor of differences among business establishments in how, and how much, they contract out work, and whether and with what success they adopt labor-management problem-solving committees).[28] Thus, the former Census Bureau economist Timothy Dunne and two of his colleagues have conducted rigorous econometric modeling of the growth patterns of these two different categories of facilities in the U.S. manufacturing sector.[29] They have shown that, for the period from 1969 through 1982, the plants of large multiplant firms had both lower failure rates and higher employment growth rates than did smaller or less complex businesses. Failure rates were especially high for the smallest plants. Moreover, changes of name in a company showing up in the Census Bureau's new longitudinal data file were predominantly associated with the incorporation of what had been a stand-alone firm into a generally larger multiunit company. In other words, on balance, spin-offs were less prominent during these years than were acquisitions. (Since 1982, Kelley's and my data show a reversal of this trend, at least in the metalworking sector, which accounts for about one-quarter of all U.S. manufacturing jobs.)

The European research from the early 1980s is also being reassessed.

Methodological criticisms similar to those appearing in the American literature have been published in British journals, monographs, and research reports for some time, especially by David Storey and his colleagues at the University of Warwick. Their results, some of which were discussed in chapter 1, strongly resemble the most recent American results. Moreover, Storey and his colleagues have shown that during the worldwide recession of the early 1980s, the OECD countries with high proportions of their manufacturing firms employing fewer than 100 workers did not systematically experience lower economywide unemployment rates or more rapid rates of manufacturing sector growth.

Again looking toward Europe, the German organizational theorists Josef Bruderl and Rudolf Schussler recently conducted a cohort survival analysis of a January 1980 sample of West German firms, tracked month by month through March 1989, and concluded that "the more employees a firm has [at any point in time], the longer it will survive." To be more precise, it is in what Bruderl and Schussler call the "adolescent" years of a business organization that the "hazard" of firm failure reaches its maximum. Companies seem to be most vulnerable when they are just starting out. What organizational sociologists call the "liability of newness" accurately describes the American experience, as well. In short, most small business start-ups fail—and quickly.[30]

WAGES AND WORKING CONDITIONS IN SMALLER FIRMS

As discussed in chapter 1, policy research comparing wages, benefits, and working conditions between small and large firms in the United States was pioneered in the late 1970s by the economist David M. Gordon and was developed more fully a decade later by the economists Charles Brown, James Hamilton, and James Medoff.[31] They and others who have looked at the question conclude without exception that, after accounting for interindustry (and, in some studies, even intercompany) differences in the availability of capital equipment, education, working conditions, age, and union status, the employees of the big firms enjoy on average systematically higher wages, better benefits, and greater job security than their counterparts in small firms. These findings are replicated in a detailed study on the Canadian economy, conducted by Statistics Canada, the official agency of the federal government.[32]

At the IILS in Geneva, Loveman and Sengenberger undertook an especially close comparison of differences in wages and employee benefits, by firm and plant size, between the United States and Japan. Average wages in the very smallest workplaces fall short of those in the very largest in both countries. The big firm–small firm employee benefits gap is even wider.[33]

Thus, quite apart from the ambiguity of the statistical story about small firm job generation itself, there is cause to worry that in the abscence of policies to change these conditoins, an economic development program that *did* target small firms for special treatment could, in the absence of other regulations or quid pro quos, actually contribute to a worsening of the national average standard of living. Such is the danger of too formulaic an approach to policy making, based on a misguided faith in small business, per se.

THE MAKING OF A MYTH

Let us take stock. Over the past quarter century, there has been no upward trend whatsoever in the small firm share of employment in either Japan or West Germany—the two most successful national economies in the world. In the United States, a (modestly) growing relative share of smaller units of production in recent years is discernible in the census data only for the manufacturing sector, and only among individual plants, not for entire firms. Among the leading international trading countries, only in the United Kingdom is there uniform and unambiguous evidence that small is becoming increasingly bountiful, for both individual plants and whole companies. Knowing what we do about the long, sad history of British deindustrialization, and how a collapse of big companies and the closure of large factories can make small business look relatively more important than it really is, these numbers do not necessarily offer a propitious sign for that beleaguered island.[34]

Moreover, during any short period of time, the vast majority of new jobs are created by a handful of successful firms. Larger companies have better chances of surviving over longer periods of time than do smaller ones. And the largest plants of the most complex multiplant companies display the fastest employment growth rates over time, at least in U.S. manufacturing.

So, small firms may not be the dominant job generators that the first generation of researchers—especially Birch—imagined to be the case. The rather extravagant claims made over the last decade about small firm–led economic growth were at best exaggerated, and at worst misleading.

And in any case, even if smaller units of production—factories, offices, stores, warehouses, whole corporations—*were* coming to dominate economic growth and employment creation in the industrialized countries, no one has ever found a way to distinguish among the many different explanations for why the size distribution of economic activity might have changed over time (if there was ever a prime example of unfinished business for policy researchers, this is it). Indeed, few decision makers even recognize that there *are* all these possible explanations.

Finally, we now know that small firms tend systematically to pay lower wages, to offer fewer and less ample benefits, to provide less job stability, and to be more likely to undermine the health and safety of their employees and the quality of their physical environments. The research on such factors is more complete for the United States than for Europe or the Far East, but virtually all reputable scholars believe that this generalization probably holds worldwide.

In chapter 1, I suggested a number of reasons why I thought the myth of small firm dominance might have become so persuasive to so many people in so short a time, including wishful thinking. But whatever the explanation, it is certain that policy makers across the industrialized world need to be reeducated on this subject. It is going to be a hard task, indeed, to convince them that, as Sengenberger concludes, "a substantial part of the restructuration that has taken place [in this era], including the expansion of the small business sector, has evolved under the control of larger firms, thus raising doubts about claims for the viability of independent, innovative and dynamic small enterprises, whose rise is due to their superiority over large corporations."[35]

In short, small, per se, is neither unusually bountiful nor especially beautiful, at least when it comes to job creation in the age of flexibility.

3

ARE SMALL FIRMS THE
TECHNOLOGY LEADERS?

How firm size is related to the ability and propensity to innovate is one of the oldest questions in political economy. During the last decade, popular writers including David Birch and George Gilder offered a portrait of the small firm as the wellspring of creativity and technical innovativeness.[1] Respected scholars such as Giovanni Dosi, Michael Piore, and Charles Sabel called particular attention to the advent of certain technologies, such as computer-controlled factory automation, as tending to systematically privilege small firms, thereby contributing to a shift toward smaller units of production.[2] Taken together, these ideas have greatly influenced economic development practitioners throughout Europe and North America, who in turn, are encouraging their policy makers to promote small firm development as a key part of any strategy for achieving local and even national economic growth.

This was not always conventional wisdom. In the 1950s, John Kenneth Galbraith was generally greeted by an affirmative nodding of heads when he wrote, with his typical light-handed sarcasm: "There is no more pleasant fiction than that technical change is the product of the matchless ingenuity of the small man [sic] forced by competition to employ his wits to better his neighbor. Unhappily, it is a fiction."[3]

Even earlier, during the 1930s and 1940s, another Harvard economist, Joseph Schumpeter, offered a series of pathbreaking observations about the sources of long-run economic growth. But where Galbraith has sustained a single position on the role of big firms over his entire career, we have *two* Schumpeters. The Depression-era economist is the philosopher of those much-quoted "gales of creative destruction," whose instrument is the innovative entrepreneur pulling the rug out from under entrenched oligopolistic power. But then there is the wartime Schumpeter, who reluctantly and unhappily saw an economy's largest corporations as having both the greater motivation for investing in innovative activity and the deeper pockets with which to finance it.[4]

Most contemporary students of industrial organization and the economics of technological change would agree with Harvard's F. M. Scherer that any effort to try to identify a "best" size of firm from the point of view of innovative capability is bound to be a waste of time.[5] In modern economies, innovation stems from all manner, sizes, and shapes of business enterprise. The economist Sidney Winter's hypothesis that big and small firms probably pursue qualitatively different innovation strategies resonates well with this era's students of the subject.[6]

More recently, Wesley Cohen and Steven Klepper, two Carnegie Mellon University economists, have posited a subtle tradeoff between the advantages to society from having an industrial structure populated by many diverse sizes of firms, on the one hand, and a more concentrated structure with a relatively few large companies, on the other. Cohen and Klepper reason that an industry containing many competitors will foster a multitude of approaches to innovation, thereby increasing the chance that at least *some* change will be forthcoming. On the other hand, because information does not flow perfectly or costlessly, and given that large firms have greater resources with which to acquire it (and generally greater opportunities to make use of it), the big firms have an advantage in appropriating the returns from innovation and are therefore likely to invest greater effort in those innovative projects they *do* choose to pursue.[7]

The upshot is that there is a tradeoff between the greater number of innovative projects likely to be attempted by the firms within an industry populated by many firms, and the fewer but more intensely pursued projects generated by the companies within an oligopolistically organized industry. Since industry structure affects the outcome, it is impossible to say, a priori, that one size class of firm—or form of industry—is unambiguously more socially desirable than another, in terms of promoting, or of being responsive to, changes in technology.

And yet, especially in the policy realm, we continue to encounter the widespread assertion that small firms really *have* become the most important sources of innovative activity and should be treated accordingly by policy makers. Even such respected scholars as the American economists Zoltan Acs and David Audretsch, who acknowledge that small and large firms may have different relative innovative advantages under different market conditions, choose to emphasize the idea that changing conditions—especially the diffusion of computer-programmable automation— now systematically favor smaller companies.[8]

Just as with the myth that small firms now create most of the jobs, so with this insistence on the special technological dynamism of the small firm sector. As I show in this chapter, little empirical evidence exists that small firms generate more product innovations than do large firms, and the available evidence is technically suspect. Moreover, there is a prodigious body of evidence from Europe, North America, and Japan showing that even the most "flexible" programmable factory automation is now being adopted to a far greater extent by, and has penetrated more deeply within, larger establishments and enterprises than by and within smaller ones. This is true even in the leading-edge high-tech manufacturing industries, where sales and jobs are even more disproportionately the province of the biggest firms than is the case for the rest of the American economy.

The new conventional wisdom about the technological superiority of small firms has it wrong in other ways, as well. Thus, it is easy to point to any number of highly innovative industries that are becoming dominated by big companies, and to many high-tech firms that are being absorbed into information, financial, and production networks with powerful lead firms, major hospitals, universities, and government agencies at their nodes. The movement toward modularized computer software offers an example of big company domination, especially dramatic in the form in which it is being pursued by the leading Japanese corporations—the "software factories." Biotech offers an example of the network absorption of high-tech firms. I will present cases on both.

To repeat: it is not that small firms are incapable of innovative activity. In many sectors—including those just named, and others that are explored throughout this book—small firms populated by highly skilled technicians and professionals play a vital role. But in the study of economic development, as in anything else, one wants to distinguish the forest from the trees. For all the currently fashionable talk about industrial capitalism's radical break with its past, certain organizing principles continue to be operative—even in the age of flexibility. Among these principles are the necessity for the creation of technical standards, so that risky new innovation-embodying investments will be worth making; firms' need to have access to the resources with which to finance, research, market test, and advertise the commercialization of new products and processes; and the competitive pressure to economize by reusing the products of previous efforts, rather than starting every new project from scratch.

The importance of standards, the desire to commercialize inventive ideas, and the pressure for reusability (which reinforces the engineering of interchangeable parts) are just some of the major underlying logics that continue to drive capitalist economic development. In some cases, these principles reinforce a classic tendency toward concentration and bigness. In other settings—especially when information and knowledge are widely dispersed—the organizational outcome will be networks of producers. What we can say with some certainty is that these principles surely do *not* systematically privilege small business, per se.

PRODUCT INNOVATION AND SMALL FIRMS

The most serious effort to empirically measure the relative importance of small firms to the innovation process in the United States is by Acs and Audretsch. For almost a decade, in books and papers published in both America and Europe, they have attempted to identify the sources of the "rate of innovation" across industries, isolating the particular role of small firms. My quarrel is not with their interesting multidimensional analysis of the market conditions under which innovations are more likely to emerge from small or large firms (for example, their finding that highly concentrated industries tend to be less innovative). Rather, I am worried that their underlying measure of innovation, taken from a private consulting firm working on contract to the Reagan administration, is sufficiently suspect as to raise the question of whether we can say anything at all—even descriptively—about the incidence of small firm innovation.

In 1982, the Futures Group delivered a consulting report to the U.S. Small Business Administration (SBA), consisting of the results of scanning more than a hundred trade publications, engineering magazines, and advertisements for announcements of new developments, mostly on product rather than on process innovations.[9] The consultants were able to identify 4,476 innovations associated with manufacturing industries. A random sample of 600 firms (with 375 responses) were successfully interviewed, in order to determine the size of the parent enterprise with which the innovation was associated. Large firms were designated as those with 500 or more employees; small firms were those with fewer than 500 employees. Innovations were differentiated by whether they "established whole new categories," were "first of their type on the market in an existing category," constituted a "significant improvement in an existing technology," or "modestly improved or updated an existing product." According to the consultants, 85 percent of the "large firm innovations" and 88 percent of the "small firm innovations" in the data base were of the last variety: modestly improving an existing product.[10]

From these data, Acs and Audretsch wanted a measure of the rate of innovation for 282 manufacturing industries (as well as for a smaller number of nonmanufacturing industries). Since the consultants estimated that innovations announced in 1982 had been invented some four to five years earlier, on average, Acs and Audretsch constructed their indicator as the number of innovations appearing in each industry in 1982 per 1,000 employees as of 1977. For manufacturing as a whole, they found that "while large firms in manufacturing introduced 2,445 innovations in 1982, and small firms contributed slightly fewer, 1,954, small-firm employment was only about half as great as large-firm employment [in 1977], yielding an average small-firm innovation rate in manufacturing of 0.309, compared to a large-firm innovation rate of 0.202."[11] It is on the basis of this calculation (which in turn rests crucially on the consultants' ability to assign each innovation to a particular firm of known size) that Acs and Audretsch offered their by now widely cited conclusion that small firms are, on average, more innovative than large firms.[12]

The authors are certainly correct that policy makers need measures of the *outcome* of the innovation process—such as published announcements of new products—rather than continued reliance on such conventional measures of *effort* as research and development (R&D) expenditures. Moreover, existing outcome indicators such as the number and interindustry distribution of patents are problematic. For example, not all products and processes are equally patentable, and some firms can more easily obtain patents from the government than can others.

But this does not mean that any measure is better than none. Product announcements or mentions in trade journals are a notoriously uneven source of information. Some firms expend much greater effort on announcing what they make than do others, depending on how strongly they need to attract customers or sources of venture capital. At least some degree of double counting is unavoidable (although the Futures Group claims to have checked for this). Identifying the true parent firm of the entity making the actual announcement (let alone its employment size) can be hair-raisingly difficult.[13]

Most important, as the Futures Group readily acknowledges, between 80 percent and 90 percent of the innovations in its data base entail the "modest improvement of an existing product." While different observers might agree on what constitutes a truly unique product, "modest improvements" are something that companies are engaged in all the time. Some get reported, while others do not, and two different observers of any particular case may not agree that something "new" has actually occurred.

At least part of society's interest in technological change is the desire to identify the sources of significant breakthroughs, in order to promote economic growth. In that policy context, a tweak in a piece of existing computer software or development of a new kind of paint thinner simply is not in the same league as the creation of automatic transmissions for cars, or the advent of fiber-optic cable capable of carrying hundreds of times the number of signals as old-fashioned copper wire.

Finally, the journal and magazine announcements or mentions in the consulting firm's database do not constitute what scientists call a probability sample, so we have no objective basis for generalizing from them. And little is done with the 375 interviews that did allegedly constitute a representative sample. Between these technical drawbacks and the admitted triviality of so many of the innovations in the "sample," one simply cannot know what to make of the claims put forward by those who either assembled or studied these data. We must look elsewhere for more convincing evidence on the relationship between innovation and firm size.

DO COMPUTER-CONTROLLED MANUFACTURING PROCESS INNOVATIONS PRIVILEGE SMALL FIRMS?

An influential school of scholars of international political economy and business development argue for the emergence of a systematic relationship between changes in the nature of factory automation and the size distribution of firms and plants. Following Piore and Sabel, others have also argued that technological change, particularly in micro-electronics, has reduced or eliminated small firm disadvantages in production costs by making competitive capital goods available at prices affordable to small firms. Indeed, the new breed of "flexible" capital equipment is considered to be especially well suited to a small firm strategy favoring small batches of customized products.[14]

In another formulation, it is suggested that "we are currently observing, at least in the industrial countries, a process of change in the size distribution of plants and firms that is significantly influenced by the new flexibility-scale trade-offs associated with electronic production technologies."[15]

Acs, Audretsch, and their colleague Bo Carlsson have formally provided what they characterize as "the first empirical test[s] of the Piore and Sabel thesis that flexible production will tend to promote the relative viability of small firms."[16] The technologies that particularly interest them are industrial robots, stand-alone computer numerically controlled (CNC) machine tools, and flexible manufacturing systems (FMS) of such tools. The stand-alone tools fall within the price range of many more small firms than during the era in which production was dominated by heavy and generally expensive dedicated machinery such as transfer lines. In addition, programmable automation (PA) reduces the expense of resetting the tools when a job with particular specifications needs to be rerun. With CNC, the operator retrieves a computer program from a library and reruns the job. In business school parlance, "break-even" is achievable at lower levels of output. This is how smaller firms are supposedly privileged by the diffusion of this technology.

What Acs, Audretsch, and Carlsson claim to have proved is that the steady growth over time in the share of all tools in use that are computer programmable substantially contributes to the measured decline in mean firm and plant size in manufacturing industries, and to what they believe is a secular increase in the share of manufacturing sales contributed by the small firms.[17] I am not persuaded.

For one thing, their statistical models yield extremely weak results, by conventional standards for social science research of this sort.[18] And even when the authors *are* able to draw the inference that the association of high average employment or market share with high rates of PA utilization tends to occur in industries containing many small firms, my confidence in the results is shaken by the suspect quality of the data.

These data were obtained from a survey conducted every five years by the McGraw-Hill Company's *American Machinist* magazine, the principal trade publication of one of several trade associations in the machine tool industry. In 1983, the response rate to the questionnaire sent out by *American Machinist* to all known metalworking plants with twenty or more employees was about one in four. In response to the questionnaire sent to 10,000 firms with fewer than twenty employees, only one in ten replied. Nowhere in the published (or, for that matter, in the unpublished) writing of Acs, Audretsch, or Carlsson are we told whether the establishments that did not respond to the surveys—and obviously, the great majority did not—are significantly different (larger, smaller, more and less intensive users of PA) than those that did respond. In other words, we should properly be concerned about *bias* in a data set with such low response rates.[19]

Even were there no reasons to suspect the quality of these data, statistical analyses based on industrywide average penetration rates (which is the level of analysis at which Acs and his colleagues choose to work) can be misleading, to the extent that businesses in different size categories are more and less responsible for the bulk of the purchases of new capital equipment in that industry. As it turns out, this is precisely what is occurring in the United States—which we can see by looking at the same data that Acs, Audretsch, and Carlsson have employed.

Specifically, a comparison of the McGraw-Hill numbers for the period 1983 through 1989 simply does not show what Acs and his colleagues (or, apparently, the editors of *American Machinist*) think they show. Figure 3.1 is based on data compiled from *American Machinist*'s surveys and presented in a recent paper by Carlsson and his colleague Erol Taymaz. In 1983 and 1989, the share of a plant's tool stock that consists of PA equipment is systematically larger, the *larger* the size of the plant. Moreover, in the course of the period, while the penetration rate rose in every plant size category, it rose proportionately *more*, the larger the plant. Thus, between 1983 and 1989, the PA penetration rate within the class of very small plants increased by 75 percent, but among plants with 500 or more employees, it rose by 142 percent.

Commenting on aggregation at the level of the "industry," the economist Bela Gold has said: "One can, of course, survey the topography of the earth from distances great enough to minimize the significance of mountain chains. But one would not do so if the purpose of such observations were to build roads."[20]

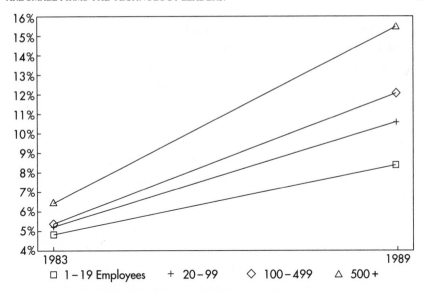

FIGURE 3.1   PERCENTAGE OF MACHINE TOOLS IN
U.S. PLANTS THAT ARE COMPUTER CONTROLLED,
BY SIZE OF PLANT, 1983 AND 1989

Source: Data from *American Machinist,* as reported in Bo Carlsson and Erol Taymaz, "Flexible Technology and Industrial Structure in the U.S." (Department of Economics, Case Western Reserve University, Cleveland, Ohio, August 1992), table 1.

In fact, it has become quite unnecessary to rely on industry averages—a procedure that lumps together one-person garage shops with the branches of a multinational giant such as the General Electric Company. Data are now available from Europe, Japan, and the United States on the adoption and penetration of programmable automation, based on government, private, trade association, and university-based surveys of individual companies and their constituent establishments. Without exception, these surveys show what figure 3.1 shows: that programmable machinery is more likely to be found (and to be found at greater relative densities) in larger business units than in smaller ones. Consider the evidence.

In the United States, in a size-stratified random sample survey of more than a thousand plants in twenty-one machining-intensive industries whose production managers were interviewed between the fall of 1986 and the spring of 1987, Maryellen Kelley of Carnegie Mellon and

Harvey Brooks, a professor emeritus at Harvard, found a significant differential between large and small establishments in the likelihood of their having adopted—or in their plans to adopt—computerized machine tools. A 1991 follow-up survey of the survivors from the earlier cohort, conducted by Kelley, showed that penetration rates, too, were strongly and positively related to plant size.[21]

Researchers at the U.S. Bureau of the Census have reached the same conclusions. As they learned from a 1988 national survey of a much larger sample (some 10,000 individual establishments were interviewed):

> technology use is positively correlated with plant size ... plants that are owned by multi-unit firms and plants engaged in defense related production are generally more likely to use advanced technologies. ... Plants using [or not using] one advanced technology tend to use [or not use] other [advanced] technologies.[22]

Evidence from other countries confirms the existence of a systematic gap between small and large businesses in their use of programmable factory automation. According to surveys conducted in the mid-1980s by the Policy Studies Institute in London, the incidence of both product and process innovation in the United Kingdom is substantially lower in plants with fewer than fifty employees—and especially low in those with fewer than twenty workers.[23] More recent survey research on Britain, covering the years through 1989, not only confirms a positive relationship between firm size and both the likelihood and the extent of adoption of CNC machinery, but also reveals that smaller firms are less likely to fully utilize the new equipment (more generally, that the rate of utilization of these machines, once acquired, is positively correlated with firm size).[24]

A technological gap between large and small *Japanese* firms in the rate of adoption of programmable automation has been uncovered in research conducted in Britain, at the Science Policy Research Unit of the University of Sussex.[25] Statistics from the Japanese government's Ministry of International Trade and Investment (MITI) also indicate that the penetration rate of CNC machine tools increases with plant size (although MITI does not collect—or at any rate does not publish—numbers on plants with fewer than fifty employees).[26] Research has also shown that in the former Federal Republic of Germany, computer-aided technologies were adopted much more often by large firms than by small ones.[27]

Even in Italy, a country that has become a major producer of computer-controlled automation, a 1983 government survey of firms with 20 or more employees revealed a 25 percentage point gap between the incidence of product or process innovation in large (500 or more workers) and small (20 to 49 employees) firms.[28] Subsequently, the economists Fabio Arcangeli, Massimo Moggi, and Giovanni Dosi found that small firms were much less likely than large firms to have adopted even one programmable machine by 1985.[29]

Given Italy's prominence in stories about small firms as technology leaders, let us dwell here on that country for a moment longer (the following chapter is devoted entirely to examining the Italian experience). A sophisticated survey-based analysis of the rates of adoption and diffusion of flexible automation in Italian industry was conducted recently at the *Politechnico* of Milan. It was discovered that, as of 1989, plants with 500 or more employees accounted for less than 1 percent of all Italian metalworking facilities in the country, but for 43 percent of all installed PA. The researchers, Sergio Mariotti and Marco Mutinelli, concluded (as do the Americans, Kelley and Brooks) that their findings imply that "the impediments to adoption of flexible automation systems by smaller companies are related not so much to strictly economical and financial factors as to insufficient human, organizational, and managerial resources." They went on to conclude: "The correlation between adoption rates and company/plant size thus discredits the illusory and hurried models that some scholars and economists proposed in the past, maintaining that the new flexible technologies were creating a favorable climate for smaller enterprises by increasing their competitive potential."[30]

The survey research I have been reviewing focuses for the most part on process innovations such as PA. But rebuttals have also been offered to the claim that small and medium-sized enterprises (SMEs) are the most likely sites for product innovation. Thus, recent evidence on Germany[31] indicates product diversity to be considerably greater in large than in small establishments, even after adjusting the data to account for differences between small and large business organizations in the resource capacity to develop a variety of products. And in their interviews with managers in the U.S. metalworking plants in 1986 and 1987, Kelley and Brooks found that large firms (defined as having more than

	Pct. of Companies With		Pct. of U.S. Employment in Companies With		Pct. of U.S. Sales in Companies With	
	<100 Employees	>500 Employees	<100 Employees	>500 Employees	<100 Employees	>500 Employees
All Industries	98.4	0.2	42.5	38.7	39.6	46.4
All Manufacturing	93.9	1.2	18.8	67.0	12.8	75.7
Selected "High Tech" Industries						
Drugs	87.0	4.3	4.9	89.6	4.1	89.0
Computers	84.9	4.7	3.8	91.2	3.4	91.4
Communications Equipment	82.2	3.6	5.1	75.8	4.6	76.1
Electronic Components	88.0	2.5	19.5	59.1	15.3	75.6
Aerospace	86.8	4.3	1.4	92.8	1.0	95.1
Instruments	91.2	2.2	8.3	87.5	6.9	85.8
Scientific and Measuring Devices	90.4	2.4	6.5	86.6	5.5	88.1
Medical Goods	91.8	2.0	18.2	(Suppressed)	13.4	(Suppressed)

Source: U.S Department of Commerce, Bureau of the Census, 1987 Enterprise Statistics,
Company Summary, ES87-3 (Washington, D.C.: U.S. Government Printing Office, 1991), Table 3.

TABLE 3.1   SMALL AND BIG FIRM SHARES
OF NUMBER OF COMPANIES, EMPLOYMENT,
AND SALES IN THE UNITED STATES IN 1987

Source: U.S. Bureau of the Census, 1987 *Enterprise Statistics, Company Summary*, ES87-3 (Washington, D.C.: Government Printing Office, 1991), table 3.

500 employees across all of their constituent plants) were just as likely to be pursuing a strategy of producing a diverse range of highly customized products as were firms with 1 to 19, 20 to 99, and 100 to 499 employees.[32]

If there is any one corner of the American economy where we have been led to expect a dominant innovative position for small firms, it is in high technology. For example, Birch identifies a series of what he calls "high-innovation sectors" (not all of them in manufacturing), which he believes to be especially fertile ground for the kind of small start-up firms about which he writes so enthusiastically.[33] I have turned to the Census Bureau's *Enterprise Statistics* volume for 1987 (the most recent available) and have examined the numbers on a few of precisely those industries named by Birch as likely to be dominated by small firms.

Even considering only the manufacturing industries on his list, the numbers just do not add up for the "small is innovative" thesis. As table 3.1 makes clear, in 1987, sales and employment in these highly innovative industries were even *more* disproportionately concentrated in firms with 500 or more employees than was the case for manufacturing as a whole—let alone for the economy as a whole. For example, almost 85 percent of all the individual establishments in the computer industry in 1987 employed fewer than 100 workers. Only about 5 percent had as many as 500 employees. Yet those 5 percent accounted for fully 91 percent of all employment and of all sales in computer manufacturing.

So much for statistics. The claims that new technologies systematically favor small firms, let alone that smaller businesses are becoming the major *producers* of such technology, cannot stand up to the evidence from case studies on particular technologies and industries, either. Let us examine two such cases: the emergence of "biotech" as a new frontier activity, and the ongoing efforts by some companies to introduce greater standardization into the writing of computer software.

BIOTECHNOLOGY NETWORKS

"The practice of altering genetic structures is as old as the art of breeding crops and animals, and fermentation techniques have been used in beer brewing and bread making for at least eight centuries." So begins one of the definitive studies of the evolving organizational form of the industry (really better understood as a family of technological processes) called biotechnology.[34]

As the sociologists Walter Powell and Peter Brantley point out, genetic alteration and fermenting as planned industrial techniques began soon after World War II: first, with Watson and Crick's description of the DNA molecule—the famed double helix—in 1953; then with the announcement of the method of "recombinant DNA" (gene splicing and regrowth of recombined elements in bacteria) in the early 1970s. Other scientific developments quickly followed, accompanied by the appearance of the first commercial companies, Cetus and Genentech, over the same period. No industrial activity in modern times has so profoundly changed the way we think about the world—and our possibilities for (literally) transforming it—as has biotechnology.

One might have expected that the new techniques would have emerged from the laboratories of the existing pharmaceutical companies, but that was not the case. The knowledge base of the giants—Merck, Hoffman-La Roche, SmithKline, and Mitsubishi—was in organic chemistry and its applications. The knowledge base of gene splicing is molecular biology and immunology. The repositories of these sciences are the research laboratories of the major universities and hospitals—*not*, for the most part, the existing pharmaceutical corporations. Because it is built on a qualitatively different theoretical foundation that threatens to undermine the dominance of older conceptions, biotechnology is what experts call a "competence-destroying" innovation.

As such, during the 1980s, biotechnology became the subject of explosive growth for hundreds of new start-ups. Partly, these new biotechnology firms (NBFs) grew so quickly because of the ready supply of scientist-entrepreneurs in the universities' laboratories. Partly—some think mainly[35]—it was because of changes in federal government regulations in the late 1970s, which greatly deregulated pension funds and other fiduciaries, freeing up a flood tide of new money for the super-profitable and seemingly well-diversified venture capital industry, which sought out (indeed, according to the University of California–Davis social scientist Martin Kenney, often literally caused to be created) new biotech firms, especially in California, Massachusetts, and New Jersey.[36]

As the industry expanded during the 1980s, the organizational form of the biotech industry began to change. From the perspective of conventional product cycle theory,[37] some observers predicted that the extremely small, science-based firms of the 1980s would turn out to be only a "bridging" institution, "specialized suppliers of esoteric knowledge," which would soon give way either to outright acquisitions by the older pharmaceutical firms with the deepest pockets, or to vertically integrated big firms that would emerge from the pack of recent start-ups.[38] Because the federal government's regulatory process can hold up approval of a new biotech product for many years, and because product development is in any case highly research intensive and protracted, expectations were heightened that big firms would be better capable of organizing the industry.

But by the early 1990s, neither the acquisitions nor the emergence of the big firms had occurred. Pharmaceutical firms have invested in or partnered with the start-ups, but they have generally not tried (or have been unable) to acquire them outright. Thus, as of 1990, the largest of the NBFs, San Francisco's Genentech, still had fewer than 2,000 persons on the payroll.[39]

Instead, what has emerged are *networks* of organizations. The members of these "federations" include the generally very large pharmaceutical/health/chemical companies, the small scientist-based biotech engineering firms, the major research universities (led by M.I.T., Harvard, Johns Hopkins, and Stanford), the big teaching and research hospitals (such as Massachusetts General Hospital and Memorial Sloan Kettering), and powerful government agencies, especially the National Institutes of Health. Outright acquisitions of the NBFs were effectively blocked by the extensive cross-licensing agreements among different partners that the small firms had already entered into in their earliest years.[40]

Moreover, since the kind of knowledge on which this family of products draws is so widely diffused and dispersed among other institutions, especially the universities and research laboratories, it is almost impossible for any individual firm, however large and resourceful, to gain effective control over it. In the past, there has been a tendency for the big companies in an industry to maintain their own R&D departments internally, and to contract out only for lower-order research. Losing control over the latter would not be so threatening to the big firm.

But with biotechnology, the leading actors such as Genentech and M.I.T. are embedded within a research and production system in which even core knowledge is so widely dispersed among so many actors that, while inside R&D departments remain important (if only to maintain the technical capability to "read" what is going on, outside),[41] technical progress for any particular player is impossible without continually renegotiating relations with other firms and universities. Thus, professors take sabbaticals at the NBFs, and research scientists devote periods of time at government labs.

Some experts have predicted that these network arrangements will prove to be unstable, given how many of them are "based on a love-hate relationship," that is, on the absence of trust.[42] In the next chapter I introduce evidence that the small firm production networks of north-central and northeastern Italy may indeed be undergoing a transformation away from the ideal typical form of cooperative competition that arguably existed in the 1970s. An important difference is that none of the Italian industrial districts that caught the fancy of Sabel, Piore, and others was ever based on *such* advanced (yet widely dispersed) scientific knowledge as that which Powell and Brantley (and the economists Ashish Arora and Alfonso Gambardella) recognize as constituting the "sea" within which the NBFs are swimming.[43]

Powell and Brantley see no reason why all of these forms—small science-based companies, giant pharmaceutical corporations with deep pockets, and networks that link them, to one another and to the financial, government, and academic communities—cannot coexist, possibly stimulating a faster rate of innovation than would occur otherwise. In

any case, the lesson for *our* inquiry is that the interorganizational networks that have evolved in biotechnology, and that Powell and others
expect to survive because neither markets nor hierarchies readily accommodate the unusual nature of what constitutes "information" within
this industry, are certainly not manifestations of a systematic privileging
of small firms by a new technology, in the sense meant by Gilder, Birch,
Acs, Audretsch, or Carlsson.

Toward the Greater Standardization of Computer Programming: Japan's Software Factories

One "high" technology that is perhaps more commonly associated with
a craft form of work organization than any other is the writing of computer software. The popular media, from movies and "cyberpunk" nov-
els to television, revel in depicting software writers as "lone rangers," or
"cowboys": independent, idiosyncratic, working at their own pace
(sometimes in their own homes), highly mobile between employers.
These traits, it is said, best promote the creativity that constitutes the
secret ingredient underlying America's competitive success in the international software industry.

What could be more unlike the allegedly rigid, formal, bureaucratically structured systems of work organization associated with the supposedly bygone era of mass production? In a factory system of organization,

> product and process innovations [make] ... goods available to
> masses of customers by systematically dividing large tasks into
> smaller and less-skilled operations, establishing standard proce-
> dures and controls to help de-skill tasks, apportioning [tasks]
> among different individuals, building tools and other equipment to
> mechanize or automate aspects of design and production, and cre-
> ating designs based on reusable or interchangeable components.[44]

The history of technology and work organization teaches us that
managers' search for more orderly, less idiosyncratic structures in the
age of automobiles, chemicals, and steel was driven by the competitive
pressure to get out from under the high labor costs (and worker control)
associated with skill-intensive production, to economize on resources by
reusing interchangeable components, and to modularize assembly operations wherever possible.[45] As it turns out, the same forces are at play in
the contemporary software industry, as well.

The leading expert on the subject, Michael Cusamano, an M.I.T. management professor, realizes that software does not readily lend itself to a factorylike form of organization, especially given the very short life cycles of most software products. This pace of change, together with the desire of high-tech companies to avoid, if possible, the need for centralized coordination and to recruit the best engineering talent from the universities, would seem to favor an organization of software production based on job shops, teams, and the cultivation of the sense of craft.

But programmers are scarce and therefore expensive, budgets get tighter, competitive success increasingly turns on timely delivery of "product" to customers, and the length and complexity of programs increase along with advances in the hardware (the computers themselves), putting a premium on quality control. Writes Cusamano: "These conditions have created huge incentives for managers to reduce skill requirements and systematically recycle key factors of production (methods, tools, procedures, engineers, components) in as many projects as possible."[46]

Some of the industry's biggest companies are in fact trying to move programming beyond the arts-and-crafts stage, to become a more structured, factorylike activity. The initial attempts in the United States were led, in the 1960s, by Xerox, TRW, System Development Corporation, and General Electric. By the 1980s, IBM was the acknowledged American leader. But, as Cusamano has documented in prodigious detail, the transformation has proceeded further than anywhere else in the country where the "creativity" needed to produce profitable software was always said to be most lacking: Japan.[47]

Four giant corporations have led the way: Hitachi, Toshiba, NEC, and Fujitsu. Between the late 1960s and the late 1980s, following careful study of and reflection on their own past inefficient practices, software managers in these companies (and, gradually, among their suppliers and subcontractors, as well) made great progress in introducing into software production the principles of long-term planning; modularization of design; reusability of parts; standardization of design, coding, documentation, and testing; the use of databases from previous projects to inform new activities; and extension of the division of labor and specialization of tasks as sometimes a substitute for, sometimes a complement to, small, integrated teams. As they have done in so many other applications, the Japanese have tested their techniques, exploited economies of scope (making multiple products with the same basic inputs), and moved down the "learning curve" (cutting unit production costs in the course of accumulating experience) by writing modularized, more standardized programming code not only as a final product in and of itself, but for their own factories, machinery, and consumer products.

To be sure, most of the employees in software factories are engineers, technicians, and other well-trained analysts. Facilities are modern and state-of-the-art, and the factories pay good wages. As in other activities within the core of the Japanese economy, employment security is part of the package, with programmers tending to remain for long periods (if not literally for a lifetime) with the same firm, in contrast to the high turnover and job hopping in the Silicon Valley system (examined in chapter 5). Some jobs continue to integrate both conceptual and implementation skills. Expression and representation move back and forth between the mind, the computer screens, sheets of paper, and the chalkboards of group meetings.

What has changed—or rather, what managers are continually and consciously trying to change—is the conceptualization and design of new software as idiosyncratic, with each new project built from scratch by one programmer or by a small team of specialists. And without question, firms are trying to economize on the employment of extremely scarce, very expensive skilled programming specialists.

American firms presently still dominate global markets for software. But Cusamano urges them to pay attention to the successful track record of Japanese companies in progressing from the attainment of production efficiency in a field to a gradual establishment of competitiveness in quality design. In some ways, he argues, the Japanese software factories today are like the Nissans and Toyotas of two decades ago: at least as, if not already more efficient than, their foreign competitors, employing higher-quality manufacturing processes, but not yet able to offer the most attractively designed products.[48]

American and European companies are pursuing these organizational methods, as well, partly under the pressure of competition from the Japanese model, and partly in order to facilitate the further development of "user-friendliness" in the form of so-called object-oriented programming. This entails the use of icons—symbols that, when activated, call into use entire subsystems of commands ("file this into the appropriate location," "compute the present discounted value of a potential stream of future returns to this investment," "send this message to everyone on a particular mailing list"). Those cute little icons (images of file folders, trash cans, and mail boxes) that appear on the screens of Unix, Apple, IBM's OS2, and Microsoft Windows–based operating systems are fast becoming the protagonists in a fierce battle for market share. They are also becoming an indispensable part of the business of using computers to assist the designers of everything from new clothes to the highest-tech automobile braking systems.[49]

Yet behind this most *au courant* technology are the most old-fashioned principles of modularization, standardization, and interchangeability. Even off-shore production—another development commonly associated with the making of cars and clothes—is now part of the industry's morphology. For example, Texas Industries, Citicorp, Borland, and Sun have established software facilities in such places as India, Hungary, and Russia.[50]

The implications for job creation in the United States are potentially far-reaching. While the U.S. Bureau of Labor Statistics forecasts continued rapid growth of domestic demand for programmers, the industry consultant Edward Yourdon predicts "massive unemployment" by the end of the decade for programmers, systems analysts, and software engineers—"the same fate that befell American autoworkers in the 1970s and 1980s at the hands of the Japanese." Why? Because

> many applications that once required customized programs can now be done with standardized modules, much as pre-fabricated units simplify the construction of a house that once had to be built laboriously by a master carpenter. Different modules encompass a variety of functions, so that a programmer sometimes merely links pre-existing modules. And assembling modules requires fewer programmers than creating a program from scratch.
>
> For another thing . . . while attention is still given to the young hotshot programmers of Silicon Valley and Route 128 around Boston, the vast majority of American programmers now work for large corporations, mostly maintaining the programs that run everything from airline reservations to inventory and accounting processes.[51]

The need for specialized *maintenance* is yet another of those underlying organizational principles that link tomorrow's high-tech industrialization to yesterday's factory-based production systems. Some things in the age of flexibility really are new—but many are not. We would do well to heed both.

TECHNOLOGICAL CHANGE AND THE CONTINUING TENDENCIES TOWARD STANDARDIZATION, SIZE, AND SCALE

Whether the future job prospects for fledgling computer programmers collapse quite so rapidly and catastrophically as Yourdon predicts, and whether scale, scope, and finance eventually come to dominate the presently decentralized organizational form of the biotech industry, only time will tell. But these cases, together with the story of how in fact computer-controlled factory automation is diffusing within and among firms across the industrialized countries, all indicate that it would be a serious mistake to see technology, per se, as systematically privileging small firms. The cases also reveal certain underlying principles of business organization in market economies that new technology by itself, however "flexible," has not negated.

Richard Gordon, a political scientist at the University of California at Santa Cruz, puts it this way:

> Contrary to prevailing assumptions, contemporary technological change does not arbitrarily reinforce a[ny] specific configuration of size advantages. New technologies do not necessarily disadvantage large corporate organizations in comparison with small and medium-sized firms [or vice versa]. Investment in flexible automation frequently exceeds the resources of smaller enterprises and . . . [m]arket differentiation does not eliminate the continued salience of scale economies (finance, design, marketing) and global distribution in a broad range of product areas, nor does it necessarily preclude large firm production of differentiated products with variable standardized inputs [or the desire by management for] reduced design times [or] savings on circulating capital.[52]

There are economies of scale and scope to be attained in at least some of even the most advanced (and the most small-batch-oriented) production processes. Smaller, stand-alone machines and customized operations may be the wave of the future—although, as we have just seen, the organization of software production could be moving in the opposite direction. But big firms have the resources to implement "small" technologies. It is not equally clear that small firms have the resources to implement "big" technologies—unless they enter into networked alliances with other companies and sources of finance capital.

In fact, writes the technology expert Yves Doz, "changes in product and process technology have *increased* the minimum efficient size of production in . . . industries such as cars, chemicals, consumer electronics, semiconductors, and machinery. . . . New product development costs have also risen considerably in a range of industries, the best-known of which are aircraft, telephone switching, cars, and semiconductors."[53] In the same vein, the British technologist Keith Pavitt, who directs the Science Policy Research Unit at the University of Sussex, notes:

> It is often argued, on the basis of either Schumpeter's notion of creative destruction or the so-called product cycle theory, that technological discontinuities are associated with the emergence of new small firms that exploit them, given the conservatism, obsolescence, and bureaucracy in established large firms.
>
> [But] the evidence does not [always] confirm this view. . . . Some of the most revolutionary business applications of information technology today are to be found not in new technology-based firms, but among the oldest, largest, and most conservative of capitalists: banks, financial services, and large-scale retailing. [Thus,] *technological discontinuities can co-exist with institutional continuities.*[54]

The cases examined in this chapter bring out a number of underlying principles that, in one way or another, operate even in the most highly innovative sectors and tend to move the economy *away* from any systematic shift toward smaller firms. I mentioned the need to commercialize innovations, the cost-reducing attractiveness of modularization or reusability of components, and the necessity for standards (governed by the public sector or at least by collective trade associations) to "incent" new risky private investment. We might also have considered the classic problem that, the greater the expense to the firm of installing new capital equipment embodying the latest technologies, the greater the pressure on management to operate it nonstop or to otherwise increase volume or diversity of products, in order to amortize the investment. The catch is that this also puts additional pressure on the economy as a whole to create sufficient aggregate demand to absorb the additional output.[55]

This underlying tendency toward ever-increasing volume does not absolutely require ever-larger units of production, but it certainly does nothing to privilege small businesses. Jeffrey Williams, a Carnegie Mellon management professor, makes essentially the same point. In commenting on Cusamano's analysis of the Japanese software factories, Williams remarks that the effort to modularize and standardize programming code "would, paradoxically, if successful, reduce product sustainability [that is, the shelf life of any particular design], shorten cycle times, and induce greater (more direct and intense) competition in the software industry."[56]

None of these points means that there is no room for smaller firms— or smaller units of production, generally—in modern industry. Indeed, in the next two chapters I build on the biotech example by exploring the argument that small firms can achieve the advantages of scale and scope by federating into spatially clustered networks, or industrial districts. But the cases presented here do, I think, seriously call into question the belief that new technologies actually give small business an inherent *advantage* over larger companies in market economies.

As with the myth of small firms as the source of nearly all job creation, so with the popular image of the dynamic small entrepreneur driving (or being the most likely to respond to the potential advantages afforded by) new technology. It is, alas, only myth.

4

THE EVOLUTION
(AND DEVOLUTION?) OF THE
ITALIAN INDUSTRIAL DISTRICTS

[A] distinction has to be drawn between those parts of corporate activity where spatial proximity is important and those where it is not. The view of writers . . . who have studied this is that commercial (sales, strategic, or financial) and basic scientific networks can work well at a long distance. However, dealing with practical, production-related issues, such as designing software or making product adjustments or applications, tend to be geographically a clustering phenomenon. Trust is built between lower managers, and the networks that they build are kept going for as long as possible until they are destroyed by mergers or acquisitions.

—PHILIP COOKE AND KEVIN MORGAN, SOCIETY AND SPACE,
OCTOBER 1993

So far in this book, we have seen that the expansive job creation claims of the advocates of theories of small firm–led economic growth have been exaggerated, sometimes markedly so. Moreover, new technological innovations do *not* systematically privilege small and medium-sized firms, as has been so strongly claimed by many writers. Beyond the frenetic period when a new industry or a new family of products is being born, during which creative small enterprises seem to be starting up almost daily, a sustained capacity to innovate and—of equal importance—to *commercialize* the new products and manufacturing processes that the new innovation makes possible tends to require the financial, market, and political power of the *big* firms and their networks.

In this chapter and the next, I turn to yet a third notion on which so much of the fascination with models of small firm–led economic development has rested: the *spatial* relationship among companies. For many years, economists, sociologists, planners, geographers, and urban historians have been documenting the continued spatial *decentralization*, or dispersal, of production. Sometimes, companies are searching for cheaper labor or raw materials, trying to escape congestion in older built-up places, or seeking to dodge trade union influence. Perhaps the pressure to relocate or branch out is driven by complaints or demands coming from distant governments, insisting that "the price for selling here is to produce here."

Whatever the combination of motives, companies across the industrialized world have been moving or expanding their production facilities from central cities to the suburbs, from metropolitan to exurban and even rural areas, and out of their home countries altogether, into new production zones on the periphery of Europe (in Ireland, Spain, Portugal, Greece, and Turkey), to the Far East, and into the Third World. Thus, American companies build plants in Taiwan and Haiti, European firms invest in North Africa and India, the Japanese set up shop in Thailand and Indonesia, and everyone establishes manufacturing operations in the *maquiladoras* along the Mexican border. Western (and Japanese and Korean) direct investment in the newly opened countries of Eastern Europe will only be a logical continuation of a long-standing trend.[1]

But the advocates of small firm–led growth perceive a countertendency to all this. According to what has come to be known as the theory of the "industrial districts," a new wave of economic growth in a number of regions in Europe, North America, and eastern Asia, from Silicon Valley in the United States to Emilia-Romagna in north-central Italy, is said to be led by highly spatially concentrated networks of mostly small and medium-sized enterprises, often using flexible computer-controlled production technology and characterized by substantial local interfirm linkages. We are told that such proximity or even co-location is fundamental to the reproduction of the conditions under which managers and entrepreneurs are more likely to share information, to exchange specialized assets, and to form and re-form projects as local or outside—especially foreign—markets

signal new possibilities for profit. The embedding of these clusters of producers within local social, cultural, and political institutions—institutions
that are inherently place-specific, from religious organizations and political
parties to trade associations—further reinforces the willingness or inclination of the small firms to subordinate their competitive tempers to the
need for cooperation and collaboration. Nowhere are these developments
thought to be more pronounced—and certainly nowhere more celebrated—than in north-central and northeastern Italy.[2]

From the market towns of medieval Europe and much of contemporary
Africa to the industrial parks, wholesaling districts, and shopping centers
of any large urban area on every continent, there is no doubt that similar
economic activities *have*, to some extent, always clustered together in
space. The phenomenon of what regional economists, planners, geographers, and business strategists call "localization" is alive and well.

But are such localized industrial districts really something *new*—and
have they become increasingly characteristic of the post-modern age?
And whether they are novel or, in other cases, simply holdovers from an
earlier era, how likely are these districts to survive as centers of more or
less symmetrically distributed economic *power* within locally oriented
networks of mostly small firms? That is, to what extent does the very success of the districts invite an invasion by outside corporations seeking to
pick off and redirect the energies of the best of the area's indigenous
firms? Alternatively, do the most successful small firms from within the
districts gradually evolve into leaders that then flex their economic muscle, imposing hierarchical rules of behavior onto what used to be a collaborative milieu? Finally, even where such districts have been successful
and have been able to compete with the powerful multinational corporations of the world, are the benefits created by this local success equally
shared among men and women, natives and foreign-born workers?

These are some of the questions I want to investigate in the next two
chapters. Here, I focus on the case of Italy. In the following chapter, I
turn to the example of California's Silicon Valley. But I might as well
announce my conclusions up front. Through statistical analysis, a
rereading of other people's case studies, and my own field work in Italy, I
have concluded that networks of mostly small and medium-sized enterprises have important functions, and supporting them belongs in any
well-balanced regional and national economic development policy mix.
But however locally embedded, these networks are *not* by themselves
likely to play a lead role in the economic development process, any more

than are the individualistic entrepreneurs that populate the visions of
Zolton Acs, David Audretsch, David Birch, Bo Carlsson, George Gilder,
and the other celebrants of the small business form.

SMALL FIRM–LED GROWTH MODELS AND THE
NORTHERN ITALIAN INDUSTRIAL DISTRICTS

The paradigmatic industrial districts are those located in that region of
north-central and northeastern Italy known as the "Third Italy"—a term
first coined by the Italian sociologist Arnaldo Bagnasco in the mid-
1970s and popularized in the English-speaking countries in the 1980s by
two M.I.T. social scientists, Michael J. Piore and Charles F. Sabel.[3] In
the wake of the intense management-labor conflicts around such big
firms of the north as Fiat, in the late 1960s, the traditional heavy industries concentrated in the northwest went into a protracted period of
decline.[4] But a quite unexpected burst of economic growth became visible in the regions to the east and southeast of the automobile center of
Turin and lasted into the mid-1980s. This growth has been traced to the
ability of the small firms in these regions—some of them spin-offs from
Fiat, Montedison, and Olivetti; others, ones that evolved from older
rural family businesses—to act as

> an integrated network of producers relying upon . . . economies of
> scale and scope arising from the division of labor between specialist
> producers; an entrepreneurial ingenuity arising from the combination
> of petit bourgeois traditions and the skill and technical advantages of
> artisan production; the accumulation of appropriate skills and innova-
> tive capability in the milieu as a result of [geographic] clustering;
> active support from local authorities and other local institutions in
> the provision of infrastructure, training, finance, and other collective
> services; the development of marketing agents to facilitate exports; a
> reputation (for example: "made in Prato") that attracts buyers; and
> the consolidation of trust and consensus-based local subcultures
> which permit social collaboration and the exchange of ideas.[5]

In the ideal typical ("canonical") industrial district, each small firm
typically specializes in one or a few phases of a complete production
process, although it may change its specialty in response to signals from
buyers or from what, in the district of Prato, near Florence, are called the
impannatori—the middlemen who bring local buyers and the small

manufacturing firms together, by organizing specific production projects to meet specific demands of the buyers (who, in turn, roam the rest of Europe and beyond in search of customers). On any particular project, the small firms will often cooperate with one another, sharing tools, information, and even skilled personnel, only to compete fiercely for a share of the next new contract or market opportunity. Some of these small firms have even become clever enough to make connections to several production networks at the same time, thereby reducing their vulnerability to the economic fortunes of any one group.

The widespread adoption of relatively small-scale stand-alone computerized automation, such as the computer numerically controlled machine tools and personal computer–based computer assisted design (CAD) systems analyzed in chapter 3, helps to make these networks of "flexibly specialized" businesses capable of rapidly reconfiguring themselves to meet the continually fluctuating demands of the world market. And because these tools can be reset quickly and inexpensively, it is argued that the collection of firms within the district may enjoy economies of both scope *and* scale.

Moreover, at a time when much of the Anglo-American world was embracing the Reagan-Thatcher philosophy that government involvement in the economy only erodes economic efficiency, local and regional governments in Italy actively supported their industrial districts with a variety of infrastructural and business services—at least prior to the national economic and political crisis of the 1990s, during which government spending at all levels has been sharply curtailed, in the wake of the inflation in the exchange value of the lira and of an unprecedented national reaction against corruption in government and business.

Italian economists in particular—notably, Giacomo Becattini of Florence and Sebastiano Brusco of Modena—have developed intricate theories of the evolution and likely future success of the industrial districts.[6] In these theories the economists draw freely on a set of ideas first promulgated in the late nineteenth century by the British economist Alfred Marshall, in the course of his observations on the development of industrial districts in Sheffield and Lancashire.[7] Thus, many contemporary writers and policy makers in Italy and elsewhere conventionally refer to the modern districts as "Marshallian."

But outsiders have taken notice of the district phenomenon in Italy not because of these theoretical constructs but because the regions in which the districts are situated fared particularly well during an era of general overproduction and declining corporate profits across the industrialized world. For example, in the Emilia-Romagna region, by the 1980s wage rates were twice the national average, a growing share of national exports were originating in the districts of this region (Carpi, Modena, Bologna, Reggio Emilia), and per capita income rose from seventeenth among Italy's twenty-one regions in 1973 to second in 1986.[8] Even as recently as 1991, the unemployment rate in Emilia-Romagna was only a third that of the country as a whole.[9] Could it be, then, that small firm–led industrial districts *have* played a leading role in the economic development process, as their advocates claim?

To get at these very complex and hard-to-measure questions, I turn now to three case studies in which I participated, or about which I had an opportunity to learn more from meetings with their Italian authors, during the summer of 1989 and through subsequent correspondence. The cases come from industrial districts located in three north-central and northeastern regions: Emilia-Romagna, Veneto, and Tuscany. (A short lesson in Italian geography and civics: In 1970 the Italian government formally divided the country into twenty such "regions," whose governments have extensive legal powers. These regions are in turn divided into "provinces" and, further, into "communes" or municipalities. As a matter of historic fact, as in the rest of southern Europe, the principal municipalities—in these cases, Bologna, Venice, and Florence—actually came first; the nation state was created around them, only in the last century.)

Why *these* three cases? All started out as part of cooperative small firm networks in their respective regions. And all have been (or are in the process of becoming) transformed into something else. Thus, the case of the Sasib Company shows what can happen when an outside conglomerate moves into an industrial district and establishes (or, in this case, reinforces the emergence of) a "lead firm." Benetton exemplifies the evolution of hierarchy—and, with it, unequal power among firms—from within an initially local network. Finally, Prato provides a case study of how an entire production system—a whole regional economy—can become vulnerable to heightened global competition when the local system becomes *too* fragmented.

Others have written about the unexpected effectiveness of small firm districts in the 1970s and 1980s. These three case studies offer some useful balance, by reminding us of the potential *instability* of production systems organized around locally oriented small firms.

PENETRATION OF THE INDUSTRIAL DISTRICTS BY OUTSIDE FINANCIAL CONGLOMERATES: THE CASE OF SASIB

A major highway heads southeast from Italy's financial capital, Milan, in the direction of the Adriatic Sea. Clustered halfway between the two poles lies a string of medium-sized cities: Parma, Modena, Bologna, and the smaller town of Reggio Emilia. We are in the heart of the region of Emilia-Romagna, home of one of the most famous—and, during the 1980s, among the most competitively successful—concentrations of metalworking activity in the world.

Some of the small and medium-sized firms in the region—the *piccole imprese*—were founded in the nineteenth century. Others came into existence in the years after World War II, especially in the wake of the big strikes that hit the auto and related industries in the tumultuous "Hot Autumn" of 1969, during which thousands of skilled craftsmen were permanently displaced from Fiat and the other large companies. As Richard Locke, an M.I.T. political scientist and professor of management, has documented, many of these craftsmen started up their own shops, some of which evolved into world-class operations within only a decade.

Italian observers including Bagnasco, Becattini, and Brusco, and foreign admirers such as Locke, Piore, Sabel, the economic geographers Michael Storper and Allen Scott, the British scholar Ronald Dore, and the architect/planner C. Richard Hatch have been understandably impressed with the technical wizardry of these highly specialized, often astonishingly automated companies, which make everything from machine tools and farm implements to ingenious food processing equipment—a natural enough development, given the rich farm country stretching out in all directions around the Emilian cities and towns.

By all accounts, during this period of the 1970s, these individually owned, independent enterprises divided the work to be done among one another, and shared information and—through the rotation of apprentices and craftsmen—skilled labor, as well. The most successful were those that had learned to develop contacts outside of the region and beyond the borders of the country, exporting custom-crafted equipment that they had manufactured to the specifications of the industrial customers with whom owner-managers and engineers regularly interacted.

Inside the metalworking districts of Emilia-Romagna, individual firms made the decisions about which local specialized subcontractors they would use on any particular job. Often, as Brusco has observed, businesses would take turns playing lead firm and then supplier, depending on their particular expertise and how busy they were at the moment. Thus, through both the fluidity of the local labor market and their unique practice of changing partners when it came to meeting a new order from an important outside customer, the metalworking firms of Emilia-Romagna built a truly nonhierarchical spatial production system that worked marvelously well for them, and caught the interest and invoked the imagination of an entire generation of scholars and policy makers all over the world who were concerned with regional economic development.

But by the 1980s, things were beginning to change in Emilia-Romagna. Perhaps it was inevitable that such highly productive, skillful enterprises would become takeover targets for outside multinational corporations and financial conglomerates. Nor is it certain that what many Italian researchers and policy makers themselves refer to as the "invasion" of the industrial districts by these powerful outside interests will turn out to be a bad thing for the Italian economy's long-run competitiveness. What *is* clear, however, is that the invasion is qualitatively transforming the nonhierarchical, collaboratively competitive nature of interfirm relations within the districts. Take the case of Sasib.[10]

The original business, which involved the design and manufacture of railway signaling equipment, began operations in 1915 in the city of Bologna, the capital of the region of Emilia-Romagna. In 1933 it was established as the Sasib Company. By the 1970s, Sasib had become one of the most important producers of tobacco-packaging machinery in the world, selling its high-quality equipment in many countries (a second division continued the company's original line of business). With an average of more than a thousand employees during the 1970s, Sasib was already one of the single largest companies in the district. Because of its extensive connections with the outside world, it frequently played the role of lead firm in landing orders for specialized machinery and systems, then assembling a local production network to carry out the various stages of the work. On other occasions, Sasib performed subcontracting work for other Emilian companies. Such reciprocal relationships were, after all, the way business was normally conducted in the industrial districts.

Enter Carlo De Benedetti, since the late 1970s the principal owner of Italy's leading producer of office equipment, the Milan-based Olivetti. Founded at the dawn of the twentieth century, Olivetti had been managed by its original father and son owners as a model of progressive industrial relations and social services for its employees. But, as happened to many companies undergoing the transition from mechanical to electronic technology, the shift from typewriters to computers threw Olivetti into great difficulty, from which De Benedetti's takeover rescued what has come to be known as the IBM of Europe. This success made him a power within Milanese financial circles.

From his base in Milan, De Benedetti had created a financial holding company—what we would call a conglomerate—in which the strategy was to combine three broad divisions, engaged in finance, per se, in synergistically interrelated domestic manufacturing, and in the management of overseas operations. In the course of its scanning of the industrial terrain for potential takeover opportunities, this "holding" focused in on the food processing equipment sector. After doing so, the holding quickly discovered, coveted, and then acquired Sasib.

The modern processing and packaging of food products, from tobacco and beverages to pasta, fruit, and vegetables, has become a capital-intensive, high-tech business. In 1977, the manufacturing division of the De Benedetti empire, Compagnie Industriali Riunite (CIR), acquired Sasib, which was by then already a highly sophisticated, high-tech producer. From its new base of operations in Bologna, the conglomerate then proceeded to reorganize an important segment of the food processing machinery sector in north-central Italy.

First, Sasib was listed on the Italian stock market in Milan. Then it was designated head of the engineering and mechanical sector within CIR, and legally became a holding company itself. Once these arrangements were completed in 1984, Sasib—fortified with the tremendous financial resources of its parent conglomerate—began to acquire several of the *piccole imprese* in and around Emilia-Romagna, some of which it had formerly traded with as one among many collaborative competitors. Between 1984 and 1988, the Sasib Group (as it was now officially known) wholly acquired eight previously independent Italian firms, and acquired or established operations in Greece, the United States, Switzerland, Brazil, and the Netherlands.[11]

According to Giuseppina Gualtieri of NOMISMA, an industrial economics research institute in Bologna, one by one, the acquired firms were absorbed into the Sasib Group: Manzini (with almost 200 employees), Simonazzi (with nearly 700), and Sarcmi (with 70) located in the city of Parma, to the west of Bologna; Comaco (also known as Montecchio)(with some 140 workers in the town of Reggio Emilia); Orlandi (with about 50 employees) located to the north, in Verona; Ricciarelli (with almost 100 workers) in Pistoia, to the south; and Idrosapiens (with 100 workers) in Turin, to the far northwest. With its 10 percent share of the world market in machinery for canning fruits and tomatoes, Manzini was an especially attractive prize for the De Benedetti conglomerate.

The NOMISMA administrator Patrizia Faraselli and I met with Sasib's chief executive officer and chief engineer Vacceri, in his modest quarters, air-conditioned to shelter us from the heat of a July 1989 Bolognese afternoon. He explained to us how he had come to this position and how Sasib had been transformed from one (albeit unusually large) independent Emilian engineering firm among many into the core holding company within an increasingly rationalized, conglomerate-managed "quasi–vertically integrated" (Vacceri's characterization) system of companies. Vacceri had been an electrical engineer with Olivetti during the early 1960s. For a short time, he worked for General Electric and then (when that corporation abandoned the computer business) with Honeywell, for whom he became the director of manufacturing in Italy and later, the head of strategic planning and marketing. (Not your typical small entrepreneur, nor a displaced Fiat factory worker.) With such experience, Vacceri was an obvious candidate when De Benedetti tapped him to become CEO of the newly formed Sasib Group.

Vacceri asserted that Sasib's explicit objective was continued expansion, both within Emilia-Romagna and beyond its borders, through the strategic acquisition of customers, competitors, and suppliers, financed by De Benedetti's own merchant bank. Consistent with contemporary best-practice thinking, Sasib, too, would rely increasingly on the outsourcing of secondary production activities that had previously been performed inside the Bologna facilities ("but of course we retain all design work in-house," Vacceri commented).[12]

Once one of the previously independent *piccola impresa* is acquired, the Sasib Group typically (although not invariably) replaces the former managers with technically sophisticated staff brought in from the outside. The central staff members in Bologna (called "technical operatives") study lists of all of the small subcontractors formerly employed by the newly acquired firm, search for commonalities and potential synergies, and impose criteria for what it takes to be a qualified supplier to the Sasib Group—much in the way that Fiat, Ford, and Toyota have established such tiered (prioritized, hierarchical) subcontracting systems (I closely examine such management practices in chapters 6, 7, and 8). The firms that make the list are then rank ordered, again by the Sasib Group. From this point on, the newly acquired *piccola impresa* is expected to work wherever possible with subcontractors on that list.

It is not only the individual member firm's freedom to make its own subcontracting decisions that is curtailed after acquisition. The extensive technical sophistication of the region's producers gives any individual company many potential choices for either performing high-level computer-controlled machining operations in-house or at the site of a neighboring small firm. Now, however, in an explicit quest for intrafirm (in contrast to regional) economies of scale, Sasib has acquired an expensive, sophisticated flexible manufacturing system (FMS) for its Bologna facilities (as noted in the previous chapter, FMS is an interconnected system of versatile machine tools, coordinated through complex computerized controls). In order to amortize its investment in this new equipment by keeping it running as continuously as possible, Sasib now requires that *all* complex machining work required by any of its newly acquired subsidiaries in the regional network be sent to Bologna for processing. A similar rule has been introduced with regard to any manufacturing processes involving stainless steel; that work, too, must now be performed in the Bologna plant. Finally, there is an inclination by management to curtail the acquired firms' freedom to sell part of their output to old and new customers of their own (including for export), with Sasib Group insisting on controlling all contacts with the market.

Each of the businesses within the Sasib Group must prepare a three-year rolling strategic plan for examination by the group's auditors. These plans are then approved by Vacceri and his technical operatives or returned for modification. The Group uses these plans to inform its control of the budgets of each member firm, and thereby their ability to modernize.

Highly specialized engineering companies such as Sasib, Manzini, and the others have always strongly emphasized the customization of machinery, tools, and systems for particular clients. They pride themselves on providing extensive after-sale support services to these clients, including installation of the transfer lines, training of the customer's engineers in how to use the new machinery, and the ready provision of spare parts. Because the kinds of machinery made by Sasib (and by other flexibly specialized Italian engineering firms such as Turin-based Comau, also formerly an independent company but, since 1982, a wholly owned subsidiary of Fiat, which acquired it in order to have access to a dedicated manufacturer of industrial robots) involve close integration of electronics with mechanics, Sasib's member firms work more and more with computers—hence, the synergy with Olivetti.

Yet Vacceri bemoans this historical orientation toward customization. The problem, he suggests, is that it makes sales growth too uncertain. In order to provide a more reliable, steadier flow of profits to the parent conglomerate, a greater degree of standardization of the product line is needed, he thinks. Sasib is attempting to move in that direction, by learning which features and services are most in demand by its past customers, building these into new product lines, and then displaying them at trade fairs and in catalogues. The Sasib Group's explicit objective is to reduce significantly the historic orientation toward customized production. Why? In order to better meet the *financial* targets of the parent conglomerate in Milan.

Franceso Garibaldo is the charismatic head of the metalworkers' union in Emilia-Romagna. Well known to visiting American scholars and trade unionists, he has long experience in the Italian labor movement. Garibaldo has a reputation as a strong advocate for the continual technological upgrading of manufacturing companies—provided that they and the regional governments continue to invest in the comparable development of the skills of the workers who must install, program, operate, and repair the new machinery in the coming era of what they (like the Japanese) call "mechatronics": the wedding of mechanical engineering with microelectronic controls.

Gualtieri and I interviewed Garibaldo in NOMISMA's Bologna offices. As Garibaldo sees it, what the De Benedetti family has done is to "buy into the district" by "buying up an entire production network." From the union's point of view, it is now facing a genuinely oligopolistic sector, where before it was dealing with a highly decentralized set of independent firms.[13] Naturally enough, Sasib still wants to negotiate with its unions on a plant-by-plant basis. Recognizing the new reality—the penetration of the district by a major conglomerate, and the systematic rationalization of production under the hierarchical control of the Sasib Group—the metalworkers' union and its parent national confederation of workers want to establish collective bargaining at the level of the holding as a whole.

Gualtieri has conducted research showing that the De Benedetti–Sasib conglomerate acquisition story is by no means unique in Emilia-Romagna. Other food processing machinery operations long part of the Bologna district have been acquired by outside Italian financial holdings, with a similar subsequent process of quasi-hierarchical rationalization. Moreover, major foreign multinational food products corporations, such as Nestlé, Kraft, and Unilever, have been "invading" Emilia-Romagna as well, acquiring the best of the *piccole imprese* and reorienting them toward the parent corporation's interests. In fact, during the 1980s a wave of both domestic and cross-border mergers and acquisitions, and a general growth in the economic importance of *big* firms, swept through the Italian economy as a whole.[14]

There are distinct advantages to belonging to a well-financed, technologically sophisticated global corporation. As has been said repeatedly by many observers of the Italian scene, the *piccole imprese* are superb at design and production but generally have limited financial resources and (with the occasional exceptions of such unusually successful global competitors as Sasib and Comau) spotty knowledge of foreign markets. Affiliation with a De Benedetti, a Kraft, or a Unilever brings with it access to "deep pockets" and worldwide marketing capability. Unified management control offers the potential for enforcing the implementation of such productivity-enhancing practices as just-in-time inventory management on otherwise recalcitrant local managers who may have become comfortable with what Vacceri calls "the old ways of doing things." And certainly there is no question that conglomerate acquisition of Sasib has brought with it a measurable, observable increase in what was already an extremely high degree of technological sophistication.

But while absorption into the orbit of a multinational conglomerate may be good for the individual acquired firm, and even for the international competitiveness of the Italian economy as a whole, the long-run impact on the metalworking districts of Emilia-Romagna remains to be seen. Managerial priorities on achieving economies of scale have reemerged as a specific corporate objective, as has the subordination of production arrangements to meeting financial targets set by the parent conglomerate. Decisions about where to produce, which technologies to employ, and which subcontractors are qualified; the freedom to deal directly with customers; and the locus of industrial relations have all been disrupted by the penetration of the districts.

Perhaps even more important, the big firms are becoming increasingly dominant in *distribution* and in *finance*. As the British economic geographers Philip Cooke and Kevin Morgan put it, following their own fieldwork in Emilia-Romagna:

consistent with . . . economic integration and the globalisation of
production [big firms gain access to one another's markets] through
the formation of complementary strategic alliances or joint ventures
with other large firms. [Small and medium-sized enterprises, or]
SMEs find this form of competition very difficult to combat
because, in many cases, they are price-takers and product-followers,
i.e., they react to the innovations of others. When price and product
leaders leave space, this works. But if large firms are seeking to introduce oligopolistic rule—by exerting stronger control of distribution
chains, for example—this makes life far more difficult for SMEs.
This is the phase in which the Emilian SME sector now finds itself.[15]

Thus, as the regional planners Flavia Martinelli and Erica Schoenberger
express it in the title of a recent paper: "Oligopoly Is Alive and Well" in
Italy.[16]

Whether cooperative competition, trust, and the other underlying
principles that have been identified as the glue holding the distinctive
Italian geographically based production networks together can withstand
the intrusion of hierarchical, concentrated economic power into the
region is an open question. In the 1990s, the regional government of
Emilia-Romagna is not resisting—indeed, is actively encouraging—closer
integration into the rest of Europe—as it should, of course. The planners' goal is to make the region an even more attractive source of production subcontracts, or a desirable site for companies that are relocating
from elsewhere in the European Community (EC). Such regional centers for technology transfer as the Center for Information on Textiles
(CITER) in Modena are being actively encouraged to support such integration, even by subsidizing outside firms (partly with EC funds). The
Catch-22 is that, to the extent that this strategy is successful, "the possibility arises that centers like CITER might find that their interests begin
to diverge from the local districts which they were set up to sustain."[17]

The penetration into the Emilian metalworking districts of financially
powerful corporations from the outside and the subsequent introduction into the districts of elements of oligopolistic market power constitute one type of transformation away from the ideal typical model of
cooperative competition. Alternatively, some of the small firms in the
districts are evolving internally into more powerful business entities,
transforming the previously cooperative and collaborative nature of the
relationships among the firms within the area into arrangements that

start to look suspiciously like the core-ring form of control currently favored by the Japanese, American, and northern European industrial giants—a form discussed at length in chapter 6.

A particularly dramatic example is offered by the famous Italian clothing manufacturer Benetton, which has evolved in less than thirty years from being one of many small "coordinating" firms embedded within the Veneto region into the financial, marketing, and production planning center of a worldwide system of franchise clothing stores at one end of the spectrum, and low-wage—in some cases, sweatshop—manufacturers at the other. Let us examine this type of district transformation more closely.

THE EMERGENCE OF HIERARCHY WITHIN A DISTRICT-BASED PRODUCTION NETWORK: THE UNITED COLORS OF BENETTON

Everyone knows Benetton: that's the company whose delightful little retail outlets feature those brightly colored sweaters, T-shirts, and jeans in the window. At last count, there were some 650 of those shops all across the United States. During the 1980s, 5,000 Benetton stores sprouted up in 79 countries—more than 300 in Japan alone.

Students of manufacturing systems certainly know Benetton. That's the company with the far-flung production system with its headquarters in Treviso (near Venice, in the Veneto region of northeastern Italy), with 8 family-owned factories and warehouses and another 500 "independent" subcontractors, 90 percent of which are also located within the Veneto (in a manner reminiscent of Japan's "Toyota City"), and with an uncounted but large number of lower-tier contract shops and home workers distributed throughout southern Italy, in Turkey, and elsewhere.

And business school researchers know Benetton as the first Italian integrated textile and clothing maker—perhaps the first of *any* nationality—to distribute its products exclusively through franchises, much like McDonald's popular fast-food restaurants. Moreover, it was the first Italian company to join the New York Stock Exchange. Clearly, the maker of the "United Colors of Benetton" (the company's slogan and internationally recognizable logo) is well on its way to becoming, in the words of the managing director, Aldo Palmeri, "a real multinational."[18]

This was not always so. In her extensive studies of the history and evolution of Benetton, the Italian trade union researcher Fiorenza Belussi documents that Luciano Benetton worked during the 1950s as an assistant in a small textile shop in Treviso.[19] His sister, Giuliana, worked in a small knitwear factory. In 1957 they joined forces: she had a talent for designing knitwear; he would collect orders from manufacturers, and she would produce them at home. What would later become one of Italy's most internationally competitive firms thus began as an arrangement—part of what historians call the "putting-out system"—that placed it squarely within the realm of traditional small Italian enterprise.

In 1965, the brother and sister started up a small factory in nearby Ponzano and hired sixty employees. With Luciano continuing to focus on marketing and Giuliana on design, their brother Gilberto joined the new family business as the head of administration and finance, and their brother Carlo took over direction of the day-to-day production. For the next decade, the family owned and operated Benetton as one of many small knitware producers nestled within the textile and clothing district of Treviso, relying for much of its production on a network of small contract shops of just the sort that had formerly employed the founding brother and sister.

Between 1965 and 1975, the Benettons began putting in place the innovations in organization and technology that would eventually prove to be so successful internationally. Their timing could not have been better. Domestically, they were able to take advantage of a series of social contracts between labor, business, and government that had been erected throughout central and north-central Italy in the years following World War II, contracts that gave rise to a shared commitment to what Italians call the "progressive local model" of activist municipal government economic development policy.[20] Through this channel, Benetton received extensive financial subsidy and technical assistance during its formative years. Not all small enterprises in the industrialized world (or even elsewhere in Italy, except for the neighboring region of Emilia-Romagna) had access to such resources.

At the same time, the rich and powerful countries of Europe and North America had begun to experience the decimation of their domestic cotton (and then, later, their woolen and synthetic fiber) textile industries under the competitive assault from cheap imported goods from the Third World. To arrest this process of deindustrialization, the governments of the rich countries effectively forced on the international trading system a set of agreements for managing trade in textiles through bilaterally negotiated quotas. The first of these was negotiated in 1961. A far more comprehensive version was put in place in 1974, under the new name of the Multifiber Agreement. Suddenly, Italy, Germany, France, and the United States were protected from unlimited competition from the mills of India and Sri Lanka. Into that vacuum, companies such as Benetton moved quickly. By 1978, fully a quarter of all the revenues earned by the four siblings was derived from exports to other industrialized countries.

Today, Benetton is a major industrial enterprise, turning out woolen sweaters, cotton shirts and slacks, and jeans. More recently, the firm has begun to produce leather products, as well, such as shoes and outerwear. Each line constitutes a division, with its own hierarchical tiers of closely dependent subordinate suppliers. Whereas suppliers and subcontractors in Brusco's and Becattini's ideal typical industrial district gain part of their much-vaunted flexibility from "changing partners," the price for being a preferred supplier to Benetton seems to be a willingness to give up doing business with other companies in the district (or elsewhere). According to Belussi, about three out of four of Benetton's Venetian subcontractors in 1989 worked exclusively for Benetton.

Instead of merchandising to department stores and clothing shops that carry many different brands, Benetton sells its products only to the franchises (typically, independently owned) that bear its name and that arrange their displays and package the product according to strict standardized procedures and formats designed in Treviso. Each individual outlet is connected to regional and world headquarters, and to highly automated warehouses, through a sophisticated computer network. In this way, the company is able to continually receive up-to-the-minute information on which product designs—and especially which *colors*—are selling in which markets, at what prices.

And what colors! All colors of the rainbow, especially the lightest and brightest, appear in Benetton's sportswear. If the franchising of manufactured goods was Luciano Benetton's great innovation, the decision to perform the coloring process as the *last step* in the production process, just before shipment to the stores, was Carlo's biggest idea. Holding off the dyeing of the goods as long as possible and then using computer-programmable equipment to enable the producer to shift from one color-specific order to another with minimal machinery adjustments, in order to meet market preferences as quickly as *they* shift, was a production-planning innovation that other European and some American firms were eventually forced to emulate (notably, Britain's Marks & Spencer department store chain). In a sense, Benetton embellished the Toyota system, by extending the principle of just-in-time delivery from the "upstream" supplier-to-final-assembler stage of production to the "downstream" manufacturer-to-retailer stage of distribution.

These innovations and the internationally competitive success they have engendered have helped transform Benetton into a kind of company different from those we have been taught to associate with the industrial districts. Indeed, in its evolving role as a major oligopolist, rather than a "cooperative competitor," Benetton has measurably undermined at least part of the fabric of the Veneto's small firm network. It has done so by imposing an extreme form of hierarchical control on a production system that had previously been characterized by far more fluid interfirm arrangements. The headquarters of the business remains in Treviso. Eight directly family owned, partly unionized, relatively high wage factories and looms operate within the immediate region. This is where most of the design work and the grading, marking, cutting, and dyeing operations take place. Extensive use of computer-assisted design/computer-assisted manufacturing (CAD/CAM) technology allows the company to quickly and efficiently translate designs into the direct production of sleeves, collars, bodices, and legs for different items of clothing. Benetton also owns its own spinning mills but often retains subcontractors to perform the weaving—a mix of vertical integration and nineteenth-century work organization that belies any simple characterization of the evolving production network.

On the other hand, nearly *all* of the labor-intensive assembly, pressing, and embroidery work is contracted out. Here again we see an analogy to Toyota (a company whose paradigmatic supplier model is closely examined in chapter 7). There is a first tier of midsized firms that work closely with the family-owned factories at the top of the pyramid. Although legally independent of the family or the corporation, these key subcontractors tend to be owned by former Benetton managers. For example, one owner is the ex–head of personnel for the corporation, while another had been the technical director of the jeans division. These contractors, who interact daily with Benetton itself, and who in many cases receive finance from and regularly exchange technical experts with the plants belonging to the four Benettons, are in turn expected to manage successively lower-order tiers of suppliers—situated within the Veneto region and farther away, in southern Italy, Turkey, and other low-wage regions.

These lower-tier suppliers typically are very small (fewer than twenty employees) and highly specialized (making or doing only one thing for their clients), are almost never unionized, and are owned and run by small-town or rural *men* employing a workforce that consists predominantly of women under the age of twenty-five. In the districts where they are powerful, the unions have in the past been able to negotiate roughly equal rates of pay regardless of firm size, and government inspectors have been vigilant in enforcing the generally equitable labor laws. Still, Belussi shows that within Benetton's production chains, wages in the lowest-tier workshops are indeed lower than at the top. Other recent Italian research documents a significant size differential in average annual wages across all manufacturing enterprises in the country.[21]

Moreover, employees at this level commonly avoid paying any taxes (and managers commonly evade paying into social security). The labor laws on unjust dismissal do not apply to firms employing fewer than twenty workers. And the artisanal laws grant such microenterprises various subsidies, such as a 1984 law's awarding of tax breaks to firms that employ "youth" workers under the age of twenty-nine on temporary employment contracts. To the extent that such arrangements cheapen production costs all the way up the line, they ultimately constitute a public subsidy to Benetton.

As for skills, at the higher levels of the pyramid, occupations such as that of computer programmer and cutter entail a great deal of skill and are compensated commensurately. At lower levels in the hierarchy, Belussi reports from her extensive interviews that managers expect newly hired workers to be able to carry out assigned tasks "without loss of productivity" within anywhere from just a few days to, at most, six to twelve months on the job. Most observers agree that the pace of production in the small contract shops is generally very intense.

This is not unique to Benetton. From his studies of the knitwear industry in Modena, in Emilia-Romagna, the State University of New York–Stony Brook sociologist Mark Lazerson concluded that the decentralized putting-out system, while attractive to small-shop owners and to the big firms to whom they subcontract, was problematic for the workforce's social development:

Its limited skill [requirements] discourage educational attainment; the average education of both employer and employee is between five and eight years. Some of its job tasks test the limits of human endurance; the young girls who begin work on the steam presses at

16 years of age have only five years before physical exhaustion will send them in search of less demanding work. Indeed, the steam-press position silently exacts its toll, at times exposing its young female operators to early sterility. Older home workers are some-times subject to premature cataracts because of the fine needles, threaded by hand, on the special machines used to attach collars and cuffs to sweaters. But if knit wear work does not sap the body, its monotonous routines of buttoning and folding sweaters day in and day out often dulls the mind.[22]

Finally, although we have very little hard data on this aspect, many Italian researchers and visitors from elsewhere report that, at the very bottom of the textile/clothing production system hierarchy, in Italy no less than anywhere else in the world, stand both registered (docu-mented, legal) and illegal *home workers*—unskilled (or, at any rate, pow-erless), receiving the lowest wages and no legislated health and safety protection. Such home work appears to be more prevalent in the south of Italy than in the more urbanized north, and may even be declining in the latter, as Lazerson suspects. But even on this question, the visitor gets contradictory stories.

At any rate, no one questions the fact that the putting-out system seems alive and well—and that it has contradictory implications for eco-nomic development and social policy. It would not be fair—let alone scientifically rigorous—to simply draw a box around the higher-wage, union- and community-regulated core regions of the north and call that "the system." Without the lower-level subcontract shops and home workers of the south, the Benetton production system could not survive. As discussed later, this is also true for other regions around the world whose economies are organized as industrial districts.

So what does the case of Benetton represent? Several times I have suggested an analogy between the evolution of this originally quintessentially small artisan-based Italian company, embedded within a region of industrial districts, and that of giant Toyota, with its core of closely owned large firms and its rings (or tiers) of highly dependent subcontractors and suppliers. A number of European social scientists, notably Kurt Dohse and his German colleagues, have explicitly pro-posed such an analogy, invoking the term "Toyotism" to describe what they see as the emergence of new forms of oligopolistic power.[23]

Belussi, Martinelli, and other Italians call this evolving form "decentralized Fordism"—a construct very similar to what I earlier named *concentration without centralization*. Whatever the academic jargon, the point is that, all at the same time, we are witnessing the introduction of state-of-the-art technology and the most innovative methods for reorganizing the structure of a company, occurring side by side with (indeed, facilitating) the segmentation of the firm into what Belussi calls a "propulsive 'core' and an 'adaptable' periphery." In the process, the industrial district is transformed from the ideal of cooperative competition among equals into a space that, at least in part, has been restructured to serve the needs and interests of a powerful lead firm.

Yet a third story of how the Italian industrial districts may be changing concerns the erosion of cooperative competition in the famous woolen textile district of Prato. A complete accounting of the crisis tendencies in Prato would take us far afield. Here, I wish to report only on two aspects of the problem. The first is the story of how the local production system has fallen prey to the unequal power of the middlemen—those who, until recently, were the principal source of the region's profitability because of their contacts with the buyers representing industrial customers throughout Europe and America, and also their function as systems integrators. The second problem concerns the difficulties now being encountered by the tiny manufacturing firms in the district because of changes in the global demand for their products—difficulties that are a direct result of these firms being *so* very small and of the districtwide production process being *so* very "disintegrated."

THE NEGATIVE COMPETITIVE CONSEQUENCES OF EXCESSIVE FRAGMENTATION: THE CASE OF PRATO

Prato has come upon hard times. In the 1980s, according to research conducted by the Neapolitan regional planner Martinelli, for the United Nations, the district was experiencing a dramatic decline in its exports of thread and fabric.[24] The few big firms in Prato were sourcing growing volumes of components from branch plants they had built (or bought) in Portugal, Greece, and Turkey. Cheap items were coming into the region (and into the country) from Asian producers, while high-level textiles were increasingly entering Italy from Germany—the same country that, up to this point, had been a major customer for the knitted woolens and other items manufactured in Prato's industrial district.

This turnaround—which many Italian researchers whom I interviewed call the "crisis of Prato"—represents a serious challenge to the received representation of the industrial district. Perhaps no other region in the contemporary world (with the possible exception of Silicon Valley, which is examined in the next chapter) has attracted so much interest and been so closely studied as have the several small towns centered on Prato, located in central Italy in the vicinity of Florence, in the region of Tuscany. It is easy to see why. Since at least the 1960s, Prato, with its production system dominated by true microenterprises, was the wonder of the industrialized world.

To be sure, the region has been an important producer of woolen textiles (yarn, fabric, and, more recently, knitwear) since the medieval era. Rather more recently, in the half century prior to World War II, a cluster of medium-sized and large firms produced mainly for the low end of the market, using rags and pieces discarded by firms from other regions of the country as their basic raw material. With the deep economic slump that set in after the war, lasting until the mid-1950s, most of the big firms underwent a process of vertical disintegration, shutting down many mills and factories and selling off or renting machinery (especially the looms) to their recently laid off workers.

These ex-employees became the owners of their own tiny workshops. By the estimate of the NOMISMA researcher Silvano Bertini, in 1990 the average size of a Pratese workshop was 4.3 persons—typically a father with his brothers or sons and perhaps a few hired apprentices. Into this mix of elements there came, after the mid-1950s, a prodigious number of independent merchants—some from old, rich families, but a surprising number of newcomers—who proceeded to organize in the valleys and villages of the area a set of production arrangements that would come to be known around the world as the "Prato system."[25]

For a time, they were immensely successful. From 1970 to 1981, textile exports from Prato increased in value by 137 percent, after inflation (compared with a rate of 93 percent for the entire Italian textile sector as a whole). The number of local firms grew by 35 percent, and employment by almost 20 percent. In 1970, Prato employers had accounted for about 33 percent of all the wool industry's jobs in the country; by 1981, that share had risen to well over 40 percent.

The textile producers of Prato have for many years had close access to a textile machinery industry that manufactured computer-controlled looms and was situated in Prato itself and to the north, in the adjacent region of Emilia-Romagna (the home of Sasib). For a time, at least, there seemed to be what Martinelli describes as significant "synergistic interactions" between the users and the makers of these machines, favoring a "continuous process of technological innovation."[26]

But the most important source of the district's remarkable success as a global competitor during the 1970s was said to be its exceptional degree of disintegration and specialization at the level of production, combined with a system of coordination or governance that emphasized the region's ability to respond expeditiously to even the most unusual, smallest demands from the market. The customers for Prato's wares consist of domestic and foreign (principally German, British, and American) buyers for clothing manufacturers. These buyers and sales agents from Prato meet regularly in a game of continual recontracting. These agents are representing those merchants or middlemen of whom I spoke previously: the *impannatori*.[27]

At the base of the system, working almost entirely on short-term, project-by-project contracts to the coordinating merchants, are more than 8,000 tiny direct producers—those typically family-run microenterprises, many of them owned by skilled but now aging refugees from the earlier era of the large mills. Their tiny size qualifies them for the designation of "artisanal" workshops under Italian national law—which, as noted earlier, effectively (if not always legally) frees them from paying social security and from honoring restrictions on arbitrary dismissal of employees. These shops typically perform only one very narrow and specialized phase in the entire production cycle.

The actual production process is straightforward. As each new contract with a buyer is executed, the merchants import and store the raw materials (especially the wool), assemble a network of small firms for various stages of manufacturing, collect the product, and see that it is delivered to the customer. At each stage of production, storage, and physical transportation of goods-in-process and finished articles, there is a prodigious amount of interfirm subcontracting (from sample data collected in Tuscany and nearby Emilia-Romagna, mostly on the clothing and knitwear parts of the sector, Martinelli estimates that perhaps half of any firm's cutting operations, and four-fifths of its knitting operations, are typically contracted out).

In general, in Italy as elsewhere, such extensive fragmentation of industrial organization and reliance on putting-out is usually accompanied by a regime of low wages, low unionization, patriarchal social relations, infrequency of social security payments, and the continued existence of unregulated home work performed by mothers, wives, and sisters.[28] Indeed, family members are sometimes impressed into working for no wages at all.[29] Prato is said to be an exception to this generalization, although I have found it surprisingly difficult to obtain hard data specific to the area or to this question.

Knowledgeable local observers such as the sociologist Vittorio Capecchi believe that at least some of the more informal working arrangements—especially those involving women, immigrants, and youth—really are voluntary, providing much-needed supplementary income, training opportunities, or bridges to more permanent jobs later on in the worker's life. Writing about Emilia-Romagna, Capecchi calls this "complicit informalization."[30] I think that observation misses the point. Whether voluntary or not, such work arrangements contribute to a feedback process that, in times of trouble, may actually undermine the long-run competitiveness of the district as a whole, for example, by reducing the incentive for firms to modernize. This problem of a low-wage, low-labor-standards periphery supporting otherwise often skillful, high-wage industrial districts will come up again, in the following chapter, when we look at the "California model" of regional development. In chapter 9, I focus mainly on these arrangements: what I have been calling the dark side of flexible production.

Whatever the degree of labor market dualism within Prato, per se, the district's ability to quickly ("flexibly") combine and recombine participants in order to fill even the smallest market demands served the region well for more than a decade. This was a golden age, about which so much has been written. But by the mid-1980s it was apparent that the local economy was in trouble, and that some of the very structural elements that had contributed to its earlier successes were now impeding its ability to respond to the crisis. What were the sources of this crisis?

Throughout the 1980s, with economic recovery in the United States and an economic boom in Western Europe, not only was there a tremendous increase in the aggregate demand for textiles and clothing, but buyers were insisting that their suppliers deliver an ever-greater diversity of designs and products, well beyond the carted woolens that had been the mainstay of the Pratese producers. Customers were also demanding ever-shorter delivery times, forcing the merchant systems integrators to find ways of shortening the cycle time from the receipt of a new order to delivery of the final output. And all of these pressures were compounded for the microenterprises and middlemen of Prato by the resurgence of the big textile and clothing companies, in Italy and elsewhere: freshly rationalized, equipped with a whole new generation of automated machinery, and with highly trained and youthful teams of dedicated sales agents operating in both domestic and foreign markets.

Pressed simultaneously by the need to diversify continually, to speed up delivery times, and to find new market niches that would enable them to coexist with the multinationals, what could the Pratese do? Could it be that the system had become *too* fragmented, *too* disintegrated? Detailed time and motion studies conducted by the researchers from NOMISMA reveal that, over the course the 1980s, the fraction of total contract cycle time during which goods were either in transit or in storage and during which machines were idle—in other words, the percentage of time that actual production was *not* taking place in Prato—rose steadily. Minor traffic accidents among trucks and carts carrying goods-in-process from one small shop to another were on the increase. In short, economic success in the context of extreme fragmentation was taxing coordination at the level of the district as a whole.

Moreover, at a time when the region might have benefited from the retirement of the oldest or least efficient of the small entrepreneurs, from the shuttering of their workshops, and from the aggregation of at least some of the smaller enterprises into larger organizations better able to grasp new opportunities and to coordinate the manufacture of a more diverse product mix, such an adaptation was being blocked by some of the very conditions that had propelled the district in earlier periods. In the formal language of industrial economics, the low wages and labor standards, the market worthlessness of the older machinery in light of the emergence of the newer vintages, the family basis of so much ownership, and the strong craft tradition of the place all combined to erect powerful "barriers to exit." Put simply, it just did not pay to leave the industry.

In such a situation of growing excess capacity, it was inevitable that the small owners would resort to mutually destructive price cutting; they had no choice. But then, so long as they and their families could absorb the resulting deterioration in their incomes, such behavior had the indirect effect of sustaining the profits of the *impannatori* and sending them a signal that all is well, when in fact it most certainly was not. To make matters worse, if anyone *was* getting out, it was the younger, most innovative owners: those who had borrowed during the 1970s to buy new machinery, who had continually modified it (mainly to speed up the pace of production), and who, in a regime of falling prices, could no longer make their payments and so chose to drop out of the industry. Thus, a process of natural selection has been leaving Prato with an increasingly aging and less efficient mix of producers.

In this increasingly fragmented and untrusting environment, even an otherwise sensible plan for using high-tech telecommunications to enhance the region's collective productivity was unable to take hold. The plan was project PRATEL, put forward by the trade association SPRINT, by which name the project became popularly known. SPRINT was first proposed in the early 1980s by RESEAU, a private center in Milan specializing in new information technology, and by ENEA, the national government's energy agency. In 1983, the three main business associations representing employers in Prato, Florence, and Tuscany; the Florence Chamber of Commerce; and even the regional secretariat of one of the unions signed a formal agreement. Under its terms, ENEA, with national funds and with a supplement from the European Economic Community in Brussels, would install networked computer terminals on the desks or workbenches of every small and large producer and merchant middleman in the Prato district.

The idea was that if the middlemen could get information on potential and actual market contracts into the hands of the direct producers even more quickly, and if the latter could give the merchants even more and speedier detail about the small firms' technical qualifications and availability for any particular project, then the productivity of the entire district would be enhanced. By 1988, three hundred terminals had been installed, a subcontracting exchange service had been set up, and advertising campaigns and training sessions were being established. In typical Italian fashion, the government, the companies, the universities, and the unions were all involved.

But walk into the office of any of the artisanal small firms, and what you see is that the terminal is usually turned off, even in periods of low demand. Why? Because the actors, the participants, and the partners in the cooperatively competitive Prato regional production system have been too *distrustful* of one another to reveal the details of their operations (in the view of the researchers from NOMISMA, the *impannatori* have been especially uncooperative). And again, it is possible to locate the problem in some of the same structural characteristics of the system that had formerly made it so profitable.

To be fair, some observers believe that it was not mutual distrustfulness that was the source of the problem, but rather the bureaucratic attempt to substitute remote for face-to-face communication—a cornerstone of the district system. Perhaps. But even so, this only provides further evidence of how "trust," when it becomes a force for defending old ways, can actually *suppress* innovation.[31]

In any case, interviews with a number of the middlemen revealed that they see themselves as being continually played off against one another by the buyers, as the latter try to get prices down. Everyone sees this as a normal part of the game. What each individual *impannatore* has going for him in this game is his *information*—about the different buyers, about *their* forecasts of market demand and style trends, and of course about the technical capabilities of the different producers within the district. Each of the small producing firms will work at one time or another with anywhere from two to as many as ninety different *impannatori*, another structural aspect of the existing system which, as noted earlier, has always been thought to be a strength.

But networking can be a two-edged sword. If a middleman reveals to the small firms with whom he is presently contracting valuable information about (say) future market opportunities that that middleman has picked up from his buyers, the very density of the production network means that this information is likely to get around. The NOMISMA thesis is that the *impannatori* simply cannot afford to let that happen; privileged information about markets—especially foreign markets—constitutes probably the most important specific asset that the merchant coordinator possesses. Information is his stock in trade. He hangs onto it with his life. And so data on future market opportunities simply do not ever find their way into the computer network. Drawing on the firsthand experience of Carlo Trigilia, an Italian trade union researcher, even Sabel acknowledges that the middlemen/merchants feared that SPRINT would "ultimately make their vast knowledge of the area's productive capacity superfluous."[32]

And so we see the contradictory nature of a highly fragmented, highly disintegrated production system such as that of the Prato industrial district. Much writing on the districts, for example, from Becattini in Italy and from Piore and Sabel in the United States, emphasizes the communal character of information; hence the frequent quoting of Alfred Marshall's poetic characterization of nineteenth-century Sheffield, according to which the secrets of industry are "in the air."[33] In Prato, information on *production* techniques does diffuse widely and rapidly, because owner-operators talk to one another, because families live in the same social context, and because apprentices move easily from one workshop to another. Moreover, the private banks, the cooperative associations, and the credit unions all stand ready to make it relatively easy for the small firms to borrow in order to finance investments in new technology.

The problem is that firms—whether three-person workshops or branches of multinational corporations—cannot judge whether it *pays* to invest in new technology unless they can forecast the level and character of future demand. And *that* kind of information does not seem to be in the air in Prato. It is privatized, rather than collectivized. The problem is not the presence or absence of the means of communication; telenetworking the small firms within a region—with one another and with their big-firm customers—still seems an eminently sensible economic development project. The problem lies in the *excessive fragmentation* of the production system, in the face of heightened foreign competition.

FURTHER EVIDENCE OF CRISIS AND CONCENTRATION IN THE "THIRD ITALY"

I have presented three case studies on the recent evolution of the Italian industrial districts, based on firsthand observation and on research conducted by both Italian and other foreign scholars. In a private communication, Alejandro Portes, the eminent Johns Hopkins University sociologist, has offered his own interpretation of my findings from the Third Italy:

> The three examples . . . have one thing in common: they all show the difficulties brought about by *the very success* of the model in an earlier period. Sasib becomes so successful that it is taken over by a Milanese conglomerate; the little Benetton shop becomes a multinational; and the Pratese system could not cope with the demand created by its own past achievements. . . .
>
> Perhaps then the story is that these socially embedded productive experiments *can* work and be viable for a while, but they cannot sustain themselves indefinitely against the logic of global capitalism. Either they are taken over, become MNCs themselves, or are torn apart by demands that their social fabric can no longer accommodate.

But in all fairness, could it not be that these stories are merely "outliers"—exceptions to a record of continuing success with and refinement of this much-publicized form of small firm–led regional economic development?

I think not. Other American visitors to Italy who have personally conducted detailed case studies on other districts are now returning home with stories of the "superexploitation" of wives, children, immigrants, and elders in order to keep labor costs as low as possible, given growing competitive pressures from Third World exporters of increasingly better made shoes, clothing, furniture, and even machinery. Or these other observers are reporting that, as in my three cases, even where a district continues to be competitively successful, economic power relations are changing, with a definite reemergence of concentrated corporate control that threatens to replace—or at least to challenge—the model of cooperative competition among small and medium-sized independent enterprises that had fired the imaginations of so many.

For example, a year after I returned from Italy, a visit to that country was made by Stuart Rosenfeld, a veteran American regional economic development policy analyst who favors Piore and Sabel's vision of a local development strategy based on networks of socially embedded small firms. Rosenfeld, the founder of the Southern Technology Council, and a leading advocate of the small firm–led development model, went to study regional economic development policy making in action in the industrial districts of Italy and Denmark.[34] In the course of his visit, he encountered Marazzi, perhaps the single most important manufacturer of ceramic tiles in Italy—and therefore in the world (among the firm's many global customers is the McDonald's chain of fast-food restaurants, for whom it makes floor tiles). With factories in the United States and Spain, Marazzi is based in the town of Sassuolo, located southwest of Modena within the region of Emilia-Romagna. The area around Sassuolo certainly fits the picture of a geographic "cluster" of interrelated businesses—an economic agglomeration. Some companies specialize in designing new tiles, others make the equipment to form them, and still others provide engineering and marketing assistance. And Marazzi is the largest of the actual manufacturers.

By the time Rosenfeld arrived, Marazzi was well into a process of wholly acquiring and vertically or horizontally integrating ten other previously independent local firms. The firm's managers told Rosenfeld that, to increase market share in the future, it was going to be necessary to gain "more control over phases of production for which [the firm] once depended on others." Thus, Marazzi was buying up shares of other suppliers, and (as I saw in Sasib) proceeding to designate the other firms its other subcontractors can work for. Rosenfeld left Italy, confident that the ceramic tile district was alive and well. But whether it would continue to function as a decentralized production network characterized by symmetric power relations among its own *piccole imprese* remains to be seen. To the extent that Marazzi's corporate strategy deforms the previous structure of the area, there will be cause for concern.

Another observer is Michael Blim, who teaches sociology and anthropology at Northeastern University in Boston. Three times during the 1980s, Blim conducted extensive ethnographic studies of the shoe, furniture, and musical instrument districts of the Marche region, immediately southeast of Emilia-Romagna, on the Adriatic Sea.[35] He focused on the fashion boot and shoe industry in and around the town of San Lorenzo Marche. The Marche emerged from World War II the most agricultural of all the regions of northern Italy. By 1981 it had become significantly more urbanized and industrialized. Between 1951 and 1987, employment in the shoe factories of the Marche increased by 600 percent. New firms sprouted up, exporting increased, and profits grew.

But since the early 1980s, average firm size, production, employment, and exports from the San Lorenzo Marche industrial district have all been falling. The utilization of the undocumented labor of children, women, and pensioners has, if anything, increased—a trend that Blim associates with a retardation in the rate of adoption of modern management methods (definitely *not* the problem in Sassuolo or Bologna). By the late 1980s, he reports the appearance of a "specter of economic decline,"[36] as suggested by several trends: heightened pressures from foreign buyers; growing competition from Third World producers; the necessity of (but social limits on) squeezing the *lavoro nero*—those working "off the books," in the "underground economy"—still further; exit from the region of well-educated young people repelled by the thought of becoming crafts workers (or, at the other extreme, frustrated by the inability of the small firms to any longer guarantee them the kind of steady work that their fathers had always enjoyed). All of these developments seem similar to tendencies observed in the case of Pratese textiles.[37]

Finally, to make matters even more difficult for the original claims about the competitive superiority of the small firm–led district model in the Third Italy, the Italian regional economist Roberto Camagni has challenged the very premise that—whether because of the district form of productive organization or not—the regions that comprise north-central and northeastern Italy *are* in fact any longer performing more successfully in international competition than the industrially mature northwest—the home of Fiat, Olivetti, and other giant multinational corporations. Camagni believes that since the early 1980s there has been a reversal in the interregional trends in manufacturing productivity growth, with the northwestern Piedmont now growing *faster* than the Veneto, Emilia-Romagna, or Tuscany.[38] The reason, suggest the British economic geographers Ash Amin and Kevin Robins, may be that the decisive defeat of the industrial unions in the northwest—beginning with the failed Fiat strike in the fall of 1980 and culminating in the entire labor movement's inability to win the 1984 national referendum on preserving the indexing of wages for inflation—successfully slowed the growth of labor costs, giving the big firms breathing room to restructure and to develop new approaches to achieving flexibility on their own terms, by creating their own "lean and mean" production networks.[39]

Evidence in support of Camagni comes from an extremely technically sophisticated survey-based analysis of the rates of adoption and diffusion of flexible automation in Italian industry, conducted by Sergio Mariotti and Marco Mutinelli, colleagues of Camagni at the *Politechnico* of Milan. They discovered that, by 1989, the northwestern region of the country had the greatest number not only of installed flexible production systems, such as computer-controlled machine tools, but also of installed flexible design systems, such as CAD-CAM (as noted in the previous chapter, this same survey revealed a strong positive correlation between size of plant and the rates of adoption and penetration of flexible automation, in every region of Italy).[40]

But might it then be that at least some of the industrial districts are coming to constitute what Amin and Nigel Thrift, another geographer, call "Marshallian nodes within international networks"?[41] Nodes, they most certainly are. But how "Marshallian"—how *cooperatively* competitive—is precisely what the evidence from all of these cases calls into question.

5

IS SILICON VALLEY
AN INDUSTRIAL DISTRICT?

In the last chapter, I showed that the industrial districts of Italy are undergoing transformations away from the locally oriented, cooperative form so widely believed to have held the key to their past success. Whatever these changes may hold for northern Italy, and for other regions like it, the larger implication is that small firm–dominated districts are probably as unlikely to lead a new wave of economic growth as are the individualistic entrepreneurs celebrated by the apostles of laissez faire, in the United States and abroad.

Another region of the world whose economic successes have been attributed by some observers to its constituting an industrial district on the Italian model is located here in the United States. This is Silicon Valley, the computer and microelectronics capital of America. Consisting of four counties (Santa Clara, San Mateo, Alameda, and Santa Cruz) situated southeast of San Francisco and west of San Jose, the Valley has been described as a full-fledged industrial district, made up of a dense thicket of mostly small and medium-sized (but also some large) firms that alternately cooperate and compete with one another. These networks of producers are said to be embedded in a local political economy that provides job training, finance capital, and an incessant flow of ideas and information about the latest design and production techniques. Although well connected to the rest of the world, the Valley has a dominant orientation toward "flexible specialists" that is said to be "Marshallian," in the sense discussed in chapter 4. That is, the district may trade with the rest of the world (and quite successfully, thank you), but *production* relationships are thought to be highly localized. Perhaps a set of arrangements that could not be sustained in north-central Italy are working well in this region of the United States. If so, then perhaps locally oriented industrial districts *are* the wave of the future, after all.

In this chapter, I show that some aspects of the industrial district metaphor fit the morphology of Silicon Valley nicely. But a deeper analysis of both the post–World War II origins of this region and its subsequent evolution and current form reveal that, at bottom, Silicon Valley was created by, and remains profoundly dependent on, major multinational corporations and on the fiscal and regulatory support of the national government—especially in the shape of the U.S. Department of Defense. As either world headquarters of (or as an important node within global networks of) big firms and their small firm subcontractors and suppliers, the organization of production is subject to many of the considerations that operate in the car, steel, and chemical industries: high-volume production, the importance of standards, and the centrality of big, resourceful institutions. These considerations are resulting in more and more of the actual manufacturing work—including the more labor-intensive, high-value-added work—being performed elsewhere. And growing direct foreign investment in the Valley, especially by Japanese companies, seems aimed more at tapping into U.S. scientific brain power (or overcoming anti-Japanese political sentiments in Washington) than in adding to the long-term productive capacity of the region, per se.

Moreover, as in the Italian case studies, there is evidence that high-tech production in Silicon Valley was, during that area's evolution, erected in part on a base of low-wage, sometimes dangerous work, which originally took place inside the Valley itself (and in the export zones of Southeast Asia). Since then, however, such work has been farmed out to the homes of women workers in what some observers call the "new sweatshops" of Los Angeles and to new factories distributed in a region extending from the Mexican border through the American Southwest and up into Oregon and Washington.

Moreover, some of Silicon Valley's most prestigious, visible companies such as Cypress Semiconductor are again outsourcing chip assembly and testing work to Southeast Asia.[1] As with our reassessment of Prato and of Benetton's production system, so here it seems only fair to complete the popular image of what makes Silicon Valley work as a regional economy by incorporating into the discussion these geographically dispersed extensions of the Valley's production system.

In sum, while networks of small high-tech firms certainly do constitute an important part of the regional production system of Silicon Valley, they are not what ultimately drives its economic growth and development. It is impossible to understand what really made the Valley take off in the first place, and what makes it a viable region today, without acknowledging the crucial role of concentrated public and private capital—however decentralized its individual units of production may be, both inside and outside of the area.[2]

SILICON VALLEY AS AN INDUSTRIAL DISTRICT: THE DEBATE

One of those periodic international crises of overproduction occurred in the merchant semiconductor industry in 1985 and 1986. At the same time, it became clear that the principle of distributed processing was decisively eroding the use of large mainframe and so-called minicomputers in all but the largest business and government applications. Since then, the Boston/Route 128 region, whose leading companies had bet on those minicomputers, has never been the same.[3]

Yet, in the face of these profound disjunctures, what some call the Silicon Valley production system hardly lost a beat. Between 1982 and 1987, sales of Valley high-tech manufacturers and software firms grew by 60 percent. During the 1980s, employment in the sector increased by 45 percent. And the number of computer firms rose from 69 establishments in 1975 to 246 by 1985.[4] Since 1987, both employment in Silicon Valley and the annual number of new firm start-ups increased without letup, even as the superficially similar microelectronics complex in southeastern Britain went flat and the roof fell in on the Boston economy.

Why Silicon Valley so successfully adjusted to the new conditions and why other high-tech regions have not are the big questions that AnnaLee Saxenian, a Berkeley planning professor and political scientist, has set out to answer. In a series of papers, and in a forthcoming book, she offers her conclusion: that, in all the important particulars,

Silicon Valley is [now] best viewed as an American variant of the industrial districts of Europe—technologically dynamic regional economies in which networks of specialist producers both compete and cooperate in response to fast-changing global markets. In these districts, technical skill and competence are widely diffused, small and medium sized firms achieve external economies through complex supplier and subcontracting relations, and the region (not the firm) is the locus of production. The result is a decentralized system which is more flexible than the traditional vertically integrated corporation.[5]

While these firms serve global markets and collaborate extensively with foreign suppliers, their *key* (her word) relationships tend to be local. In other words, an important segment of the Valley's producers have eschewed vertical integration and standardization. Instead, they have gradually formed small firm–led production networks operating on principles similar to those claimed to be operative in Italy's Emilia-Romagna, France's Lyons district, and Germany's Baden-Württemburg.

Saxenian argues that, by embracing such flexible manufacturing systems and factory designs as modular wafer fabrication, which relies on small equipment islands scattered throughout the plant rather than on giant assembly lines, many of Silicon Valley's small semiconductor producers have been able not only to afford to set up shop (because the cost of entry *for their specialty* is so much less), but to attain profitability even in the face of excess capacity in the industry as a whole. The latter achievement results, as well, from a highly competitive search for new market niches, a search that has produced a stream of new or continually improved products, from ever more powerful microprocessors and logic chips to disk drives, networks, power supplies, software, and all manner of computers and workstations.

As in the European districts, so in Silicon Valley an important aspect of the system's performance rests on a dense thicket of supportive institutions, from universities and technical institutes to trade associations and consulting groups (but *not* labor unions; the Valley's companies are fiercely antiunion, and have been so since the beginning). The sense of community is maintained by frequent trade shows, conferences, seminars, and other forums. Overall, "by socializing costs and risks and pooling technical expertise," writes Saxenian, "these institutions allow Silicon Valley's specialist firms to continue to innovate and react flexibly" to new pressures and opportunities from the market.[6] Venture capitalists play a particularly important role in this regard, acting in many cases not only as the source of funds for new projects but as systems integrators, bringing otherwise independent firms together to join forces in making some new product or pursuing some new process development.

Saxenian readily acknowledges that, at least among the engineers and programmers, loyalty is ultimately to the craft rather than to any particular firm, as such. But unlike other observers, she sees this as an arrangement favorable to the Valley's economic development, promoting the spirit of cooperative competition. After all, says Saxenian, "as individuals move from firm to firm in Silicon Valley their paths overlap repeatedly; a colleague might become a customer or a competitor, today's boss could be tomorrow's subordinate. Professional respect, loyalties, and friendships transcend this turmoil."[7]

To be sure, not all Silicon Valley companies play by these rules. Indeed, since the early 1980s the very survival of the region has, in Saxenian's judgment, been threatened by certain inclinations of the merchant chip manufacturers such as Intel and Advanced Micro Devices (AMD). Specifically, they have tended to further vertically integrate and standardize; to seek federal government protection from imports; to shift manufacturing operations to low-wage areas beyond the Valley; and to initiate lawsuits against former licensees, rivals, and what they see as footloose employees whose interfirm mobility these big firms consider to be the agency for the theft of the big firms' intellectual property.

These efforts to restore concentrated economic control over the sector (and the region) are, in Saxenian's view, precisely what has made high-tech industries in Boston, England, and other areas unable to respond flexibly to rapidly changing market conditions, partly by undermining the growth of an independent network of suppliers in the region. Were it not for the vibrant, locally oriented small firm–led networks that have taken root in Silicon Valley, that area, too, would be endangered by the reactionary behavior of the big firm merchant producers.

Saxenian's greatest concern is with the implications for system governance of the lack of self-consciousness by the managers of the region's small and medium-sized enterprises. While these managers *behave* like the cooperative competitors of industrial district theory, the managers Saxenian interviewed almost always failed to see themselves as part of a collective organization with mutual responsibilities. The inclination to revert to going it alone, if only the opportunity arose, seemed all too widespread.

Saxenian's provocative thesis—that Silicon Valley has become a de facto (but not yet a self-conscious) industrial district on the northern Italian model, and that therein lies the key to understanding its resiliency and the source of its creativity—has attracted critics. None of these has mounted a more elaborate attack than two social scientists, Richard Florida of Carnegie Mellon University (CMU) and Martin Kenney of the University of California at Davis. In a series of papers and book chapters, at professional conferences, and in the business press, Florida and Kenney have articulated the opposite concern: that excessive fragmentation of the U.S. electronics industry, generally, and of Silicon Valley, in particular, are gradually eroding the international competitive position of American firms in this sector.[8] In this respect, their concern echoes that of the NOMISMA research group in Bologna, over what the latter consider to be the excessive fragmentation of the small firm economy of Prato.[9]

Based on her reading of the business press and on her own interviews with managers of some one hundred high-tech firms in the Valley, Saxenian identified what she called a "definite trend" toward greater collaboration among the companies. Florida and Kenney challenge this claim, arguing from their own evidence that "there is little burden sharing between companies; contracts are broken and suppliers let go when a better deal can be had elsewhere."[10] They point to the continued existence of stockpiling and hoarding of components, in violation of the principle of just-in-time inventory delivery, as further evidence of a lack of trust. They argue that the tone for the entire region is dominated by the behavior of the biggest firms, such as Hewlett-Packard, whose managements (according to Florida and Kenney) continue to play their suppliers off against one another.

The evidence that most strongly calls the assertion of widespread interfirm collaboration into account, however, concerns the tremendous growth in the frequency with which Silicon Valley electronics firms now bring lawsuits against one another (or against one another's employees) for the alleged violation of intellectual property rights. For example, Intel's total litigation expenses increased tenfold between 1985 and 1988.[11] Most of the objects of the lawsuits by these big firms are the small entrepreneurial companies within the region. In his pathbreaking work on the theory of the social embedding of economic relations, Mark Granovetter, a sociologist at Northwestern University, had explicitly offered the tendency of firms in a cooperatively competitive network to *avoid* formally suing one another as a marker of the existence of strong social embedding.[12] Thus, the exponential growth in lawsuits among Silicon Valley companies in recent years may well constitute evidence of an erosion in the social basis for the reproduction of the region *qua* industrial district.

Of course, up to a point, lawsuits and other legal actions may be seen as normal institutional approaches to regulating conflict. The question is the extent to which the proliferation of lawsuits over intellectual property in such high-tech districts as Silicon Valley has come to inter- fere with system reproduction, for example, by discouraging the inter- firm circulation of mobile software programmers and systems designers. As discussed earlier, such mobility is thought to be an important mecha- nism by which information and skills are diffused throughout the region.

Florida and Kenney's critique of Saxenian also turns importantly on their argument that her emphasis on the proliferation of new firm start- ups as a measure of regional economic success is misplaced. For one thing, they suggest, statistics on start-ups only prove that the regional economy is organized to maximize the opportunities for technological rent seeking. Following an argument first put forth some years ago by the Harvard University science policy expert Harvey Brooks, Florida and Kenney believe that U.S. companies are particularly inept at translating technological "breakthroughs" into new commercial products. Saxen- ian's long list of computer-related products developed initially in Silicon Valley and now sold worldwide goes some way to reply to this charge. Still, some of the most profitable recent high-tech products, such as flat panel active-matrix video displays, have until now been taken up entirely by Japanese companies, even when the initial innovation was made in America.[13]

But Florida and Kenney are actually making a deeper point. They note that the explosive growth of high-tech U.S. start-ups in the 1980s was fueled mainly by the venture capital industry, as part and parcel of the general speculative financial boom that also gave us the savings and loan crisis. In what *Business Week* (and the British political economist Susan Strange) first named the "casino society," small firm start-ups themselves became commodities, to be bought and sold in order to gen- erate capital gains. It is difficult, say Florida and Kenney, to reconcile the idea of local embedding with a climate in which whole companies are "grown" purely for the sake of asset appreciation and ultimate resale.

As discussed in the last chapter and again in the review of Saxenian's argument, an important cornerstone of the theory of industrial districts is the hypothesis that, as a result of ever-shortening product cycles and the necessity for constant improvement in product quality and perfor- mance, firms in an industrial district are likely to contract out a major share of the production activity associated with any particular project to other producers located within the region. Moreover, the theory predicts that it is generally the most customized or technically sophisticated parts, components, subassemblies and services that are the most likely to be sourced from within the district, in order to maximize mutual access to information about the problems encountered in designing and manufacturing high-tech products.

On precisely this question, Richard Gordon, a political scientist at the University of California at Santa Cruz, has provided some answers, based on his own surveys of Silicon Valley firms. What Gordon has discovered is that nearly two-thirds of all component inputs into new product development by Silicon Valley companies are being sourced from *outside* the Valley. And of the remaining third procured from within the region, most comes from departments or divisions located inside the firm doing the final assembly.[14]

As for the relatively small volume of genuinely local interfirm component sourcing, it tends to be concentrated in such non-technology-intensive products as cabinets, casings, power supplies, raw materials, process materials (for example, photo masks and chemicals), and documentation (however, relatively complex disk drives *are* still produced within the region—for the moment). The more technically sophisticated and specialized inputs, such as CAD design equipment, wafers, and customized integrated circuits, are sourced mainly from outside the Valley—and from predominantly *large* firms, to boot. Since these are the inputs that are most likely to require a high degree of customer-supplier collaboration, Gordon concludes that, at least in Silicon Valley, collaborative manufacturing is *not*, despite the theory, strongly associated with *local* interfirm linkage, per se.

Still another characteristic of the Silicon Valley production system that calls into question its status as an industrial district is the nature of its insertion into the global economy. During the 1980s, and with increasing frequency since then, the process by which the Valley's companies are articulating with the outside world is, in important ways, progressing less and less on the region's own terms. The 1980s saw a steady increase in direct foreign investment in the area, as well as indirect foreign investment in the form of venture capital funds structured so that, after a period of years, the limited partners obtain ownership shares in the Silicon Valley companies in which the funds have invested.[15] Fujitsu and Kubota are among the many Japanese corporations that have been buying into the microelectronics, workstation, and supercomputer companies in the Valley. The acquired companies are nontrivial names in American high tech: Wyse, Fairchild, Akhasic, MIPS.

The acquisitions and portfolio investments by foreign multinational corporations are not limited to microelectronics. In 1990, for example, the Swiss conglomerate Roche acquired a 60 percent share in the pioneering American biotechnology start-up Genentech, based in South San Francisco on the northern fringe of Silicon Valley. Mergers and acquisitions have been occurring regularly *within* the Valley, as well. For example, in 1989 Hewlett-Packard (H-P) acquired workstation maker Apollo for an enormous sum, making H-P—itself a Fortune 500 company—the world's largest supplier of this kind of equipment.[16]

What the Japanese, Taiwanese, South Korean, German, and Swiss corporations provide is, of course, the "deep pockets" and "patient capital" so badly needed by American companies, especially in the wake of the bursting of the domestic financial bubble after October 1987. But according to David J. Teece, a business strategy theorist at the University of California at Berkeley, what the foreign (especially the Japanese) companies get out of locating in the United States, generally, and in Silicon Valley, in particular, are testing grounds for adapting their own technologies to the particular conditions of the American market, a vantage point from which to "monitor and assimilate U.S. technological and scientific advances," and the chance to "shore up their historic weaknesses [in software development] by hiring U.S. talent."[17]

The foreign companies bring engineering talent, a commitment to manufacturing excellence, and (in the case of the Japanese parents) access to the internal Japanese market. What they do *not* yet show signs of bringing is a commitment to local economic development, per se. Thus, foreign-owned or -acquired firms and branches outsource at least as much, if not more, of their overall production as do the mostly domestic (let alone the truly local) companies.[18]

INEQUALITY AND DUAL LABOR MARKETS IN THE SILICON VALLEY PRODUCTION SYSTEM

There is a tendency in much popular (and even in the scholarly) writing about Silicon Valley to emphasize the glamour, creativity, skill, and high pay associated with the technical, engineering, and scientific occupations in the Valley's many high-tech companies. Yet as early as the 1970s, it was becoming apparent that the workforce employed inside the semiconductor companies at the heart of the area's economy was in fact highly stratified.

As Saxenian herself has documented, in her earliest published research, at the top of the hierarchy were the highly educated, extremely well paid managers, engineers, and other professionals.[19] At the same time (and often within the same factories and laboratories), nearly half of all workers in the Valley's high-tech companies performed production and maintenance tasks, four-fifths of which were officially classified as semiskilled or unskilled. Wages in these jobs were dramatically lower, and benefits often nonexistent. The remaining jobs consisted of clerical and secretarial work.[20]

During the 1960s and 1970s, immigration into Santa Clara County reflected this stratification. Well-educated engineers and scientists moved into the western foothills near Stanford University, in the far northwestern corner of the Santa Clara Valley, to be near their offices and labs, as well as to the more expensive luxury homes, golf courses, and tennis courts.

At the same time, the industry's demand for production workers stimulated an equally large in-migration of unskilled, predominantly minority workers . . . displaced agricultural workers from California and the Southwest (primarily Chicanos and some Filipino-Americans), foreign-born Mexicans and Filipinos, and smaller numbers of U.S. blacks and Native Americans.[21]

Subsequently, the in-migration came to consist mainly of Southeast Asians.

These unskilled foreign workers were shunted off to new residential areas situated far to the southeast of the zone around Stanford, in and around the explosively growing city of San Jose. The marked differences in income and wealth were quickly reflected in differences in the quality of life and the availability of social services between the elite northern and the working-class southern reaches of the Valley. It is only now, in the 1990s, that city planners in San Jose are making headway in installing public services, housing, and local service sector job complexes geared to accommodating the needs of the several communities of color living in the area.

But during the 1970s, this geographical stratification of Silicon Valley's population—itself a reflection of the internal labor market segmentation within the electronics companies themselves—was giving rise to a host of problems that imposed new, much higher costs of doing business on those very companies. Housing prices in the affluent north skyrocketed, requiring the companies to come up with wage increases, subsidies, and perks in order to attract new engineering talent. Production workers forced to commute long distances from their homes in the southern Silicon Valley to the high-tech plants in the north contributed to levels of congestion and air pollution previously unheard of in the region. As is almost inevitable, with this sort of linear spatial configuration, there popped up seemingly overnight a mass of poorly regulated strip development: shopping centers, industrial parks, parking lots, and more freeways, all of which only exacerbated the impending environmental crisis. Finally, the rising cost and difficulty of commuting within the Valley created shortages of production workers available on a reliable, daily basis to the high-tech firms in the north. Turnover rates rose to unacceptable levels. The costs of doing business within Silicon Valley rose still further.

By 1980, the semiconductor and other electronics companies' solution was to shift new direct production operations to locations outside the Valley. This was not, in fact, an entirely novel development. As early as 1961, Fairchild had established chip assembly and testing facilities in Hong Kong. These were production activities requiring (at that time) the least skill and entailing the most dangerous working conditions. Moreover, they were discrete tasks, comparatively easily separated off from the rest of the production process and therefore easily relocated to distant sites without disrupting overall coordination.

The operations that Silicon Valley companies erected in Utah, Arizona, New Mexico, Texas, Oregon, and Washington during the 1980s tended to involve middle-level production and, as time went by, research and development, as well. In this evolving spatial division of labor, the lowest-skilled, most standardized production work was relegated to plants on the Mexican border and throughout Southeast Asia. Subsequently, many of these Asian facilities began to take on higher-level functions, themselves, as industrial and consumer products markets grew rapidly in that part of the world.

New operations sited within Silicon Valley have tended to focus on higher-level functions, especially R&D and management of these increasingly far-flung networks. The many new highly specialized start-ups about which Saxenian has written more recently tend to engage in relatively little direct manufacturing themselves. Indeed, it is precisely this separation of design and development from actual manufacturing in the Valley that so concerns such observers as Florida and Kenney. Manufacturing has been dispersed to such an extent that a popular trade magazine columnist on the computer industry remarked recently that chip making has become so widely scattered that "nowadays, Silicon Valley is a virtual reality."[22]

In sum, this quintessential "postindustrial" high-tech industry and region have been characterized by rather classic dual labor markets. Technical and managerial jobs are found at the top (although, as noted earlier, with much greater interfirm job changing and fewer long-term "careers" with a single firm than was ever the case in steel, autos, or chemicals). At the bottom is the generally lower wage, more expendable labor force made up predominantly of Latino and Asian immigrants performing the more mundane, standardized, sometimes dangerous tasks— dangerous because of the chemicals that have to be physically handled, or because of chronic eye strain resulting from having to stare constantly into high-powered microscopes.

This, too, is part of the Silicon Valley production system.

CONCENTRATED POWER AND THE ROLE OF THE MILITARY IN SILICON VALLEY'S ECONOMIC DEVELOPMENT

Quoting the anthropologist Karl Polanyi, the development economist Alice Amsden ironically reminds us that "laissez faire was planned."[23] In his own history of the origins of Silicon Valley, Stephen Appold, a Carnegie Mellon sociologist, reminds us of just how much the Valley is a construct of a small number of powerful institutions, and *not* a "cluster" to which independent firms were attracted by market forces.[24]

For all the myriad legends of the free-wheeling entrepreneurs who founded the Valley's earliest firms, two powerful institutions—the U.S. Department of Defense and Stanford University—arguably played the greatest role in transforming this region of fruit orchards into a world center of microelectronics-based high-tech industry. Actually, the heroic and the more sober, big institutions explanations for the origins of Silicon Valley are inextricably interwoven.

The first of the heroes, Frederick Terman, a professor of electrical engineering at Stanford (and later a dean, provost, and vice-president there), is often credited with having launched the concept of the university-industry partnership. Following his service at Harvard during World War

II, managing military research activities, Terman returned to Stanford to reorganize its research and teaching efforts in these fields. He recruited faculty, brought in Defense Department contracts and grants, launched the Stanford Research Institute and the Stanford Industrial Park, founded the nation's first high-tech industrial park located on university property, and managed the transition in military-related university R&D from radar and microwave communications to control systems for ballistic missiles. It was during Terman's years that Sylvania, Fairchild, General Electric, Philco-Ford, Westinghouse, ITT, Admiral, and IBM all bought or constructed facilities adjacent to Stanford University. Hewlett-Packard and Varian Associates, two other key players in the subsequent evolution of the Silicon Valley production system, were there already.

Because of its location as the country's western-most outpost during the Asian war, California was by 1945 already the base of operations of much of the U.S. aircraft, radar, and atomic energy industries. As the Rutgers University regional economist Ann Markusen and her colleagues document in two recent books, aerospace, missiles, and advanced microelectronics followed over the next quarter century. The production of military-related equipment was centered in southern California, while electronics research and development were, from the beginning, located in and around Stanford, four hundred miles to the north.[25]

Enter the second of the great heroes of Silicon Valley legend, William Shockley, the co-inventor of the transistor. In 1955, Shockley returned to Palo Alto from his stint at New Jersey's Bell Labs to found his own firm. Two years later, a group of Shockley's engineers spun off their own start-up with backing from Fairchild; by 1980, almost every semiconductor firm in Silicon Valley could trace its origins in some way back to that company. From the inception of this new technology, growth of the semiconductor industry within the Valley could be linked in some important way to the fact that the unusually high concentration of aerospace and missile producers located in California provided a ready-made and virtually insatiable market for chips.

From such historical materials, Gordon and Joel Krieger conclude that

> the embryonic microelectronics industry emerged in the U.S. as a result of combining the autonomous innovative capabilities of a small number of existing firms with the driving force of a highly centralized and technologically sophisticated state demand. Con-

trary to almost universally accepted mythology . . . in its infancy the microelectronics industry proceeded less from ineffable scientific genius and heroic entrepreneurialism than from a conjunction of innovation in established firms and extensive state intervention. . . .

Advanced military and aerospace demand provided the principal market for microelectronics, established research priorities in product and process innovation, stabilized high profits and underwrote the risks of new product development. The vast majority of scientists, engineers and technicians in the microelectronics industry acquired state-of-the-art theoretical and practical knowledge in government-financed university or corporate research and development programs. The performance specifications of the world's hegemonic military power pushed development continually beyond existing technological frontiers into more advanced miniaturization and higher levels of performance. Government funding significantly extended to development of the manufacturing equipment and to technologies facilitating the transition of device R&D into commercial production capability. Defense support was also critical in the establishment of the learning economies which governed the subsequent evolution of the industry; since market growth and expanded production volumes resulting from military sales accelerated cost reductions, and since military requirements meshed broadly with civilian market demand, U.S. firms were able to move faster and earlier than their foreign competitors down the learning curve towards more consistent quality, higher yields and lower prices. Government policy also provided a bridge between traditional industrial leadership and new entrants, for while established firms did receive substantial military funding for microelectronics R&D, defense contracts were also awarded to smaller start-up companies, helping to diversify the structure of the emergent semiconductor industry. Military demand, therefore, was instrumental in creating and maintaining U.S. leadership in electronics.[26]

Research grants were awarded to universities and private laboratories, helping to fuel the creation of pools of scientists, engineers, and computer programmers. Small start-up companies also received procurement contracts, which helped to create what would later become the basis for networks of diversified suppliers and subcontractors to the big firms. Military financing permitted companies in Silicon Valley (and throughout California) a degree of trial-and-error exploration of new technologies that private sources of finance would never have allowed.

The same effect was produced by the Pentagon's allowing premium prices and offering guaranteed long-term contracts. Thus, even when the direct *volume* of contracts was small for any particular firm, the *value* of these kinds of contracts for launching a risky new industry was inestimable. While Silicon Valley's direct dependence on the military has declined over the intervening years, as recently as 1984, Santa Clara County—home to such major military contractors as IBM, Ford Aerospace, FMC, and Lockheed's Missiles and Space Division—still ranked third among all U.S. counties in the receipt of Defense Department prime contracts, surpassed only by Los Angeles and St. Louis.[27] Indeed, in 1992 Lockheed was still Silicon Valley's largest single producer.

Gordon and Krieger, as well as Markusen and her colleagues, agree that the formative role of the major American universities in all high-tech regions is inextricably bound up with government spending influenced directly or indirectly by military/geopolitical considerations. In the case of Silicon Valley, of course, the dominant academic partner to the Defense Department was—and continues to be—Stanford University. Indeed, ever since World War II, Stanford has been by far the single largest academic recipient of federal electronics-related research funding in the United States. As one of the country's premier universities, Stanford was vital to the growth of Silicon Valley. This was not so much for its research strengths per se as for its world-class graduate engineers, and for its readiness to encourage and underwrite commercial activities by its faculty and students.

Thus, with the help of the federal government—especially the Defense Department—Stanford was, and continues to be, able to provide a steady supply of well-trained technical talent for the region's emerging high-tech companies. Other regions may contain the campuses of one or another engineering school. But no one else has the networks of a Stanford.

THE THREE FACES OF SILICON VALLEY

As noted earlier, small entrepreneurial start-ups make up only one face of Silicon Valley. This has been true from the Valley's inception. Networks of alternately cooperating and competing small and large firms, manufacturing semiconductors and other electronics equipment in the manner of an industrial district, do exist in Silicon Valley and offer still another of its faces. A third face involves the deep interconnections between multiregional, often multinational American firms, Japanese and European corporations investing in the Valley, and such fundamentally outwardly directed, powerful institutions as Stanford University and the Pentagon.

But if all three faces are observable, there seems little doubt that it is the third—the Silicon Valley whose insertion into the global economy is being led by the big firms and avowedly hierarchical institutions, just as the region's high-tech system was created initially by a small group of engineering entrepreneurs working hand in hand with institutions of concentrated economic and political power—that will drive the region's economic growth and development in the years ahead. Nothing in recent American economic history brings this home more powerfully than the bombshell dropped on the markets and the politicians in the summer of 1992, when the Silicon Valley resident AMD and the microelectronics giant IBM both announced that, after years of trying to compete head to head with the Japanese and Europeans for greater shares of the world semiconductor market, they would now pursue their objectives through strategic alliances with their foreign competitors. Thus, IBM has joined forces with Germany's Siemens and Japan's Toshiba to design one new kind of microchip, while AMD has hooked up (even exchanging stock) with Japan's Fujitsu to design and manufacture another type of computer memory.[28] At least initially, the first of these alliances will work out of an IBM facility in New York State. The AMD-Fujitsu partnership will, from the beginning, manufacture in Japan.

If the dominant actors in Silicon Valley are multinational corporations and their production networks, and to the extent that governmental subsidies (plus actual direct and indirect reliance on government markets, in the form of the military) do continue to play an important role in sustaining the system, then it becomes more difficult to characterize the regional economy as being largely internally self-actuating, linking to the outside world substantially on its own terms, a reasonable approximation of (say) Prato during its heyday. Moreover, if computer programmers' basic loyalty is to their craft, which they can effectively practice anywhere in the world, then it becomes more difficult to draw a logical connection between the organization of work and loyalties to the locality, as such. In other words, the production and labor systems of Silicon Valley may not be all that locally embedded, after all.

Finally, if the industrial labor markets of the region are segmented, as well, then yet another element of systematic hierarchy (as well as exploitation) enters the story, further confounding efforts to depict the Valley as a collaborative of more or less symmetrically powerful managers and other agents.

I am not suggesting that tiered interfirm contracting relations, increasingly complex linkages to the outside world, and the admixture of craft with hierarchically structured labor markets constitute a combination that necessarily threatens regional economic growth (although Teece, for one, continues to warn of the "shortcomings of the Silicon Valley industrial system—the absence of patient capital, the lack of [local] manufacturing skills, and poor capacity to access foreign [especially Japanese] markets").[29] My point is that, confronted with such institutional facts of life in Silicon Valley, depictions of the region as either an example of a textbook free market in action or as a Marshallian industrial district held together by a thickening of local interfirm connections and a common industrial atmosphere become stretched—perhaps beyond the breaking point.[30]

PART III

THE EMERGING SYSTEM
OF GLOBALLY NETWORKED
PRODUCTION

6

"FLEXIBILITY" AND THE EMERGENCE OF LARGE FIRM–LED PRODUCTION NETWORKS

In recent years, there has been a considerable amount of research on organizational practices and arrangements that are network-like in form. . . . One would need to have followed the field of international business, technology strategy, industrial relations, organizational sociology, and the new institutional economics, as well as interdisciplinary work on such themes as cooperation, the embeddedness of economic life in social structure, and the proliferation of small business units to have kept abreast.

—WALTER W. POWELL[1]

In the decade and a half following the mid-1960s, the share of corporate profit in gross domestic product fell sharply across the developed world. According to the Organization for Economic Cooperation and Development (OECD), between 1965 and 1976 the corporate rate of return fell by 37 percent in England, by 16 percent in West Germany and Canada, and by 12 percent in Japan.[2] According to the British economist Andrew Glyn and his colleagues, profit rates, averaged across the seven richest industrialized countries and defined as net operating surplus divided by net capital stock at current prices, declined in the manufacturing sector from 25 percent in 1965 to 12 percent in 1980

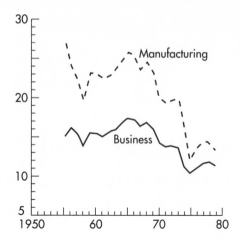

FIGURE 6.1 NET PROFIT RATES IN THE G-7
INDUSTRIALIZED COUNTRIES, 1955–1980

Note: Net profit rate is the net operating surplus divided by net capital stock at current prices. The G-7 countries are the United States, the United Kingdom, Japan, Canada, Germany, France, and Italy.

Source: Andrew Glyn, Alan Hughes, Alain Lipietz, and Agit Singh, "The Rise and Fall of the Golden Age," in *The End of the Golden Age,* ed. Stephen Marglin and Juliet Schor (New York: Oxford University Press, 1989), p. 53.

(figure 6.1). Across all sectors, the average rate of profit fell from 17 to 11 percent.[3]

Whatever one's preferred explanation for the great profits squeeze of the 1970s,[4] the crisis ushered in a veritable sea change in the nature of international economic competition. The world experienced nearly two decades of deindustrialization, the shakeout of weak competitors and older capacity, all manner of financial and real business restructuring, and the penetration of new technologies onto the factory and office floor. Now that the smoke has cleared, the number of competing firms in both developed and developing countries has greatly increased. Market conditions now change more quickly than ever before, due at least in part to the relaxation of the trading regime of fixed exchange rates, and in part to more heterogeneous forms and sources of product demand. The spread of new technologies in transportation and communications has made it possible for companies to move physical and financial capital ever more rapidly from one place or use to another.[5] The shelf life of commodities, competitively acceptable delivery times for new products, and average production cycle times all seem now to be undergoing continual compression.[6]

These changes have given rise to a sometimes dizzying array of management innovations in corporate organization that have become the stuff of daily discussion in both the professional and the popular press. Who has not encountered the practices of lean production, the greater use of outsourcing of work from the big firms to generally smaller suppliers, or the growing employment by the big firms of contingent (for example, part-time) labor? All can be thought of as experimental reactions by big business to the trauma of the worldwide economic crisis of the 1970s and early 1980s.

Of all the reactions, all the experiments, the most far-reaching may well turn out to be the creation by managers of boundary-spanning networks of firms, linking together big and small companies operating in different industries, regions, and even countries. *This* development— not an explosion of individual entrepreneurship or a proliferation of geographically concentrated industrial districts, per se—is the signal economic experience of our era. In this chapter, I explore the underlying reasons why network forms of organization have emerged as an important part of big business's solution to the challenges posed by the collapse of the old order. Then, in the following chapter, I examine examples of particular networks in the United States, Japan, and Europe, and ask whether they are likely to turn out to constitute a stable form of industrial organization into the next century.

CRISIS AND THE CORPORATE SEARCH FOR FLEXIBILITY

How have business firms coped with the dynamic, incessant, greatly magnified competitive pressures of the new world economic order? Managers everywhere have responded since the 1970s in various ways— all of which can be characterized as a search within large and small firms alike for greater *flexibility*: through reorganization and technological change, in labor-management relations, and in the reconfiguration of each firm's (and establishment's) transactional and longer-term relations to other companies and operating units. In other words, firms are becoming more *integrated* into one another's orbits. Not by accident has the British economic geographer Philip Cooke identified a worldwide trend toward "flexible integration."[7]

Mainstream economists have always believed that free markets are the best way to guarantee such flexibility. According to this view, the ideal institutional arrangement consists of decentrally organized markets, in which more or less equally powerful, independent firms voluntarily exchange resources, goods, and services mainly at arm's length, at openly quoted, widely publicized prices. Such arrangements will maximize the efficient spread of information about choices, without imposing top-down control over the freedom of individual firms and consumers to do the choosing.

An intriguing branch of mainstream economic theory known as "transactions cost economics" is helpful—up to a point—in making sense of why decentralized markets are *not* necessarily or always well suited to solving the flexibility problem.[8] In a cogent summary of the theory, Richard Walker, an economic geographer at the University of California at Berkeley, writes:

> The principal limits to the formation of stable, workable market exchanges are uncertainty (incomplete information, an uncertain future); small numbers (few parties, irregular transactions); bounded rationality (inability [of managers] to handle all information and contingencies); and opportunism (misrepresentation, reneging). As a result, commodities may not be available as needed, monetary payments may not be forthcoming, critical information may be withheld, labor processes may be poorly coordinated, and capital flows may be blocked. In short, neither linkages nor regulation among the parts of the division of labor may be sound enough for production to proceed in a stable and effective manner. . . . Competitive individualism [thus] has its drawbacks as a way of organizing social action; and markets can be a weak integument for complex production systems.[9]

According to Berkeley's Oliver Williamson, the leading contemporary architect of transactions cost theory, the greater the problem of coordination—of the "governance" of the exchange of commodities—the more likely companies will be to organize vertically integrated *hierarchies*, designed to internalize under a single management those exchanging units involving assets that are the most highly specific to the transaction. In other words, exchange will be organized through and by competing big firms.

The orthodox economics approach to transactions costs has come under attack from sociologists and institutional economists.[10] As Walter W. Powell, a University of Arizona sociologist, puts it, the key flaw is the "exclusive focus on the transaction, rather than the relationship," along with neglect of the role of government in shaping the institutional "context in which exchange is conducted."[11] Moreover, as virtually all contemporary schools of political economy now agree, while it may offer a solution to the anarchy of the market, the highly centralized, vertically integrated business organization has itself become insufficiently agile to solve the new problems posed by global economic integration. This proposition constitutes the starting point for Piore and Sabel's theory of the sea change, according to which (in Powell's words) "the disadvantages of large scale vertical integration . . . become acute when the pace of technological change quickens, product life cycles shorten, and markets become more specialized."[12]

Powell's criticisms notwithstanding, transactions cost theory and its friendly critics provide a coherent, scholarly, *non-normative* explanation for why the kind of economic growth and development posited (and devoutly wished for) by Birch, Gilder, and other celebrants of the free market is simply not viable in this brave new world of increasing complexity and uncertainty.[13] The criticism of simplistic characterizations of how markets are organized is valuable. But transactions cost theory has not made much progress in helping us understand which transformations in the form of business organization *have* been developed to cope with the need for enhanced flexibility, and why.

Before answering that question, we must first develop a more focused definition of just what is meant by *flexibility*. From the rapidly growing multidisciplinary and institutional research on corporate and industrial restructuring emerge at least three distinct (but not mutually exclusive) meanings for the term, as it applies to strategic behavior.

Functional flexibility refers to the efforts of managers to redefine work tasks, redeploy resources, and reconfigure relationships with suppliers, as in the just-in-time system of minimal-inventory delivery of parts and goods-in-process to manufacturing assembly plants. Attempts to achieve greater functional flexibility include the adoption of new technologies that facilitate more rapid product design or tool changes of the sort discussed in chapter 3, and that permit a greater decentralization of decision making and responsibility and hence accelerate the speed at which production operations can change over from making one type or design of product to another.[14]

Wage (or, as it is sometimes called, "financial") *flexibility* refers to the various efforts by managers to reintroduce greater competition among individual workers, particularly in those occupations and industries that had become substantially sheltered from direct wage competition during the long post–World War II expansion as a result of unionization and government regulation. It is these latter developments—from wage take-backs, the introduction of two-tiered wage payment schemes, and payment via bonuses for individual performance, to systematic union avoidance (even by historically unionized companies)—that have occupied the greatest attention of researchers in the fields of industrial relations and labor economics in the United States and Western Europe.[15]

By contrast, there is *numerical flexibility*, whereby jobs are redesigned so as to substitute part-time, contract, and other "contingent" workers, who (at least in the United States) receive few or no benefits such as health insurance and pensions, for full-time employees who had been receiving more or less comprehensive fringe benefit coverage.[16] A second type of numerical flexibility is evident in the tendency of managers to outsource production, maintenance, catering, clerical, and other activities that arguably were formerly (or, in the absence of heightened competitive pressure, would otherwise be) undertaken in-house.[17] That the make-buy ratio moves cyclically, with firms tending to increase subcontracting during macroeconomic expansions and bring in the work during recessions in order to keep their own core workers occupied, has long been known and well documented.[18] What some researchers now suspect is the advent of a secular upward trend in outsourcing in the last decade and a half.[19]

To the Stanford University sociologists Jeffrey Pfeffer and James Baron, and to the M.I.T. economist Paul Osterman, these forms of labor contracting represent a strategy pursued by large, high-wage companies to externalize the employment relationship.[20] Rather than continuing to rely on internal scale economies and a long-term commitment to preserve stable employment relations within the organization as the means by which more senior employees are encouraged to pass along their "tacit know-how" to younger employees, the largest firms seek to reduce costs by retaining an in-house set of core activities, while contracting out other work, either to low-cost or to more specialized external subcontractors and suppliers. Over the long run, the implications of such a strategy would be to undermine the job stability and high-wage advantage of workers employed in what, during the post–World War II era, came to be known as the "primary labor market." I return to this subject in chapter 9.

Now I am prepared to reengage the original question: How have companies tried to cope with the growing uncertainty, fragmentation, and time compression that characterize so much of contemporary industrial competition? The answer is that, in all the industrialized countries, the solution to private industry's search for greater flexibility has, in one way or another, increasingly come to entail the creation of *networks* among producers. The industrial districts described in the last two chapters are particular examples of such networks. So is the recent strategic alliance in research and development among the giant multinational corporations IBM, Toshiba, and Siemens. So, too, is the explicitly hierarchical relationship between the Toyota automobile company and its rings of more and less dependent suppliers. The group of European corporations and the four governments that together created Airbus Industrie, in order to compete head to head with America's Boeing, represents still another example of a production network.[21]

PRODUCTION NETWORKS AS A MEANS OF ACHIEVING FLEXIBILITY

Naming specific examples of networking behavior does not seem so difficult; to some extent, we know it when we see it. But is a more analytical (and therefore predictive) characterization possible? As Joseph Badaracco, of the Harvard Business School, says in a recent book on the subject, with no consistently collected data, nor even universal agreement on the definition of terms, "efforts to be precise can be frustrating." So frustrating, indeed, that, paraphrasing a paraphrase by the economist Kenneth Boulding of a comment by the eighteenth-century British philosopher David Hume, Badaracco writes: "After considering the difficulties of defining, categorizing, and measuring knowledge, [I] felt that the effort led into a philosophical morass from which . . . the only escape was to climb out, clean oneself off, go home, have a good dinner, and forget all about philosophy."[22]

In spite of the challenge, a hardy band of sociologists, economists, geographers, and students of business strategy are making a concentrated effort to bag the elusive quarry of the production network as an identifiable form of industrial organization in the age of flexibility. Some of their conclusions follow.

Williamson himself acknowledges that hybrid forms of industrial organization, situated somewhere between atomistic competition and a world of vertically integrated corporations—between markets and hierarchies—may actually be more common than he originally thought. Powell, who once took a similar view, has since become the leading American theorist of the production network as a class of industrial organizations in and of itself. Working in North America, eastern Asia, and Europe, he and others argue that network forms of organization may in fact be in the process of becoming the *signature* institutional form of this era—precisely because they offer managers the best working solution to the challenges posed by the increasing need for flexibility.[23]

To Powell, network forms give greater room for the entry into business affairs of reciprocity, altruism, friendship, reputation, and collaboration as principles of governance, in contrast with unbridled self-interest in the case of free markets and central administration in the case of the vertically integrated corporation. Even mainstream neoclassical economists and positivist political scientists have become interested in the growing importance of collaborative relations in business affairs, modeling them in terms of mathematical game theory and the theory of "tournaments." Distinguishing among these different attacks on the problem, Powell observes that, whereas "anthropological and sociological approaches ... tend to focus more on the normative standards that sustain exchange, game theoretic treatments emphasize how individual interests are enhanced through cooperation."[24]

Network forms seem especially appropriate—perhaps even necessary—when what is involved is an exchange of commodities whose value is not easily measured. For example, if one firm wants to acquire the tacit knowledge (know-how), technological capabilities, or style of or philosophy toward production that characterize some competing company, these assets are not easily tradable through markets or accessible by top-down command within individual corporations. "The open-ended, relational features of networks, with their relative absence of explicit quid pro quo behavior, greatly enhance the ability to transmit and learn new knowledge and skills."[25] Indeed, all students of network forms of organization now agree that what such systems are especially good at is diffusing the kind of information that tends to become blocked ("impacted," "sticky") within both markets *and* hierarchies.[26]

Since rapid access to information is the single most important requisite for facilitating flexible approaches to changing competitive conditions, network forms of organization are especially appropriate for attaining flexibility. For individual managers, companies, and governments, the price for obtaining the benefits from networking is the surrender, at least temporarily, of "the right to pursue their own interests at the expense of others." Unlike spot transactions in a market, relationships "take considerable effort to establish and sustain, thus they [may] constrain [the] partners' ability to adapt to changing circumstances."[27] Coordination of the organizational routines of partners with different corporate cultures may be difficult. Nor does networking eliminate the ambiguity of properly defining the implicit prices of intangible assets.[28] Contacts within networks may become points of conflict as well as harmony. Moreover, even though no formal legal merger may have been involved, partners nevertheless become indebted to one another, and it may take some effort and cost for one participant to pay off the debt and dissolve the relationship when it no longer works or is no longer needed (termination of such partnerships need not, of course, imply failure).

There is also a social cost to networking, since "by establishing enduring patterns of repeat trading, networks restrict access. Opportunities are thus foreclosed to newcomers [outsiders], either intentionally or more subtly through . . . unwritten rules or informal codes of conduct."[29] This is one reason that American companies have found it so difficult to penetrate Japanese markets.

In sum, from the point of view of Williamsonian theory, networking actually *increases* transactions costs. Yet, as will be clear from evidence presented here and in chapters 7 and 8, network forms of organization continue to proliferate. Apparently, "networks are more than the sum of their parts."[30] There is such a thing as "network externality." Or, as the school of "new institutionalists" would prefer to put the case, the social embedding of economic relationships makes the depiction of strategic business decisions as the outcome of self-interested calculation of incremental costs and benefits an inaccurate way of representing real-world behavior.

Types of Interfirm Production Networks

Production networks have evolved in many different industries and sociopolitical settings. Indeed, Powell himself admits that

> the network story . . . is a complex one of contingent development, tempered by an adjustment to the social and economic conditions of the time.

The absence of a clear developmental pattern and the recognition that network forms have multiple causes and varied historical trajectories suggest that no simple explanation ties all these cases together.[31]

Nevertheless, a typology of classes of production networks is beginning to emerge from observations of business practices around the world. One class of networks is associated with *craft-type industries*. Here, work tends to be organized around specific projects rather than around stable, sharply bounded "firms." From construction, publishing, the film and recording industries, and architecture to engineering and computer programming, a highly skilled labor force shares technical know-how among one another, even across the boundaries of the firms with which they are currently employed. While such informal trading of proprietary know-how may even be illegal—recall the discussion in the last chapter of the problem of intellectual property "theft" in Silicon Valley—such practices reflect loyalties to the underlying craft that provide the technical specialists with the means by which to more flexibly construct and deconstruct the partnerships that are most appropriate to compete in the market for new projects.

Two other classes or types of networks include those *small firm–led industrial districts* and *geographically clustered (agglomerated) big firm–led production systems*. The small firm–districts have already been considered in some detail. Examples of the big firm–led production systems include Toyota and its rings of suppliers concentrated in and around "Toyota City" in Nagoya, Japan, and the General Electric Company's Aircraft Engine Group and the many small machine shops huddled around it in northeastern Ohio and on the northern shore of Massachusetts.

In some of the big firm–led production networks, especially in Japan and Korea, the corporations actually helped to *create* their generally smaller firm suppliers when, after World War II, none previously existed or had been decimated by the war.[32] The American experience has been the other way around. Dense networks of small workshops and factories go back to the mid-nineteenth century, only to be exploited in the twentieth century by the emerging monopolies in one industry after another, from U.S. Steel and General Electric to Westinghouse and Ford, as subcontractors and as training sites for skilled craft workers whom the big firms would periodically absorb and then, during recessions, spit out again.[33]

Yet another basis for the emergence of big firm customer–small firm supplier networks since the early 1970s has been the process of the *vertical disintegration* of the big firms themselves. In chapter 2, I presented the idea that modern territorial agglomerations may result in part from the break-up of the biggest auto, steel, textile, and aerospace firms, and the spinning-off of what had formerly been units within those corporations to outside, but still localized, suppliers.[34]

Whether territorially based or not, large firms can achieve greater flexibility by resorting to the practice of *subcontracting* (also known as outsourcing or, in Britain, outworking). Apparently, since the early 1970s, a growing number of large firms have turned to both greater and more varied uses of subcontracting, and the existence of a large sector of small subcontractors has proven *not* to be an anachronism (as had been widely supposed), but rather an integral part of all modern industrialized as well as developing economies.[35] In this connection, Berkeley's Richard Walker reminds us:

"Putting-out" is the oldest specifically capitalist form of organizing production, and it predominated in many parts of Europe in the eighteenth and nineteenth centuries.... Thought to have passed from history, it is nevertheless to be found today in automobiles ... garments ... films ... computers ... agriculture ... insurance ... semi-conductors ... aerospace ... machine tools ... and so on. It is widespread in Japan ... Italy ... and Spain ... and is more firmly rooted in Germany, France, Britain, Canada, and the USA than previously recognized.... It is prevalent, too, in the informal economies of Third World cities.[36]

Finally, there is the class of production networks known as *strategic alliances*, especially those crafted among big firms striving to span a manageable but diverse range of related activities and markets without having to undertake the full expense involved in actually building new plants or acquiring existing ones, outright.[37] The M.I.T. economist Charles Kindleberger, one of the deans of the field of international economics, has noted that international alliances among distinct companies are hardly a new thing.[38] But, the British geographer Peter Dicken contends: "What is new is their current scale, proliferation, and the fact that they have become *central* to the global strategies of many firms rather than peripheral to them. Most strikingly, the overwhelming majority of strategic alliances are between compe-titors."[39]

The main lesson managers learned from the restructuring experi-ments of the 1970s and 1980s is that, especially in cross-border high technology, few big firms can go it alone.

Figure 6.2, constructed by the United Nations Centre on Transna-tional Corporations, illustrates the extent in 1987 of the various alliances erected by just one large multinational high-tech corporation whose home base is Germany. Siemens manufactures computers and software, robotics, engines and parts, materials, telecommunications components and systems, and semiconductors. Not all of the particular connections depicted in the figure are still in force today, and some links are not represented because they came into existence after 1987. Still, the linkages forged by Siemens and the global extent of those are typical of this form of production network.

Cooperation among legally distinct firms is of course not new. Compa-nies have collaborated on one or another project, often on an ad hoc basis, since the earliest days of the Industrial Revolution. The concern here is with the evolution of the contemporary *institutionalization* of cooperation as a deliberate strategy for attaining greater flexibility in the face of the heightened competition and chronic uncertainties of the new world order.

In the years after World War II, many corporations, both within indi-vidual countries and across borders, cooperated by *licensing* particular technologies and process or product innovations to one another.[40] As far back as the late 1950s, such arrangements were seen by managers them-selves as unsatisfactory, and the Stanford economist Kenneth Arrow thought he knew why. If the inventor of a particular technological asset—say, a new way of electronically storing and retrieving data—wants to attract potential investors, the inventor (the seller) must reveal information about the process or product, to justify his or her asking price or license fee.

But to the extent that the information so revealed can be absorbed and exploited by the prospective investor (the buyer), why buy? Given the Arrow revelation problem, it is therefore not surprising that, while licensing deals continue to be used by companies and do constitute one form of networking, they have been steadily supplanted by joint ven-tures and other forms of alliances that make greater demands on all of the parties to a relationship.

The Berkeley professor of management strategy, David Mowery,

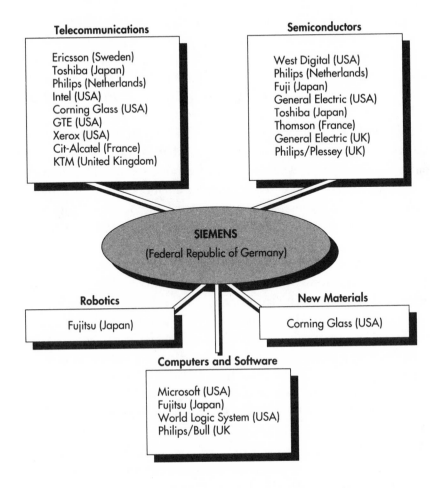

FIGURE 6.2 MAJOR STRATEGIC ALLIANCES INVOLVING
SIEMENS BETWEEN 1984 AND 1987

Source: United Nations Centre on Transnational Corporations, reproduced from Werner Sengenberger and Duncan Campbell, eds., *Is the Single Firm Vanishing? Inter-Enterprise Networks, Labour, and Labour Institutions* (Geneva: International Institute for Labour Studies of the International Labour Office, 1992), p. 20.

remarks that strategic alliances "may reduce the severity of the Arrow revelation problem," since the prospective partners will typically experiment with the collaboration before any money changes hands. Moreover, such alliances literally license the partners to *monitor* at least some aspects of one another's progress in the competitiveness game.

Still, all such technical problems of revelation, ambiguity of asset pricing, and appropriability aside, the present spate of strategic alliances among different countries' largest corporations is mainly attributable to the growing awareness of increased financial or political (for example, protectionist) barriers to market entry, the need to accelerate product delivery times, and the crucial importance of accessing know-how that is embedded in competing organizations. The Italian economist Roberto Camagni adds the pertinent observation that, in the race to set standards that will make whole classes of products profitable, individual companies' objective in entering strategic alliances is "not just the control over a given technology or a given stock of complementary assets, but rather the control over the optimal development trajectory of these assets or technologies."[41]

Each partner brings to the marriage its own specialty—technology, financial power, access to government regulators or procurement officers—and its own constellation of small firm suppliers. As the name suggests, alliances tend to be somewhat less formal, less rigidly contractually obligational versions of the market-sharing, joint-venturing, licensing, equity-partnering, R&D consorting, coproduction and even bartering arrangements that have been linking the big firms both within and across national borders since at least the 1960s.[42]

Badaracco has developed case studies of two corporate giants, General Motors (GM) and IBM, which illustrate these principles.[43] By the late 1970s, both companies were literally desperate to achieve greater flexibility in the face of unexpectedly intense competition, and, in the case of IBM, an unanticipated change in technological trajectory. For GM, the pressure was to improve the quality, diversity, and timeliness of its products, especially in the smaller-vehicle range. For IBM, the challenge was to transform the company from the world's largest maker of one class of products—stand-alone mainframe computers, employing a highly specialized ("dedicated") programming language and components—into a standard setter and manufacturer *cum* systems integrator in the brave new world of distributed information processing. In short, IBM had to learn how to design and build communications systems whose nodes would be desktop personal computers and workstations.

For both companies, an important part of their strategy for change involved the deliberate erection of new intra- and international strategic alliances. Some of these have worked out better than others. I postpone until chapters 7 and 8 a detailed examination of what these (and many other) companies have actually done in creating alliance forms of production networks. Here, I want to extract a few of the general organizing principles behind Badaracco's examples.

The starting point for this work is, not surprisingly, the phenomenon of *globalization*, or, specifically, the globalization of *knowledge*. By this, Badaracco means that more entities and places are producing potentially commercializable information than ever before. Some of this information is "migratory"; the blueprints can be bought and sold (or stolen), and passed fairly readily from one institution to another. Alternatively, one firm will purchase the new products of another, subject them to reverse engineering (disassembling them, like a child's toy, to see what makes them tick), and will then use the knowledge thus acquired either to literally clone the original or to aid in manufacturing a modified version.

By contrast, other information is embedded (impacted) within the organization(s) where it was first created or unearthed. *This* sort of knowledge "moves only slowly and awkwardly among organizations." It is not just a matter of individual employees knowing certain things (although both the explicit and tacit knowledge of key personnel is, as noted in chapter 6, certainly an important object of much interfirm rivalry). In the course of developing new knowledge about processes and products, whole organizations develop routines that both reflect and ratify the existence of that knowledge. The upshot is that, if you want to access *this* sort of information—if what you are after are *capabilities* rather than commodities—you must penetrate the outer skin of the target organization. You must, in short, become an insider. Outright acquisition may achieve this objective, but it is expensive, constitutes yet another potential drain on the resources and energies of the acquirer, and risks squeezing out of the acquired business precisely the creativity that probably contributed to creating the new knowledge to begin with. The alternative approach is to negotiate a strategic alliance.

Thus, relative to the many other forms of interfirm relation, alliances are more likely to be formed when knowledge is embedded within the organizations that created it. These alliances may focus on the design, production, or distribution of specific products, thus enabling the partners to fill out their portfolios. Such joint ventures, specific technology exchanges, and the like are what Badaracco calls "product links." By contrast, the alliance may be sought in order that one partner to the collaboration can learn from the other(s) how to do new things. These are "knowledge links." GM's many recent alliances have for the most part been aimed at gaining the ability to manufacture or import inexpensive small cars, and to increase global market share in electronic database systems and in robotics. These are product links. IBM has also erected its own product-oriented alliances, but *its* large-scale R&D and networks have been aimed mainly at helping the company learn how to do things it did not previously do. These are knowledge links.

In general, the ability to form and sustain these two broad types of production networks becomes part of a modern flexibility-seeking firm's strategic repertoire. Moreover, there is a positive feedback process at work. As new competitors emerge, they create alliances to access embedded knowledge. The greater their success, the greater the pressure on the older firms to enter (or expand on existing) alliances of their own. Thus, competition in the marketplace increasingly takes on the form of competition among alliances whose composition may itself be shifting over time. These alliances thus constitute for Badaracco a network of "core" firms, each of which, "like a Renaissance city-state," is itself embedded within a "dense web of [sometimes] long-standing relationships."[44]

The cores are then "enmeshed in a sphere of alliances that link them with other organizations." Since the individual firm can no longer single-handedly invent, own, or control all the assets, or even parse all the available knowledge it needs, managers actively seek out new alliances and are forced to learn how to manage in such a strange environment. Why strange? Because it is inherently cooperative and hierarchical *at the same time*. Within Badaracco's core, management is command driven, at least to some degree. But management across the alliance can *only* be collaborative, or there can be no alliance.

Badaracco's distinction between migratory and embedded knowledge is not in itself a new idea. He inherited it from Arrow, the Berkeley economist David Teece, and the M.I.T. strategy theorist Eric von Hippel (whom he freely acknowledges). Moreover, that organizations have limited capacity to absorb information or to make any sense out of it was first suggested many years ago by Herbert Simon, Carnegie Mellon University's Nobel laureate economist-psychologist, in his famous theory of bounded rationality.[45]

What Badaracco has done is to take these theoretical elements and bring them to bear on a specific, historically concrete problem: how the very corporations that found themselves in crisis during the 1970s have discovered a strategy for simultaneously gaining or maintaining flexibility, while reorganizing themselves to maximize their access to new knowledge. Badaracco's answer is the same as Powell's: the managers of a growing number of large corporations are doing it by creating or entering production networks.

Strategic alliances may eventually lead to outright mergers and acquisitions, although various national laws and customs may effectively rule that option out (the Japanese are especially resistant to majority acquisitions by foreigners). More important, the same desire for greater flexibility that encourages networking in the first place tends to discourage the expensive, although not irreversible, commitment to one firm's actually acquiring another in order to gain access to the tacit knowledge embedded within the target company. On the other hand, as discussed in chapter 7, merger mania is currently alive and well in Europe, suggesting either that the desire for flexibility is not the only concern motivating business managers, or that flexibility and the age-old desire to exercise total control over business relationships may not, after all, be incompatible.

TERRITORIALLY BASED PRODUCTION SYSTEMS AS NETWORKS

How geographically focused production networks emerge, what holds them together, in what ways they may be changing over time, and what vital local networks contribute to *national* economic growth are questions that have become central to the contemporary discourse on competitiveness, business strategy, and industrial policy.

The first full-fledged attempt to integrate theories of industrial organization with those of economic geography and regional economics was published in the early 1970s, by the now-deceased, highly influential American economist Stephan Hymer.[46] He imagined that it was the location decisions of the multinational corporations that were shaping a new economic geography. These corporations tended to

> create a world in [their] own image by creating a division of labor between countries that corresponds to the division of labor between various levels of the corporate hierarchy. [They] tend to centralize high-level decision-making occupations in a few key cities (surrounded by regional sub-capitals) in the advanced countries . . . confining the rest of the world to lower levels of activity and income.[47]

Thus, an emerging "new international division of labor" was coming into existence, reflecting the hierarchical social division of labor commonly associated with the big firms from the 1930s through the late 1960s. Hymer forecast a continual expansion of production work into all regions of the world, with headquarters operations tending to remain concentrated within the business capitals of the most powerful countries.

By the 1980s the Hymer theory was increasingly being criticized. The evolving spatial division of labor was revealing itself to be much more complex than what Hymer had imagined. With the benefit of hindsight, we know that the advent of production networks was responsible for confounding the story. Networked production permitted far more complicated spatial arrangements than that suggested by Hymer's simple hierarchy—even when the big firms remained at the center of the action. By the 1990s, new theories of spatial economic organization in a world of production chains were being crafted to account for these more intricate phenomena.[48]

The University of California at Los Angeles economic geographer Michael Storper and I have attempted to deconstruct and then reassemble the elements of networked production systems so as to explicitly link economic geography into the broader context of the study of organizations and of business strategy. For us, a full production network has an input-output structure (a set of transacting production units of various sizes), a structure of governance (authority and power), and a territoriality (spatially dispersed or concentrated).[49]

SOME BASIC DEFINITIONS

A *production unit* is a physically integrated set of production activities occurring at a single location. By contrast, the *firm* is a legal entity. Most firms consist of a single, usually (but not necessarily) small unit, while *multiunit firms* are either sets of functionally related units or collections of units united by ownership but not by function. The *input-output (I-O) system* is a collection of activities that lead to the production of a specific marketable output. There is no necessary one-to-one correspondence between particular firms (or even plants) and a single I-O system; just as firms may own many units of production, they may also be involved in many I-O systems.

The *branch of production* (which the French call the *filiere* and which is more commonly called the "sector") is the statistical aggregate of similar I-O systems. Finally, a *territorial agglomeration* is a collection of production units in a limited area, such as a city or region. While the industrial district story (especially in its Italian version) assumes that any given territorial agglomeration will be focused on a particular branch, this need not necessarily be the case; regions, too, many contain a variety of *filieres*.

Types of Input-Output Systems

The units of production in an I-O system differ according to their degree of potential for realization of internal economies of scale and scope. As a general proposition, the greater these economies, either separately or in combination, the larger the units of production.

Internal economies of scale exist when increases in the capacity of productive units lead to a more than proportional increase in output or decrease in unit cost. Internal economies of scope are said to exist where a single unit—a particular plant, workshop, or store—produces a range of related products or includes multiple phases of a complex production process such that the overall unit cost of production is less than the sum of what it would cost to manufacture the individual products in separate facilities. Economies of scope and scale need not occur together, but neither are they necessarily in conflict.[50] When there is a deep division of labor, so that many units of production are knit together into a system, there is a greater likelihood that *external* economies of scale or scope will be high. In this case, the production cost advantages are enjoyed by the system as a whole rather than by individual units, as such.

The Territorial Dimension

Production networks with deep social divisions of labor and, therefore, with the possibility of substantial external economies of scale or scope may or may not be territorially agglomerated. A dispersed network with large units is typified by the IBM system for producing and assembling personal computers. Dispersed networks with no large units—that is, with no units enjoying significant internal economies of scale or scope—are sometimes encountered in long-distance alliances of specialized small consulting firms. By contrast, agglomerated networks may principally involve production units with insignificant internal economies, such as the industrial district of Prato, or they may consist of a mix of small and large units, as seems to be a more accurate description of Silicon Valley, and of the Baden-Württemburg region of southwestern Germany.[51]

Obviously, it is important whether the region's firms are situated at the center of an agglomerated network, whether the region is merely a site for units in a dispersed network, or whether the region constitutes a node in a network that is partly concentrated and partly dispersed. This will have much to do with the bargaining power of the region's firms vis-à-vis the sector as a whole, and with the extent to which the region's policy makers can bargain successfully with powerful agents in the sector, especially at the international level.

GOVERNANCE STRUCTURES

A production system consists of more than just the agglomerated or dispersed I-O systems I have defined thus far. In addition, we must specify a particular system of *governance*: the degree of hierarchy and leadership (or their opposites: collaboration and cooperation) in coordinating the I-O system. It is with the specification of a governance structure that we finally get to relations of power and decision making within the system.

Following a long tradition in political theory and business history, it seems useful to distinguish forms of governance along a dimension capturing the relationship between the firms within the *core* of a system and the *ring*, or periphery, of that system. The existence of a system core implies that power is asymmetrically distributed, and that some core firms have the power to determine the existence of others. Within the ring, power is more likely to be distributed symmetrically, and the existence of one or another unit or firm is not obviously under the control of other units within the ring. (An exception is the *tiered* supplier system, in which those closest to the big firm customer are responsible for managing lower-level, usually smaller, contractors.)[52]

A rough typology of governance structures might begin with the case in which a system is essentially all ring and no core. That is, there is no enduring lead firm, or there is a rotating leader, as in systems that link firms together on a project-by-project basis. Such systems may be highly agglomerated, as in the prototypical case of Prato, or they may be dispersed worldwide, as was arguably the case through the 1980s for small, independent film producers.[53]

In core-ring systems with a coordinating firm, we begin to encounter a degree of hierarchy and asymmetric power in the system. At the very least, a well-placed "coordinator" can influence the configuration and direction of the system, and quite possibly the internal operations of at least some member firms. In the United States, the Xerox Corporation was an early leader in conducting joint design work with first-tier suppliers, as part of a general restructuring of its operations in the face of unexpected competition from the Japanese.[54] In Germany, the degree of control of lead car companies such as Porsche and BMW over their parts and systems suppliers is restrained by the unusually high degree of skill and other forms of asset specificity within those rings.[55]

When key firms become dominant in their control over their suppliers or other partners, we may say that a "lead" firm has emerged within the system's core. Power is highly asymmetrical, and there may be considerable hierarchy governing relations among units. This seems a reasonable characterization of production systems in the electric products branch, involving the core-ring systems organized around companies such as America's General Electric, Japan's Sony, France's Thomson, and Germany's Philips. It surely describes the mass market–oriented car companies, such as Japan's Toyota, America's General Motors, and Italy's Fiat. In chapter 4, I suggested that Sasib, the Italian food processing machinery group, is evolving in just this manner.

NETWORKED PRODUCTION SYSTEMS

Having constructed a language for describing the division of labor, the size of units, their interconnectedness or network character, their territoriality, and their governance, I now want to fit the pieces together.

Thus, for example, as discussed in chapter 5, Silicon Valley has customizers, systems integrators, and giant merchant chip manufacturers, each associated with a different type of governance. Moreover, the entire network is linked to other, geographically dispersed networks, as exemplified by IBM's global sourcing strategy for manufacturing personal computers and its R&D alliance with Germany's Siemens and Japan's Toshiba. The different market segments of Hollywood's entertainment industry—independent productions, theatrical motion pictures, and films for television—are governed by qualitatively different rules. Some of these networks contain mostly small units, while others are organized around large coordinating lead firms.[56]

Still another example is offered by the congeries of relationships between the core-ring systems that have evolved around Porsche, the auto maker, and Germany's premier auto parts manufacturer, Robert Bosch. Both are based in the same region of southern Germany and are tied to one another by strong I-O links. Yet within their own orbits, Porsche serves as a coordinating firm among more or less equally small units, while the "Bosch system" includes some units (starting with Bosch, itself) that are much larger than others.[57]

Analogously, strategic alliances may cross governance/I-O boundaries, as well. For example, one such alliance connects Italy's Marpos to America's IBM. Marpos, an agglomerated weak core-ring system located outside of Bologna, designs and manufactures sophisticated gauges for measuring tolerances in metalworking machinery. In my own visit to Marpos, I discovered that IBM trains Marpos technicians, equips Marpos factories, and provides both financial and technological resources, in return for which Marpos gives IBM yet another "window" into the European capital goods branch.

Another alliance between two highly localized strong firm core-ring systems connects General Electric's Aircraft Engine Group and the French public enterprise SNECMA (Societé Nationale d'Étude et de Construction de Moteurs d'Aviation). These companies collaboratively produce jet engines for both civilian and military markets. GE uses its supplier networks located around Lynn, Massachusetts, and Evandale, Ohio, while SNECMA relies primarily on a highly agglomerated supplier network centered on the "Arc de SNECMA" to the west of Paris.[58]

DEVELOPMENTAL TENDENCIES

Networks evolve—which is to say that, over time, their production, locational, and governance structures may all change. For example, I have shown earlier how, as the result of the intrusion of Milanese finance capital into the region, the Sasib network in Italy's Emilia-Romagna is evolving from an agglomerated network of locally embedded small firms into a lead firm–driven, more geographically extensive system. I also reported how Benetton, which during its early years was content to function as a loosely coordinated set of units concentrated in the Veneto region, is now moving toward a truly hierarchical form of coordination, with the establishment of totally dependent branch plants and the extensive use of home workers situated elsewhere in Italy and even outside the country.

International competitive pressures, including the search for cheaper labor or markets and the need to operate inside other countries' borders, have compelled Toyota to extend its production system far beyond the confines of its original base in Nagoya, Japan, and the same has happened to other companies. Toyota is now establishing its own overseas assembly operations in North America and Europe, and attracting both its traditional Japanese auto parts suppliers and local, non-Japanese subcontractors to follow it.[59] The much-praised agglomeration of auto assemblers and parts producers in southwestern Germany is undergoing a process of international decentralization.[60] In yet another example of system evolution (but in the opposite direction), the Cornell University urban planner, Susan Christopherson, suggests that a certain reintegration by the largest Hollywood film studios is occurring in the United States, following the reconcentration of theater ownership and the shake-out of financially troubled independent studios.[61]

So, change is common and need not imply (as has been recently suggested by the Berkeley management theorist Raymond Miles) that network organizations are failing.[62] Underneath the change, however, lies a common trend. Everywhere in the world, we can now find examples of a shift away from agglomerated, fragmented, symmetrically powerful, mainly small firm production systems to core-ring systems— some agglomerated and some dispersed, but commonly organized around powerful lead firms.[63] I think that this movement toward core-ring networks organized around powerful lead firms is becoming the *dominant* tendency in the new world order—especially when financial and marketing operations, and such other key partners as research hospitals, big banks, and key government agencies are brought into the picture.[64]

PUTTING THE CONTEXT BACK IN

A fully developed theoretical model of the morphology and developmental tendencies of networked production systems will eventually have to account for historically contingent specifics and relations *outside* the I-O nexus, within which production relations per se are embedded. The list of missing pieces to the puzzle is a long one and might include labor relations, the wage-setting process, the relative importance of public regulation, and even direct ownership (as with such national holding companies as Italy's IRI and the Canadian Crown corporations), the extent and significance of trade associations, local traditions of family business ownership, and the local content and other policies

148   THE EMERGING SYSTEM OF GLOBALLY NETWORKED PRODUCTION

that governments commonly impose on private companies to change their investment, procurement, or locational behavior. Moreover, it is now generally agreed that companies are linked not only by commodity exchanges (those I-O connections), but by the sharing of commonly available knowledge and freely supplied services—sometimes called "non-traded interdependencies." All are relevant to fashioning descriptions of how existing production networks are configured and to predicting how they are likely to evolve in the future.

Moreover, while I and Storper have presented a schema in which geographic clustering is possible—and may contribute significantly to the efficiency of production units and even entire firms within these networks—the actual extent and significance of such locational clustering remain open questions for further research. To be sure, many writers and policy makers think either that such clustering is already common or that it should be actively promoted. This is a principal theme in *The Competitive Advantage of Nations*, a highly influential book by the Harvard Business School management guru Michael Porter.[65] And of course it lies at the very heart of the theory of the industrial districts.

But in an era in which productive units in any particular location are often in constant communication with parent or partner firms located on the other side of the earth, I think that the jury is still out on the question of the extent of such geographic clustering—and especially the significance of agglomeration for the actual performance of individual firms and business establishments.[66]

Still, enough thinking has now gone into the construction of these various organizational maps that we may begin to see the outlines of the sort of industrial order that is evolving out of the breakdown of the older mass production system, in the wake of the global profits squeeze of the 1970s. In the new world economy, the atomistic competitors of the sort celebrated in conventional economics texts, and by free market ideologues such as Gilder and Birch, are socially *inefficient*, profoundly *unstable*, and tend to be technological *laggards*—precisely the opposite of the conventional wisdom. Instead, production (and distribution) *networks* are becoming the dominant organizing principle; this is how business achieves the greater flexibility demanded by chronic uncertainty, market fragmentation, and accelerating product cycles. Some of these networks have segments that are geographically clustered; some do not. Embedding matters, to be sure, but it does not follow that *locally* embedded, primarily small firm–led and inward-looking industrial districts are the dominant (or even the most energetic) form. Indeed, I am arguing that they probably are not.

Ultimately, these are empirical questions. Therefore, I now turn to an examination of the evidence of the growing incidence and importance of actual networked forms of productive organization in Japan and Europe (chapter 7) and then in the United States (chapter 8).

7

LARGE FIRM–CENTERED
NETWORKED PRODUCTION
SYSTEMS IN JAPAN AND EUROPE

In the previous chapter, I discussed why, in an effort to escape from the profits squeeze of the 1970s and faced with the heightened competitive pressures of the present era, companies across the industrialized world have been turning more and more to network forms of productive organization.

In some countries more than in others, aspects of networked production had come into existence even before the onset of the crisis period of the 1970s. Thus, the Japanese *keiretsu* evolved into their present form during the years after World War II, and their roots go back even further into Japanese history. The "habit of cooperation" within European capitalism between big firms and their national governments also has earlier historical precedents (for example, in the fascism of the 1930s), although the flowering of large-scale interfirm collaboration through networks of producers is a more recent development. Collaboration has been most episodic, sporadic, and incomplete in the United States, where both the formal legal system and informal norms have always treated interfirm cooperation with the greatest ideological skepticism. Nevertheless, American corporations and their networks of suppliers are currently engaged in an unprecedented degree of networking, too.

The most formal, deeply rooted, truly paradigmatic examples of institutionalized cooperative competition through network forms of industrial organization are to be found in Japan. I therefore begin with a brief sketch of the *keiretsu* system in that country. While there can no longer be any question that modern Japanese economic development has been promoted by an activist government, the innovative production networks of that country are best understood as substantially private in nature. For a look at collaboration that includes the public sector at a significant scale, we must turn our attention to the Europeans. There, what has been a long-standing tradition of close government-business cooperation has only grown stronger, as the impetus toward continentwide economic integration motivates an ongoing shift from the post–World War II emphasis on nationalistic collaborations between particular governments and their "national champion" firms, toward an astonishingly rapid growth of cross-border alliances involving the companies and governments of many countries.

The head start in networking that east Asian and European companies have gained over U.S. firms as a whole goes some way in explaining the much-discussed problem of American competitiveness in the global economy. American-based corporations *have* become engaged in interfirm collaborative networks, but they often do so in order to buy into the foreign webs. That is why I first want to look abroad before examining the United States.

JAPANESE PRODUCTION NETWORKS
No economy in the world depends to a greater extent on the institution of the production network than does Japan. Indeed, the Japanese models have given the rest of the world a vocabulary with which to understand how concentrated economic power can be reconciled with decentralized, cross-border production in ways that promote organizational learning, systemwide innovation, and—most of all—flexibility in the face of competitive pressures.

The preeminent examples of complex, networked production systems in the world today are the Japanese *keiretsu*. Literally meaning "societies of business," these groups of interconnected companies come in essentially two flavors. The sociologist James Lincoln, at the University of California at Berkeley, and his colleagues call the first of these types the "big six intermarket groups": Sumitomo, Mitsubishi, Mitsui, Dai Ichi Kangyo, Fuyo, and Sanwa.[1]

Sumitomo, Mitsubishi, and Mitsui evolved from the pre–World War II family-owned *zaibatsu*. Although the U.S. military occupation officially tried to break them up after the war, by the 1950s they had

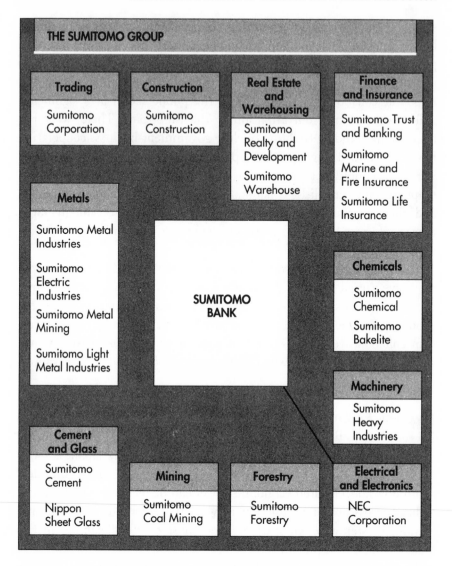

FIGURE 7.1 THE SUMITOMO GROUP

Source: Marie Anchordoguy, "A Brief History of Japan's *Keiretsu*," in Charles H. Ferguson, "Computers and the Coming of the U.S. *Keiretsu*," *Harvard Business Review* 68 (July–August 1990): 58.

reassembled themselves and have only grown stronger over the years. Thus, by the 1980s, according to estimates made by the management consultant Charles Ferguson and the University of Washington scholar Marie Anchordoguy, though the 182 companies that make up the core of the six groups account for only about 10 percent of the companies on the Tokyo Stock Exchange, more than half of Japan's 100 largest companies are group members. Virtually all of Japan's top city banks, trust banks, insurance companies, and computer, telecommunications, and semiconductor makers are group members. In the late 1980s, these six keiretsu earned some 18 percent of the total net profits of all Japanese business [and] had nearly 17 percent of total sales.[2]

In keeping with what the Japanese refer to as the "one set principle" (*wan setto shugi*), each intermarket or conglomerate keiretsu tends to operate one company in each major industry, from chemicals to computers. Each has a lead bank, a foreign trading company (the famous *soga shosha*, also a prewar holdover), and close connections to the most important government ministries. This remarkable mix of activities is exemplified by the organizational map of the Sumitomo Group (figure 7.1).

In the postwar era, other large corporations emerged and gradually built networks of suppliers and affiliates around themselves. This second kind of industrial grouping in Japan is known as the "supply *keiretsu*." In the early years, these companies faced great obstacles to competing domestically with the older and much more politically well connected conglomerates. Even the big banks were not initially supportive. Their eventual success came largely as a result of their proven ability to penetrate overseas markets.

Especially common in such industries as automobiles, consumer electronics, and machinery, where the final product is composed of many parts and subassemblies, supply *keiretsu* such as Toyota, Sony, and NEC are organized as giant, so-called parent final assemblers—what the American auto industry calls original equipment manufacturers, or OEMs—and their dense networks of more and less dependent, generally (but not always) smaller suppliers. Figure 7.2 displays the structure of the topmost tiers of NEC's supply *keiretsu*.

The two ideal types of keiretsu may be contrasted by differences in the extent and structure of hierarchical relations within them. The Dutch social scientist Eric Van Kooij distinguishes between the "pyramid," or top-down, form of a canonical supply *keiretsu* such as Toyota, on the one hand, and the "cobweb," or more conglomerate, networks that make up such intermarket groups as Mitsubishi, on the other (figure 7.3).[3] Many supply *keiretsu* are also members of one of the big six intermarket groups (note from figure 7.1 that

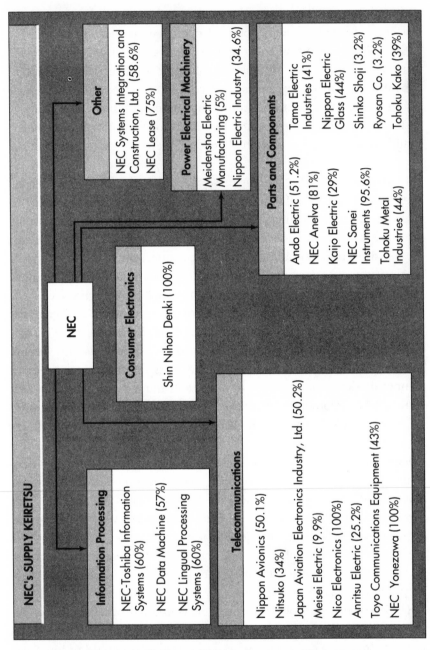

FIGURE 7.2   NEC'S SUPPLY KEIRETSU

Note: Percentages indicate share of the affiliate's stock owned by NEC.

Source: Marie Anchordoguy, "A Brief History of Japan's *Keiretsu*," in Charles H. Ferguson, "Computers and the Coming of the U.S. *Keiretsu*," *Harvard Business Review* 68 (July–August 1990): 63.

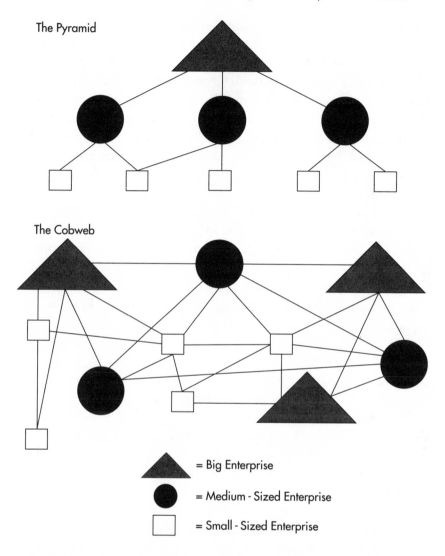

FIGURE 7.3   THE TWO CANONICAL FORMS OF KEIRETSU

Source: Eric Van Kooij, "Industrial Networks in Japan: Technology Transfer to SMEs [Small and Medium-Sized Enterprises]," *Entrepreneurship and Regional Development* 2 (1990): 298.

NEC, a major supply *keiretsu*, is itself a member of Sumitomo's intermarket *keiretsu*). Even when they are not formally part of a conglomerate family, the big supply *keiretsu* are, more often than not, affiliated with the big six through technology agreements or other strategic alliances, according to the management professor Michael Gerlach, Lincoln's colleague at Berkeley.[4]

In the course of undergoing international expansion since the 1960s, the Japanese supply *keiretsu* have entered into a growing number of alliances with Asian, North American, and European corporations. Consider Toyota's ten-year alliance with General Motors to make small cars in Fremont, California (a case examined closely in chapter 8), or Toshiba's recent alliance with IBM and Siemens to design and test a new generation of microchips in IBM's upper New York State facility. Through such alliances, what began as more or less stand-alone pyramidal chains is clearly evolving into a system of interlinked supply *keiretsu*—what Michael Storper and I call "strong core-ring production networks," a language first presented in chapter 6. These webs interconnect the clusters of activity that originally formed around particular big customer firms.

Toyota offers an especially clear example. From a core firm situated within an agglomerated supply *keiretsu* centered in one region of central Japan, the company has evolved into a node within a worldwide system of automobile design, parts sourcing, and final assembly. Honda is moving in a similar direction.[5]

SUPPLIER RELATIONS

Both types of *keiretsu* are notable for the complex relationships that have evolved over time between the parent companies and their dense networks of generally (but, as Lincoln and his coauthors have shown, by no means invariably) smaller suppliers: enterprises that the Japanese refer to as "children." There is by now an enormous scholarly and business literature in Japan on the subject.[6] One entire department of the Ministry of International Trade and Industry (MITI) devotes itself to monitoring the operations of customer-supplier relations. Western scholars including professors Ronald Dore and David Friedman believe that the extraordinary vitality of Japanese industry in world trade can be traced in large part to the inner workings of what has come to be called "relational contracting" within these networks—a concept further explored in a moment.

The essential organizational characteristic of the *keiretsu* supply system is that of *tiered* relations. For example, in 1980, Toyota had 168 first-tier subcontractors, 4,700 second-tier subcontractors, and 31,600 suppliers in yet a third tier.[7] In such a system, the parent big firm directly manages its relationship with the top tiers, whose member firms in turn take responsibility for managing those lower down in the hierarchy.

Many suppliers are organized horizontally into cooperative associations, or *kyoryokukai*. For example, within Toshiba's domestic supply keiretsu (that is, not counting the growing number of overseas suppliers that sell to Toshiba), at one point in the 1980s, van Kooij found 500 enterprises that belonged to one *kyoryokukai*.[8] The "parent" (core) firm provides financial and technical assistance to the members of the cooperative association to improve their economic and engineering capabilities, to teach them quality control, and to enable them to communicate—with one another and with Toshiba, itself—via computer. This is essential to the operation of the *kanban* ("just-in-time") system of shopfloor continuous improvement, reinforced by the minimal holding of buffer stocks of inventory. Matsushita has a similar arrangement with its tiers of subcontractors and suppliers.

With or without supplier associations to mediate the relationship, parent firms in Japan tend to practice what Dore has named relational (or "obligational") contracting with at least their first-tier suppliers. Relational contracting may have roots in prewar, even feudal principles of mutual obligation, as some writers have suggested. But it also turns out to be immensely productive, by diffusing innovations throughout a network and by sharing (and therefore encouraging the taking of) risk. Thus, to take a decidedly contemporary example, such modern organizational methods as just-in-time inventory control and total quality management demand that suppliers make significant investments in new equipment, incur debt, and undertake training of both managers and production workers in order to make themselves attractive partners (or, as the Japanese prefer to say, "associates") of the big customer firms.

The suppliers in turn expect *reciprocity* from their customers—a sharing of the risks and the burdens associated with such investments, and signs that the parent firms are willing to make long-term commitments to the nurturing of their "children." These commitments take the form of long-term procurement contracts and equity investments, plus technical assistance from customer firms and the exchange of personnel with those firms. Indeed, the extent of such reciprocal relations—the complexity of the ties that bind *keiretsu* members and affiliates to one another—is one of the things that so frustrates Western companies that are trying to break into such a substantially closed system, in order to sell to the big Japanese corporations. By this means, the Japanese seem to have found ways of reconciling seemingly contradictory behaviors of businesses; however, it would be a mistake to imagine relational contracting to "mean a cozy relaxed atmosphere—far from it. . . . Japanese suppliers face constant pressure to improve their performance, both through constant comparison with other suppliers and contracts based on falling costs and (therefore) delivery prices."[9]

Whether the big customer firms are part of the big-six conglomerates or constitute the cores (hubs) of supply *keiretsu* such as Toyota, Toshiba, and Matsushita, the more specialized the supplier firms—the more highly specific their informational, human capital, or machinery assets—the more independent they are likely to be of these big customers. For example, van Kooij has studied the Japanese color television customer-supplier network. He finds that, on average, nearly three-quarters of the value of production in NEC, Panasonic, Mitsubishi, and Sony is contracted out.[10] Highly specialized independent manufacturers of resistors, tuners, and switches sell to these OEMs but typically have other customers, as well, and even export on their own. Both the OEMs and these highly specialized independent suppliers then sit atop tiers of generally much more dependent suppliers and subcontractors.

In previous chapters, I introduced the idea of vertical disintegration as providing one of several explanations for the origins of locally oriented clusters or industrial districts. Disintegration can help us better understand the nature of power relations among the members of Japanese customer-supplier networks. For the most part, control in the *keiretsu* system flows "upstream"; customers generally control the behavior of suppliers in Japanese supply chains. Vertical disintegration is one reason why power and control flow asymmetrically:

> Size and concentration account for much of the power of downstream assemblers but not all. The purchasing firm is likely to be older . . . better known, and to enjoy higher status. [Moreover] the supplier and subcontractor firms with closest ties to a large assembler such as Toyota or Matsushita are in many cases spun-off divisions. . . . Despite the spin-off's new-found status as an independent firm, the hierarchical relationship between it and the "parent" is very likely to be preserved.[11]

In other words, what matters is not so much the downsizing of the big firms, per se, as the nature of the subsequent relationship between the original parent firm and its now ostensibly independent suppliers.[12]

KEIRETSU GOVERNANCE

A glance back at figure 7.2 immediately brings home a fundamental property of the *keiretsu* production networks' mechanisms for system governance. NEC directly owns substantial shares of stock in the companies that supply it with parts, designs, components, subassemblies, and services—as much as 100 percent, in some cases. Indeed, Lincoln's extensive research leads him to the conclusion that "more than any other criterion, this network of cross-shareholdings superimposed on financial and commercial transactions *defines* the term *keiretsu* [my emphasis]."[13]

Originating in the 1950s as a way of protecting themselves from domestic hostile takeovers and, later, from the influence of foreign capital, the cross-holding of equity among the members of a Japanese *keiretsu* has become an extraordinary source of institutional stability. Such cross-holding frees corporate managers to a significant extent from concerns about the availability of investment capital and encourages them to adopt a longer-term planning horizon than that which characterizes especially British and American corporations.

Averaging across all keiretsu members during the 1980s, an estimated 15 percent to 30 percent of their stock was held by other members of the group. Moreover, "some 60 percent to 80 percent of *keiretsu* company shares are never traded [at all], so managers do not have to worry about takeovers and can focus on long-term issues."[14] This is why the biggest Japanese corporations were to some extent sheltered from the recent Tokyo Stock Market crisis. The subject of short-term versus longer-term horizons and how they vary between Japan, Germany, and the United States has recently undergone considerable study by a group of economists, business strategists, and historians working through the Harvard Business School, on a project under the leadership of Michael Porter. In the next chapter, I review their findings at length, in the course of drawing conclusions about why American business has lagged behind its foreign competitors in the adoption of the new ways of operating that have proved so successful for others.

In Japan, cross-shareholding tends to follow the development of a preexisting economic relationship. Once such transactional relationships prove mutually beneficial to the parties, cross-shareholding helps the customer firm exercise greater control over the setting of prices by the supplier firms. But—and this is the essence of what Dore and others mean by calling Japanese contracting *relational* rather than simply transactional—it is not a matter of the big firm simply exploiting the smaller:

> The more the small firm is brought under the large firm's wing, the greater its confidence that it will continue to get the parent's business and the better its reputation in the business community. In the Japanese corporate stratification system, these translate into more resources at lower cost owing to improved credit standing, access to the parent firm's main bank, access to the technical and managerial skills of the parent, and the ability to attract higher quality labor.
>
> High-level managers in large Japanese manufacturing firms observed that . . . it was normal to maintain a minority equity position in a supplier or subcontractor.[15]

It used to be thought that especially in the conglomerate *keiretsu*, the leading merchant bank was the main governor of the entire system (indeed, Ferguson still refers to this type of *keiretsu* as "bank-centered"). But while group-affiliated merchant banks such as those of Mitsui and Sumitomo are certainly powerful and visible, and indeed have become among the largest banks in the world, Lincoln and Gerlach have shown convincingly that they no longer exert quite the degree of day-to-day dominance over the industrial firms within their groups that had been believed.

On the other hand, the Berkeley sociologists' research supports the view that power and control within these networks flows substantially from the bigger to the smaller firms. That is, these production networks really are hierarchically organized, such that power is distributed asymmetrically:

> Equity and financial capital, as well as directors, flow from large, established, big-six affiliates to smaller, younger independents, whereas industrial products (parts and materials) move the other way (except that large firms are central in the supplier network and big-six firms are as likely to be suppliers as [are] independents). Other evidence of asymmetric control of small firms by large firms is the finding that shareholding reciprocity is much stronger among the top 60 Japanese financial firms and industrial firms than among the top 250.[16]

Note Lincoln's point that large firm centrality is no longer exclusively located at the purchasing or final assembly end of the value-added chain. Some key suppliers, such as the auto parts producer Nippondenso, are themselves powerful, well-connected, multinational corporations. The existence of such partners within the network may complicate simple one-dimensional pictures of hierarchy within Japanese production networks, but we are still talking about asymmetric power relations, in which firm size remains an important parameter.

KEIRETSU PRODUCTION NETWORKS
IN A CHANGING WORLD ORDER

In recent years, a growing number of Japanese supplier firms have succeeded in making themselves more independent of any one big customer firm within their *keiretsu* than used to be the case. In the auto industry, for example, some members of the Toyota Group now routinely subcontract to Honda and, occasionally, to Nissan. Some semiconductor suppliers now also sell to customers who compete head to head with those small firms' own parents.[17] Often, with the slowing of economic growth, lead firms actually encourage their key suppliers to find additional customers, in order to keep the former alive and well.[18]

Moreover, it is undeniable that, under increasing pressure from the West to open its domestic economy to imports of industrial parts and components, Japan has been encouraging its big firms to do more business with outsiders. As noted in chapter 5, Japanese corporations are also under pressure to go global themselves, building research laboratories and assembly plants in such places as the Mexican border and Silicon Valley in order to access cheaper labor, to participate in joint technology agreements with foreign partners, or to meet the local content requirements and other nontariff barriers erected by foreign governments. As is already clear from the behavior of Japanese multinationals in the auto sector, there is also a definite sign that the big firms are encouraging (if not demanding) their long-standing child firms back home to go global themselves, building branch operations abroad in order to follow their principal parents.[19]

But we should be wary of predictions of the impending demise of the *keiretsu* system. As discussed throughout this book, global competitiveness now turns increasingly on a combination of attention to *quality* and *timeliness* of the development and delivery of products to customers. Unlike competition based mainly on price, or on the monopo-

162   THE EMERGING SYSTEM OF GLOBALLY NETWORKED PRODUCTION

lization of raw materials or access to particular markets, the kind of competition that characterizes the new world economic order requires ever-greater *coordination*—among stages of production and across spatially dispersed units—to get things done and to win and retain market share. This is especially so, the greater the degree of *complexity* of the relevant design, engineering, or manufacturing processes.

One thing on which there is general agreement among scholars and business analysts studying the Japanese *keiretsu* is that so long as markets place a premium on quality and (therefore) on the ability to coordinate within complex systems, keiretsu-like big firm–led production networks will continue to have a significant competitive advantage. As Lincoln and his colleagues conclude: "It is premature to assume that the keiretsu is an obsolete organizational form."[20] In the 1990s, the problems being encountered by the Japanese economy have to do with the high exchange value of the yen, the bursting of the real estate speculative bubble of the 1980s, and a genuine crisis of legitimacy within the ruling political party. Against these storms, the mutual supports within the keiretsu system of production have arguably kept a bad situation from becoming even worse, as regards the innovativeness and international competitiveness of Japan's big firms.

EUROPEAN INTERCORPORATE NETWORKS

The reader will surely have noticed that, in my entire discussion of the Japanese *keiretsu*, I barely mentioned the role of the State in supporting the formation of that country's production networks. That role was certainly crucial in the formative years after the war and since then has never failed to be relevant. Those academic curmudgeons who, after all these years, continue to insist that the Japanese government does not engage in capitalist planning are simply wrong.[21]

They are not, however, wrong to assert that, in the contemporary era, Japanese development is *mainly* the product of the workings of the private sector. Of course, as just discussed, those "workings" bear almost no relation to the economists' textbook models of free markets, characterized by arms'-length spot transactions. The Japanese private sector is pervaded by hierarchy, asymmetric power, and the practice of long-term planning.

European corporations have also struggled to restructure themselves in the wake of the worldwide economic crisis of the 1970s. They, too, have sought to rationalize relationships with suppliers. Licensing agreements, joint ventures, and more loosely coupled strategic alliances are all present in the European story, as well. And given the long tradition of corporatist collaboration between business and government (and, in varying degrees, with labor) in much of the European continent (although not in the United Kingdom), government policy has played a continuing role in supporting the European evolution toward networked production systems.

Of all the forms of interfirm "partnering"—a favorite word in European business and policy circles—outright mergers and acquisitions are the least novel. And they are clearly on the rise, actively encouraged by some governments, and both implicitly and explicitly supported by the European Community (EC). During the 1980s, the twelve member nations of the European Community experienced an average rate of growth of industrial (that is, not counting retail trade, banking, or insurance) mergers of 25 percent per year. As a result, the market share of the top 100 companies in the manufacturing and energy branches rose from 14.8 percent in 1982 to 20 percent in 1988.[22]

The regional planners Flavia Martinelli and Erica Schoenberger remind us that waves of concentration are not a new development in the history of capitalism, any more than are periods in which, as new sectors and technologies come into existence, industrial organization seems fragmented. But in Europe today, the wave of mergers and acquisitions is, in their words, "unprecedented,"

> spurred on by strategic repositioning in the face of the approaching . . . integration of the EC [European Community Single] Market. . . . This latest wave [is sweeping over] both mature (oil, chemicals and food) and new sectors (electronics, telecommunications and business services) . . . families of closely related projects . . . carried out not only by large multinational groups pursuing a strategy of globalization, but also, as in Italy, by smaller, national or local firms attempting to strengthen their market positions.[23]

In Italy, for example, between 1983 and 1988, acquisitions of major Italian firms by foreign companies grew by a factor of nearly six, and acquisitions of major Italian companies by other Italian firms grew by a factor of five.[24]

The merger wave in Europe began in earnest during the mid-1980s and shows no signs of letting up anytime soon. One accounting firm counted 1,200 mergers and acquisitions among major corporations in

1988 alone. In what *Business Week* describes as a "feeding frenzy on the continent," cross-border partial and complete takeovers fill the news. In 1992 alone, France's Elf-Aquitaine, a petroleum giant, acquired a chain of refineries and gasoline stations in eastern Germany, Switzerland's Nestlé acquired the manufacturer of France's famous Perrier water, Germany's Lufthansa and Air France merged their four-star hotel chains, and Italy's state-owned cement giant, Italcementi, acquired a 45 percent share of what had been France's largest cement producer, making Italcementi a $4-billion multinational corporation overnight. Nowhere is the urge to merge gathering more steam than in Germany, where the pressures created by reunification and heightened global competition pushed merger and acquisition activity in 1991 to a record level of nearly 2,000 business deals. The targets of the German takeovers are the highly productive midsized companies (the *mittelstand*), whose older founders and senior managers are nearing retirement or are worried about competing in a world of ever-larger global production networks.[25]

Many observers think that the principal motive for the timing of this wave of European mergers and acquisitions is the need of European companies to bolster their ability to compete in the more open international economy being brought into existence by the policy of the Single European Market, with its continentwide internal deregulation. This pressure has been greatly heightened by the aggressive investing of American and Japanese capital in Europe, buying up companies and building plants in order to get inside the nontariff barriers (such as local content and offset requirements) that the EC is erecting around the continent.[26]

In no sector is the invasion of the Americans and the Japanese more determined than in high-tech electronics. After trying to go it alone in the development of new microprocessor technology, and in spite of EC efforts to shore up European computer firms through government-subsidized intra-European alliances, France's Bull recently sold a small but significant stake in the company to IBM. Another French conglomerate, Matra, has sold fully a fifth of its communications division shares to Canada's Northern Telecom, to build cellular telephones in Europe. After an unsuccessful strategic alliance with AT&T, Olivetti has now sold a tenth of its shares to Massachusetts-based Digital Equipment Corporation, giving DEC a long-desired European sales network for distributing its workstations. Fujitsu has acquired the United Kingdom's biggest computer maker, International Computers, Ltd. (ICL). Holland's Philips is going both ways: joining forces with America's Motorola to design computer chips for consumer electronics products, while simultaneously merging its VCR and camcorder business with its German rival, Grundig, to fight off the invasion of its European markets by Sony and Panasonic.[27]

STRATEGIC ALLIANCES

If formal mergers and acquisitions involving European-based companies represent the continued use of an old business strategy, the recent proliferation of strategic alliances constitutes a prime example of the newer, more flexible approach to governing relationships among firms.

The number of intra-European alliances increased by 1,000 percent between 1980 and 1985, while joint ventures between European and U.S. companies nearly doubled in the decade after 1978. In the electronics branch, the floodgate of recently announced technology-driven alliances was opened in June 1989, when German's Siemens and Japan's Matsushita announced that they would collaborate in the manufacture of components used in car electronics, computers, telecommunications equipment, and household appliances. Siemens' annual worldwide sales in this niche had, to this point, averaged about $350 million, whereas Matsushita's had more closely approximated $3 billion. Said one Dusseldorf-based analyst: "The only way to improve business now is to join one of the giants."[28]

But perhaps the biggest—certainly the most well publicized—of the recent cross-border alliances was the announcement in 1990 that German's Daimler-Benz and Japan's Mitsubishi would begin engaging in what the two corporations called "intensive cooperation" in a broad range of fields, including autos, aerospace, and microelectronics. Daimler gains access to Mitsubishi's prodigious knowledge of mechatronics (the wedding of mechanical and electrical engineering). What Mitsubishi gets from the alliance is Daimler's aid in blocking French and Italian resistance to letting Japanese car makers into post-1992 Europe, and a connection to Europe's single largest and most commercially successful public-private partnership: the Airbus Industrie consortium that has made itself one of the world's three largest producers of civilian aircraft and a leader in airframe technology.[29]

Typical of the strategic behavior of these giant corporations in wanting to participate in multiple alliances or networks, even as it becomes more engaged in the European Airbus project, Mitsubishi Heavy Industries has simultaneously become a major supplier to Boeing, Airbus's biggest competitor. Moreover, in keeping with usual Japanese corporate practice,

Mitsubishi announced that it is unabashedly using its connections to Daimler and Boeing to learn how to eventually manufacture wide-bodied passenger aircraft on its own. In keeping with the competitive dynamic wherein one strategic alliance begets another, in response to Boeing's partnering with Mitsubishi, the seriously financially strapped McDonnell-Douglas has for over a year been engaged in discussions with the Taiwan Aerospace Corporation, a government-backed industrial group, about forming an alliance of their own.[30]

Strategic alliances are especially attractive to firms seeking to make progress in researching, developing, refining, testing, or commercializing new technologies. Luc Soete is the director of the Dutch University of Limburgh's Economic Research Institute on Innovation and Technology at Maastricht. His extensive research and policy advising across Europe have led him to propose eight motives for technology-oriented companies to seek cooperation via networks:

- the reduction, minimizing, and sharing of uncertainty in R&D
- the reduction and sharing of the mounting costs of conducting R&D
- the increased complexity and intersectoral nature of new technologies and the cross-fertilization of scientific disciplines and fields of technology, monitoring of the evolution of technologies, technological synergies, access to scientific knowledge or to complementary technologies
- the desire to capture another company's tacit knowledge or technology, to facilitate technology transfer, or even to know what technologies one wants to "leapfrog"
- the shortening of product life cycles, i.e. reducing the period between invention and market introduction
- monitoring of changes in the environment and of new opportunities
- internationalization, globalization, and entry into foreign markets
- the desire to expand the firm's existing product range by making new products and entering new markets[31]

All of the examples I have offered in this chapter display one or another of these conditions. The point is that alliances *are* strategic, and not simply serendipitous or opportunistic. They have an underlying logic that economic development policy analysts and decision makers will need to study more closely.

GOVERNMENT ENCOURAGEMENT OF HIGH-TECH NETWORKS: THE CASE OF ESPRIT

Prior to the onset of this wave of partnering with Japanese and American electronics giants, European industry was lagging far behind its competitors. In the early 1980s, per capita consumption of electronic components was two to three times lower in Europe than in the United States or Japan. Having experienced earlier failures at intra-European joint venturing, and with the exhaustion of the postwar policy of subsidizing "national champions" to represent individual countries in the global arena, the big European firms now turned increasingly to cross-border and overseas alliances. In the first half of the decade, only a quarter of the agreements involving a firm based in the EC were concluded with other EC companies. In fully half the cases, the new partner was American.

Yet even as they individually sought overseas partners, Europe's top high-tech companies—Bull, Siemens, Philips, and Compagnie Generale d'Electricite—were already thinking about a new round of intra-EC experiments in collaboration. Thus, in 1980, they and eight other major information technology (IT) companies—including Germany's AEG, Britain's Plessy, France's Thomson, Italy's Olivetti, and four others—were invited by the EC commissioner for industry, Etienne Davignon, to meet to draw up a collaborative R&D program for the community.

The resulting European Strategic Programme for Research and Development in Information Technology, or ESPRIT,[32] was launched in 1983. Half its funding came from the EC, and half from the participating companies. Its objectives were to engage in precompetitive (basic) R&D in advanced microelectronics, software, advanced information processing, office systems, and computer integrated manufacturing, and to develop European standards for the relevant technologies. In 1987, ESPRIT was renewed at a higher level of financial commitment that actually exceeded by 60 percent the total budget of the United States' unofficial industrial policy institution: the Pentagon's Defense Advanced Research Projects Agency, recently renamed ARPA (to accentuate its growing involvement in civilian or so-called "dual use" R&D).

From the beginning, ESPRIT's rules deliberately required that any project proposal brought to it for funding must involve at least two private companies based in at least two different EC member countries. In fact, the leading companies sought and signed up many more partners than that, all across the continent; the extent and breadth of this networking have increased over time.[33] Partner firms include major telephone operating companies, airlines and aircraft manufacturers, car makers, machinery and robotics firms, chemicals and plastics companies, banks, health care centers, and insurance providers—as well as a number of universities and *polytechnics*. If nothing else, as a goad to intra-European networking, ESPRIT must be counted as an unqualified success.

One criticism leveled at ESPRIT's managers, headquartered in Brussels, is the favoritism shown to the original "big twelve" firms in the allocation of grants. For example, while Bull's total subsidy from ESPRIT for 1984 through 1986 amounted to only 5 percent of the company's total R&D budget, that allocation constituted more than a fifth of all ESPRIT disbursements.[34] On the other hand, the rule that all participating firms have equal access to all research results might have effectively tempered this inequality, by giving small firms access to knowledge that they could not possibly have generated for themselves. A case can be made for both sides of this debate.

Moreover, the places in which the big twelve conduct most of their collaborative R&D activity are mostly located in the highest-income countries and regions of Europe. Even after accounting for the locations of their smaller network partners, it has been argued that ESPRIT systematically privileges the core regions of the continent.[35] Again, whether this comes at the expense of the periphery depends on the extent of the information spillovers, as these diffuse through subcontracting and supplier networks.

ESPRIT is not the only EC-wide government-inspired project to encourage industrial networking. Launched in 1985 by eighteen European countries plus Turkey on the initiative of France's President François Mitterand, the EUREKA program goes beyond promoting precompetitive collaborative R&D to encourage participating firms' actual development of marketable products. The more recent Joint European Submicron Silicon Initiative, or JESSI, has been less successful in countering the wave of alliances with and outright acquisitions by American and Japanese corporations, which threaten to undermine intra-European solidarity at the level of actual commercialization. On the other hand, experts have given high marks to the European Synchrotron Radiation

Facility, a twelve-nation collaborative electron particle accelerator sited in Grenoble, France.[36]

Nor has ESPRIT itself succeeded in its goal of establishing specifically European standards in microelectronics and telecommunications environments, architectures, and interfaces that might have created effective barriers to entry into the European scene by non-EC companies. The integration of European with east Asian and North American big firms and their supplier networks now seems unstoppable.[37] But if ESPRIT and its sister programs failed to achieve continental exclusivity, they have certainly added to their member companies' cumulative experience with participation in interfirm networking.

A PARTICULARLY SUCCESSFUL PUBLIC-PRIVATE PRODUCTION NETWORK: AIRBUS INDUSTRIE

Individual European automobile, machine tool, and aerospace companies have always exhibited great skill and resourcefulness. These are precisely the skills and resources necessary to build quality airplanes. But the financial and technical demands on the design, manufacture, and marketing of modern, wide-bellied jet aircraft have become astronomical. By 1970, two American companies, Boeing and Douglas (now McDonnell-Douglas), dominated the world's markets for passenger aircraft, with another U.S. company, Lockheed, a distant third. Two of the three largest aircraft engine manufacturers were also American: General Electric, and Pratt and Whitney.

Motivated by the desire to compete globally in the passenger aircraft industry, and also to use aircraft explicitly as an economic development "driver" because of its extensive backward (upstream) linkages, the British and French governments gingerly joined forces during the 1960s to try their hand at designing and building a high-tech passenger jet of their own. The initial product was the mammoth, sleek, distinctively long-nosed Concorde: architecturally stunning, but extravagantly expensive and the world's largest gas-guzzler, brought to market precisely on the eve of the decade of oil shortages and rising fuel prices. Concorde was never a profitable venture, but it did foster the habit of collaboration that set the stage for what was to follow: Airbus.

As early as 1966, a consortium consisting of the French, British, German, and Spanish governments, together with a number of leading European corporations and their suppliers, had begun negotiations that would lead to the formal launching in 1971 of Airbus Industrie. Headquartered in Toulouse, site of its core manufacturing facilities, Airbus is a cobweb production network, part public enterprise and part private. Companies such as the French-owned SNECMA and Germany's Daimler-Benz are among the major hub firms in the network. In addition to the four governments that pay the bills, partners in Belgium, Italy, and the Netherlands have subcontracts.

Since the founding of Airbus Industrie, the French and German governments alone have invested some $14 billion in the enterprise. The result has been a series of true innovations. Airbus, not Boeing, produced the first wide-bodied jet capable of flying with only two engines. Its planes feature the first all-electronic cockpits, designed to be manned by crews of only two pilots. And the innovations have paid off: as of the spring of 1991, Airbus had a backlog of orders for 1,600 new planes from private and government-owned airlines all over the world, worth an estimated $70 billion. This constitutes nearly a third of all current industry orders, still behind giant Boeing (with 53 percent of all world orders) but twice the share of new orders held by faltering McDonnell-Douglas.[38] To those (including officials of the U.S. Department of Commerce) who claim that Airbus operates from an "unlevel playing field" because of its massive subsidization by European governments, managing director Jean Pierson accurately observes that Boeing and McDonnell-Douglas have for many years also been government subsidized, via military contracts. The fact is that, given the huge capital costs involved, no airplane manufacturer in the world operates without some public subsidy.

Airbus has a French director and a French headquarters, but it is decidedly a European firm (although some of its constituent hub companies, such as SNECMA, are linked to American or Japanese firms through their own strategic alliances—again in imitation of the extended cobweb-style keiretsu). It is also an aggressive actor in foreign markets. For example, Airbus now has a North American office, directed by a man who was once the secretary of the U.S. Department of Transportation and and the chairman of Amtrak. It has a large and growing American sales force, a training center in Miami, and a parts warehouse at Dulles Airport outside of Washington, D.C. In short, Airbus Industrie is the very model of a modern cross-border production network.

The international competition in the production and sale of aircraft is thus shaping up as a battle between entire big firm–led production networks: Boeing and its suppliers and partners versus Airbus and its associated companies and governments. But consistent with the strong underlying impulse of the era toward more complex collaborations among the big firms, Boeing has now approached some of Airbus's largest member companies about forming strategic alliances "on the side." Thus, for example, in January 1993, Boeing approached Deutsche Aerospace, the aircraft division of Daimler-Benz and the firm that holds fully three-eighths of Airbus stock. The proposal was that the two giants jointly develop the next generation of super-jumbo-jet passenger aircraft. Boeing also wants the big-three Japanese aerospace giants—Mitsubishi Heavy Industries, Kawasaki Heavy Industries, and Fuji Heavy Industries—to join the project, with Boeing as the lead firm. The question now is whether the various countries' antitrust laws could ever be bent sufficiently to permit such a potentially powerful alliance among industrial giants.[39]

PRODUCTION NETWORKS AS AN EXPRESSION OF CONCENTRATION WITHOUT CENTRALIZATION

I have examined the empirical evidence on the response of Japanese and European business to the profits squeeze of the 1970s. Building on the particular historical experiences of the different regions, and proceeding (as it inevitably would) at an uneven pace from one company and country to another, a trend is occurring toward growing reliance by business on production networks of all sorts, as a method for gaining greater flexibility in all of its many meanings. The small firm–centered industrial districts of northern Italy and elsewhere are now seen as special cases of a far more general phenomenon. Moreover, the empirical evidence seems overwhelming that the evolving global system of joint ventures, supply chains, and strategic alliances in no sense constitutes a reversal—let alone a negation—of the 200-year-old tendency toward concentrated *control* within industrial capitalism, even if the actual *production* activity is increasingly being decentralized and dispersed.

American-based companies have been pursuing many of these approaches to interfirm and cross-border organization, too. But taken as a class, U.S. firms still lag behind their foreign counterparts—and their own offshore divisions—in developing extensive collaborative production systems. This issue requires a separate chapter for exploration of the reasons why.

8

INTERFIRM PRODUCTION
NETWORKS IN THE UNITED STATES

Elements of both the east Asian and the European approaches to networked production are becoming visible in the United States, too. A growing number of American firms are partnering with foreign companies and with one another.

But U.S. managers' implementation of the principles that make the most effective use of interfirm networks has been at best incomplete and often half-hearted. Some companies have made great progress in instituting these innovations, and I will name them. But, as with the still abysmally low commitment by U.S. employers to skill training, vis-à-vis the companies of other nations,[1] far too many American firms seem still inept at developing truly collaborative relationships with others.

To be sure, the leading business media and best-practice management strategy books all subscribe to the ideas of collaborative production, relational contracting, joint labor-management teamwork, and the rest. *Business Week*'s editors have even called for the advent of American-style *keiretsu*.[2] But these practices remain relatively rare, there is still more hype than substance, many managers remain almost instinctively suspicious of all such reforms, the American legal and political systems produce an institutional climate that is inherently inhospitable to collaboration, and too many private industry efforts at implementation are terminated unilaterally by management before they have been given a chance to take root.

In this chapter, I show how and why the United States still has a long way to go in following through on the development of durable strategic alliances, truly relational contracting between customers and suppliers, and the other aspects of networking that have been held out to business as offering the greatest promise for achieving flexibility in the new world order.

CORPORATE RESTRUCTURING AND PRODUCTION NETWORKS IN U.S. INDUSTRY

The Boeing example from chapter 7 makes it clear that at least some American companies *are* seeking out more collaborative, networked approaches to competition.[3] During the 1980s, a number of leading U.S. firms began experimenting with one or another element of the models of corporate flexibility that were already an integral part of Japanese and European practice. For example, the big-three auto firms—General Motors, Ford, and Chrysler—have shifted in varying degrees toward the Japanese method of supplier selection, awarding three- to five-year contracts to preferred suppliers and putting more emphasis on rewarding quality and innovation. Similar changes have occurred in the U.S. textiles and electronics industries.[4]

In the auto sector, the ongoing surveys of the Case Western Reserve University economist Susan Helper, conducted through the M.I.T. International Motor Vehicle Project, reveal that longer-term contracts from the customers (the original equipment manufacturers, or OEMs) increase the likelihood that the supplier firms will acquire and implement more up-to-date manufacturing technology.[5] The automotive division of TRW has become an increasingly collaborative supplier of air bags and seat belts to Ford, designing as well as manufacturing such subsystems. TRW is also trying out similar relationships with the smaller companies from which *it* purchases parts and subassemblies. In an interview with professors Richard Walton and George Lodge of the Harvard Business School, TRW Vice-President John Marshall described the company's new "downward-looking" initiatives this way:

In the past we sought bids from a number of suppliers, and price was the principal issue. Now we want flexible relationships with a few suppliers. . . . It is not unusual these days for two or three engineers from our suppliers to be working in our plants for a while.

We network through computers. I might call one of our suppliers and urge them—if not help them—to locate a plant near us.[6]

Nor are such experiments confined to the auto business. Motorola, the manufacturer of electronic components (including computer chips) and cellular telephones, has made substantial investments in just-in-time inventory management, training, teamwork, and technology alliances with such companies as IBM and Apple. At the highly profitable and much-praised (but, like Motorola, determinedly nonunion) Texas min-imill of Chapparal Steel, teams of engineers and workers collaborate with suppliers of steelmaking equipment to tailor the design of this very expensive machinery to fit Chapparal's particular needs. So does the Massachusetts-based Digital Equipment Corporation (DEC), the country's leading maker of minicomputers. To conduct such a relationship sometimes requires the customer—in this case, DEC—to share unusually detailed information with its suppliers about existing operations and even long-range strategic plans. Indeed, one DEC executive told the Harvard Business School's Rosabeth Moss Kanter, "We are telling stuff today [to our suppliers] which I'm sure if the old purchasing manager of two years ago knew we were doing, he would roll over in the grave."[7]

It is even possible to identify successful examples of the federal government encouraging supplier development in the defense industrial base, with the U.S. Department of Defense playing a role comparable to, if not as extensive as, that of Japan's MITI or Europe's Airbus. Since the late 1970s or early 1980s, the Defense Department has actively encouraged its prime contractors (especially those manufacturing aircraft) to make special efforts to build up the technical competence of their suppliers and subcontractors. New empirical research conducted for the Congressional Office of Technology Assessment by the Carnegie Mellon University professor of management and public policy, Maryellen Kelley, and the Lehigh University professor of business economics, Todd Watkins, shows a dramatic difference in the technical capabilities, extent of adoption of programmable automation, presence of customer-supplier collaboration, and adoption of joint labor-management problem-solving committees between shops and factories that are part of the Pentagon procurement network and those that are not. It is from this evidence that Kelley and Watkins urge that, in the course of managing the post–cold war military build-down, the government be wary of inadvertently dismantling the nation's most successful industrial *keiretsu*.[8]

GROWTH IN CROSS-BORDER ALLIANCES

Mergers and acquisitions involving cross-border deals between American and foreign firms became an increasingly important component of all merger and acquisition activity during the 1980s. In 1983, only 7.3 percent of the takeovers of U.S.-based companies were by foreigners. By 1989, that share had risen to nearly 37 percent—an astonishing increase in so short a time, attributable only in part to the undervalued dollar, which made American assets relatively cheap on world markets.[9]

More interesting than simple takeovers have been the sort of cross-border strategic alliances that have, as noted in the last section, been spreading throughout Europe. Of course, alliances initiated by American corporations are not a new invention. Some have existed for a generation and apparently have been working well for all parties. Take the case of the alliance between the General Electric Company's Aircraft Engine Group and France's Societé Nationale d'Étude et de Construction de Moteurs d'Aviation (SNECMA), whose aircraft engine and other aerospace production plants and their suppliers are situated on the western edge of Paris, in what the locals call the "Arc de SNECMA."[10]

The two independent corporations, one private and one mostly government owned,[11] originally joined forces in the early 1970s, creating a fifty–fifty joint venture called CFM International to manufacture a particular type of jet engine for medium- and short-range aircraft. The government-owned national airlines of Europe were the main target market for these engines, and both GE and Pratt and Whitney (P&W) wanted in. Moreover, SNECMA was about to become a major subcontractor to the new Airbus Industrie consortium, so that an alliance with SNECMA promised the possibility of still further networking connections to Europe. At the beginning, the U.S. Department of Defense was leery of the alliance, seeing the technology of the engine's compressor (designed by GE) as a "national security asset," not to be disclosed to foreigners. It took letters passed back and forth between the French and American presidents to eventually clear the way for the alliance to be formed.

The value of the alliance for GE was confirmed at the end of the decade. In the late 1970s, Air France was about to build a new line of aircraft in its government-owned facilities. Although GE submitted its own bid to build the engines, P&W, which had already supplied the engines for Air France's previous generation of planes, was widely expected to win the contract. P&W even offered to completely refurbish its by-then aging JT9D engines, at no cost to Air France, and to construct a maintenance facility for the new engines on French soil. The management of Air France and the French Transport Ministry favored the deal. But it was vetoed, anyway, and the contract was awarded to GE. Why? The Defense Ministry and the unions supported GE's bid because of the latter's ongoing alliance with SNECMA. Their reasoning was simple: By 1979 the French government had already sunk half a billion dollars into the joint venture. Why not build on a good thing?[12]

The particulars of the relationship between GE and SNECMA have varied over the years, from project to project. Sometimes, GE plays the role of project manager, receiving a fee for service from SNECMA, which does the production work. On other occasions, SNECMA performs subcontracting work to GE on the latter's mainly U.S.-oriented projects. Experts agree that GE remains the technologically superior company, while SNECMA has become an extremely highly qualified specialist subcontractor and parts supplier to aerospace companies all over the world.

Cross-border alliances of this sort became more common for American companies in the 1980s and early 1990s. In computer hardware, alone, IBM and Toshiba are now partnering to develop new video displays, IBM and Canon are jointly developing small printers, Intel and Sharp are working together to develop flash memories (the kind that do not clear out when the power goes down), and Motorola and Toshiba are making high-powered chips.[13]

The steel industry offers another dramatic example. Decades of gradual deterioration of the domestic integrated American steel industry, characterized by management's unwillingness to innovate and by labor's insistence on rigid work rules in order to provide much-needed job security, had culminated in the disastrous wave of mill closures of the late 1970s and early 1980s.[14] Output, employment, productivity, and profits were all in decline. Yet by the middle of the last decade, we began to hear talk about a revival of the U.S. steel industry. This was being attributed largely to the success of the so-called minimills, which use electric arc furnaces, scrap steel (rather than coking coal) as the principal input, and new flexible production and personnel management practices. Companies such as Chapparal grabbed headlines in all the business media as leading the way toward a new era of (admittedly more modest) economic growth for U.S. steel.

But there is another facet to the story of the restructuring of the steel industry within the continental United States. This is a story of a rash of joint ventures, coproduction deals, and other strategic partnerships between large-scale Japanese and American capital.[15] In some cases, such as the NKK–National Steel alliance, the two countries' companies jointly own the entire operation. In other cases, such as Kawasaki's investment in Armco or Kobe's investment in USX, control remains substantially in the hands of the American firm, which has been energized by the infusion of foreign finance. Finally, there are many joint ventures for specific products, such as the partnership between Sumitomo and LTV to make steel for auto bodies for the Japanese "transplant" car companies that have been setting up assembly plants in the United States for a decade.

This last case has been about the development of interfirm networks within an entire sector of the economy: integrated steelmaking. Other cross-border alliances were crafted by American firms in the 1980s for particular products or technical processes. By any standard, one of the most successful has been the alliance between GM and FANUC, the Japanese manufacturer of controllers for computer-programmable machinery. Together, they have forged the world's leading production system for designing and manufacturing industrial robots.[16]

FANUC is the world's largest producer of computer numerical controls: the programmable devices that are capable of guiding the operations of many different machines by translating instructions written in computer code. FANUC also makes a wide array of industrial robots. The company was founded in Japan in 1955, when Fujitsu created an engineering team dedicated to developing a factory automation business for the parent firm.[17] At the same time, in order to rapidly establish a national standard for the controlling of machine tools, MITI encouraged all industrial users to purchase their controllers from Fujitsu-FANUC and urged FANUC to build the simplest possible designs into its black boxes, so that the technology would diffuse quickly among the many smaller supplier firms that the keiretsu were already reorganizing into the vast production networks described in chapter 7, in steel, shipbuilding, cars, and consumer electronics. If ever there were evidence that the Japanese economic miracle cannot be attributed to the operations of the free market, alone—if ever we needed definitive proof that the Japanese *do* practice long-range capitalist planning—this must surely be it.

In 1972, FANUC became independent of Fujitsu. Unlike its American competitors, such as Cincinnati Milacron, FANUC quickly moved to exploit the new solid-state technology and microprocessors, so that, by the late 1970s, this one firm controlled 80 percent to 90 percent of the domestic Japanese market in controllers, and, by the early 1980s, 40 percent to 50 percent of the world market. Confronted by FANUC's success, most of the competition dropped out of the industry altogether or—as in the case of Siemens and GE—entered into joint ventures with FANUC.[18] These, however, are minor links, compared with the relationship between FANUC and General Motors.

Like all car makers, GM needed industrial robots: for assembly, painting, spot welding, and the performance of other standardizable repetitive tasks. By the early 1980s, GM was the largest user of robots in the United States. Dissatisfied with its own American vendors, GM in 1982 proposed an alliance with FANUC to "design, market, service, and develop applications for factory automation robots."[19] The new fifty–fifty joint venture, GMF, with its world headquarters in Michigan, was officially launched in 1983. While much of its sales continue to go directly to GM, the new firm also sells its high-tech products to other customers all over the world, including Japan and Germany. The alliance has exclusive rights to FANUC's robots in the United States, and nonexclusive rights in Europe. By the mid-1980s, GMF held about a third of the U.S. market for industrial robots of all types and perhaps two-thirds of the market for painting (finishing and coating) robots.

How is GM-FANUC governed? Senior executives from the two parent firms meet four times a year to plan strategy. Teams of engineers from GMF, FANUC, and GM are continually being reconstituted to work on different projects, one of which has been the development of a special programming language for linking GMF's robots to vision screens and other GM products.[20]

When he was studying FANUC in the mid-1980s, the Carnegie Mellon economist Steven Klepper remarked that, "unlike other industries, robotics has not yet been influenced by government procurement policies, nor has it been the subject of international trade negotiations."[21] That may be about to change. A domestic U.S. industry that once included such major names as Cincinnati Milacron, Unimation, and Prab is now virtually monopolized by GM-FANUC. Meanwhile, according to the U.S. Department of Commerce, in 1991 about $300 million of the $350-million estimated American industrial demand for robots and parts was filled by imports—from Japanese, German, Canadian, Swiss, and Swedish vendors.[22] This situation is prompting U.S. government officials to begin paying attention to the institutional details of international trade in robots. Moreover, the British, French, and Swedish governments are also considering industrial policy with respect to robotics. In the last two years, robotics has also appeared on "critical technologies" lists developed by both public (for example, ARPA) and private (Council on Competitiveness) agencies in this country, suggesting a growing policy concern here, as well.

GENERAL MOTORS AND THE JAPANESE
JOIN FORCES IN MAKING CARS

By suddenly and substantially pushing up the retail price of gasoline, the two oil shocks of the 1970s reinforced the growing demand by at least a segment of the American public for smaller, more fuel-efficient cars. At the time, the Big Three seemed not to know how (or to want) to build them. But eventually they had to do something to deal with a crisis not of their own making. As the industry leader, GM took the first plunge—via extensive cross-border networking.[23]

Actually, GM made its first move to gain access to imported small Japanese cars as early as 1971, when it purchased a one-third interest in Isuzu Motors of Japan. Ten years later, GM acquired equity in another small Japanese producer, Suzuki. Then, in 1983, "GM startled the automotive world by announcing that it would create a 50-50 joint venture with Toyota, a major competitor."[24] New United Motors Manufacturing Inc., or NUMMI for short, makes small and midsized cars in a refurbished GM assembly plant in Fremont, California, located not far from Silicon Valley. By most accounts, NUMMI has been a great success, especially at familiarizing American managers and blue-collar workers with the organization of production around teams, and with the development of just-in-time parts delivery arrangements with American suppliers.[25] Nor was NUMMI to be the last of GM's cross-border alliances. In 1984, GM and the Korean conglomerate (chaebol) Daewoo, announced a joint venture to produce small vehicles. Other alliances were established during the 1980s with Nihon Radiator, with Japan's largest manufacturer of brakes, Akebono, and with NHK Spring, the world's largest producer of automobile suspension systems.[26]

What did GM gain from these varied alliances? In the particularly antagonistic context of American capitalism, GM certainly intended to "mix its portfolio" in order to gain leverage over its partners.

With four Asian sources of small cars and the prospect of Saturn [the newest division of the Corporation, which builds small cars in Tennessee], GM was much less dependent on any one of its Asian allies than it would have been with only one Asian partner. Reliance upon low-cost, high-caliber labor in Asia also enhanced GM's bargaining position with the [United Auto Workers' union]... [and] the availability of high-quality, Asian-made parts gave GM the option of reducing its dependence on its in-house parts suppliers and its [external] U.S. suppliers.[27]

In 1983 General Motors had announced that it would invest billions of dollars to create a new compact car called Saturn, the first new GM nameplate since the introduction of the Chevrolet in 1918. Located in the largely rural ("greenfield") site of Spring Hill, Tennessee, far—but not too far—from its "brownfield" Great Lakes regional center of gravity, Saturn would further extend the experiments with teamwork, just-in-time (JIT) parts delivery, and "continuous improvement" (*kaizen*) that had begun in NUMMI. Moreover, the plant layout, car design, selection of suppliers, and even to some extent decisions about the pricing of different models would in principle be made jointly by Saturn's management—itself semiautonomous from GM headquarters—and the UAW. GM also sought to build longer term, more collaborative relationships between Saturn and its key suppliers, again on the keiretsu model. As Joseph Badaracco notes: "The project marked the first time in GM history that the United Automobile Workers had participated in GM corporate planning."[28]

For all its aggressive alliance building, and given the successful launching of Saturn, GM remains a corporation in confusion. In the fall of 1992, CEO Robert Stempel was accused of insufficient attention to innovation and was forced out of office by the outside members of the board of directors. Now it is the new corporate president, John F. "Jack" Smith, Jr., who will have to find a way to restore GM's global position. He has already made it clear that interfirm alliances and other cooperative agreements—along with continued downsizing and a studied effort to become leaner and meaner—will be important elements in the further restructuring of the company.[29]

WHY MANY AMERICAN COMPANIES STILL DO NOT GET IT

The lessons from abroad—and from the most successful American corporations—are clear. Successful competition in the contemporary global economy now puts a premium on the well-thought-out formation of strategic, rather than on again–off again alliances, joint ventures, and other network forms of production: with customers, suppliers, and even competitors.

Yet for all the examples of how individual U.S.-based firms have been at least dipping their feet into the stream of interfirm collaboration to test the waters, the reality is that all too many companies still do not seem to get the message. Or, having gotten it, they continue to elect to follow the low road to economic growth, seeking profitability mainly by cutting costs rather than by enhancing productivity through new mechatronic and organizational technologies—the high-road approach. In the words of the M.I.T. authors of the definitive study of the restructuring efforts of the automobile industry: "Unfortunately . . . the reforms made to date have involved pushing the traditional . . . system to its limits under pressure, *rather than fundamentally changing the way the system works* [sic]. . . . Everyone knows the words of the new song, but few can hold the tune.[30]

The problem is that, unless a sufficiently large number of companies make the shift to a high-road strategy together, the forces of market competition act to push the system as a whole back to the low road. This a classic example of what mathematicians and economists call a "low-level equilibrium trap." If I am the only one to make the shift toward a more collaborative way of doing business but my competitors do not, I will lose out in the short run, as they undercut my costs of doing business. The trick is to move the whole system above a threshold—to achieve a critical mass of participants committed to collaborative production relations and a long-term planning horizon.

Thus, for example, in a recent, authoritative, comprehensive review of the evidence on the extent of collaborative linkages between industrial customers and their suppliers in the United States, the Congressional Budget Office came to the sobering conclusion that

it is [still] longstanding custom for American manufacturers to discourage—even forbid—design engineers from developing close relations with suppliers. Direct approaches to suppliers are known as "going around the purchasing department," and are [often] against company rules. Purchasing agents themselves are frequently reassigned to different types of supplies, so they won't develop overtly cozy relations with suppliers. The ideas behind all this are, first, that maintaining arms-length, impersonal, strictly contract-based relations with suppliers is the best way to get a good price and keep costs down; and second, that it is unfair to give any supplier a privileged position and deny the others an equal chance. Some company officials even believe that they might be subject to lawsuits if their suppliers were deprived of the chance to bid for contracts.[31]

Interviews with auto parts suppliers reveal that the typical manager of a supplier firm in this country continues to be wary of being blindsided by his large business customers, whether by being forced to take a sudden large price cut or by being expected to upgrade his technology without the financial wherewithal to do so. To be sure, among at least the first tier of suppliers, such collaborative arrangements with the OEMs as long-term contracts or exchanges of information are more prevalent in the industry now than was the case even a decade ago. Nevertheless, the majority of suppliers continue to report that their customers are "not trustworthy" and that they still do not provide much assistance to suppliers to reduce costs or adopt new technologies. Most disturbingly, there is a definite trend toward a practice in which GM, Ford, and Chrysler "obtain improvements in quality and delivery by forcing suppliers to adopt methods, such as JIT delivery without JIT production [that is, stockpiling of parts at the supplier's facility, with all the associated inventory costs], that provide better service to customers at the expense of supplier [profit] margins."[32]

In June 1992, the worst fears expressed by the researchers were confirmed. The entire existing supplier network to GM—and many outside experts and journalists, who had thought (hoped?) that the world's biggest car company *was* finally getting it—were shocked and dismayed when a new management team, brought in from Europe in response to a stockholders' revolt over GM's deteriorating profitability, announced that all existing long-term contracts with parts suppliers were off, and that a new bidding war among existing and prospective domestic and foreign suppliers for GM's business was now on. Many of the small firms had borrowed heavily over the past several years, precisely in order to invest in upgrading their technical capabiltiies so that they might remain (or become) preferred suppliers to GM. One supplier firm manager offered the opinion that, with the new GM policy, "the small guy is really up the creek." Distrust and fear again permeated the GM production network.

In the Japanese keiretsu system, customers typically require suppliers to lower their prices regularly as part of a strategy of pressing the latter to innovate continually. But the quid pro quo in the Japanese system is long-term contracts and parent investment in the supplier. The new GM policy means "lower prices on [even] existing contracts," squeezing the cash flow of suppliers since they "now must finance tooling which GM used to bankroll," and the potentially devastating undermining of long-term cooperation between suppliers and GM. Apparently, GM did not get it, after all.[33]

The experience from Japan and Europe revealed that a vital aspect of collaboration within production networks is this sharing of finance within the system, usually in the form of customer firms purchasing equity stakes in (or lending money to) suppliers. But in a random sample of *all* U.S. metalworking plants in 1987 (not just automobile OEMs and parts suppliers), Carnegie Mellon's Kelley and Harvard professor emeritus Harvey Brooks found that only 3 percent reported having received any financial assistance from customers between 1985 and 1987 to upgrade their technology, and only 16 percent had received technical assistance in the form of visits from the customer's engineers.[34] Those numbers increased somewhat between 1987 and 1991, but not by much.

In some ways, it is even more important in the United States than in other countries that such financial and technical assistance to smaller suppliers come from the big firms. The reason is, unfortunately, that the U.S. government has made itself so conspicuously absent from this kind of activity. The Congressional Office of Technology Assessment estimates that, where the Japanese government spends perhaps $31 billion on technology extension and loans to smaller firms each year, all levels of government in *this* country—federal, state, and local—probably spend no more than $50 million.[35]

CHRONIC SHORT-TERMISM: EXPLAINING THE SHORT TIME HORIZONS OF AMERICAN MANAGERS

The slow pace of diffusion within U.S. industry of such innovative practices as concurrent engineering, labor-management problem-solving teams in which workers have a real measure of independent power, the systemic nurturing of suppliers by their customers, and the forging of trustworthy strategic alliances and other forms of networked production does not occur in a vacuum. The underlying problem, many believe, is the chronic myopia of American management. This impression was reinforced in June 1992, with the release by the private, nonprofit Council on Competitiveness of a timely batch of new research papers, synthesized by the Harvard Business School's Michael Porter.[36]

The participants in the Time Horizons Project concluded that the underlying factors distorting real capital formation in the United States since at least the 1960s have to do not so much with the profligate behavior of "big government" as with the rules, procedures, and customs by which the *private* sector allocates capital among competing uses. Put simply, the twenty-five economists, finance experts, business historians, and industrial relations specialists who participated in the project argued that the Germans and the Japanese inherited from their own past, and then refined in the years after World War II, a coherent set of institutional arrangements that systematically favor the long-term growth and survival of their companies over the search for short-term monetary earnings.

In one of the project's most important papers, the economists Lawrence Summers and James Poterba, of Harvard University and M.I.T., respectively, presented the results of their interviews with CEOs of the Fortune 1,000 in the United States and of the fifty largest firms in Japan, Germany, and the United Kingdom.[37] They found that U.S. investors tend to demand considerably higher minimum acceptable rates of return for their capital—so-called hurdle rates—than do their foreign competitors. This behavior on the part of American investors is way out of whack: in recent years, investors here have been rejecting projects whose expected inflation-adjusted payoff falls below 12 percent—even though the recent real cost of debt has averaged only about 2 percent, and the real cost of equity about 7 percent. Moreover, where a healthy share of the R&D portfolios of foreign corporations include explicitly long-term projects (47 percent in the case of the Japanese companies studied, and a whopping 61 percent for the Germans), only one out of five projects being undertaken by U.S. companies could be classified as long-term.[38]

As a result, in the United States, useful projects that could eventually create lasting employment, downstream profits, and opportunities for experimentation with new technologies and new forms of production organization in actual workplace settings are being systematically underfunded. In their place, vis-à-vis their German and Japanese counterparts, Americans invest disproportionately in unrelated acquisitions, which have generally been poor performers in the marketplace,[39] and in real estate. Such portfolio investments may generate short-term earnings (not to mention the fat commissions for the lawyers and bankers who handle the transactions), but they do not add to the net stock of real productive capital in an economy. Instead they merely involve the trading of [existing] assets from one owner to another.

The key to understanding *why* American corporations take such an ultimately counterproductive short-term approach to investing begins with the ways that the players in what Porter calls the *external* capital markets behave. In the parlance of formal economic theory, we appear to have a profound "principal-agent" problem on our hands. In contrast to Germany and Japan, where the major owners of shares of stock tend to be other big corporations directly or (via customer-supplier relations) indirectly involved in the running of the companies they own, and where the dominant owners are continually researching their companies' progress, drawing easily on inside information, U.S. insurance companies, mutual funds, and individual investors alike often know little or nothing about the messy details of making steel, wiring silicon chips, or communicating via satellite.

Instead, American investors employ on their behalf armies of professional market analysts, whose own careers depend on their guessing correctly more often than not. In the absence of hard, substantive inside information (the trading on which by outsiders is, in fact, against the law in this country), or with "experts" whose knowledge is so specialized that they have no common language for communicating with one another (let alone with the owners: the "principals" who own the stock), everyone tends to fall back on those indexes that can be most easily and inexpensively measured and readily communicated. Typically, that comes down to recent and projected quarterly earnings. Agents build their careers by bouncing from one investment bank and mutual fund to another, competing for status and income by how well they can predict tomorrow's share values and by how many Wall Street firms have hired them. No lifetime employment in *this* system!

Inside the corporations themselves, in what Porter calls the *internal* capital markets, things are not much better.[40] American boards of directors are made up of executives of other firms, who (again) know far too little about the substantive business they are being paid to direct. And the firms themselves have become so complex, after thirty years of conglomerate mergers and acquisitions and the widespread diffusion of the principle of breaking companies into divisions doing very different things, that even *their* managers and senior technicians have difficulty talking across what often amount to fire walls. In the effort to communicate across these barriers, even the insiders are driven toward a reliance on simple goals such as maximizing share prices. Once again, we get management by the numbers.

Moreover, more than is the case elsewhere, American managers are compensated with stock options, which of course only ties their behavior that much more closely to current share prices instead of to long-term prospects (there is apparently no correlation between short-term values and long-term business profitability, anyway). Note that Porter is not saying that incentive pay is in principle the problem, only that the American system ties incentives to a short-term rather than a long-term signal.

In any case, the predictable result of the existing incentive system is that, as with those stock analysts out in the external capital market, inside the corporation's internal market the managers' tenure with any one company is also short-lived, in comparison with their Japanese and German counterparts' tenure. But this only further compounds the problem of inadequate development of key personnel with firm-specific knowledge— which in turn further promotes the use of those easy-to-read indicators such as stock price and quarterly earnings. Half a century ago, the Swedish social democratic institutional economist (and later, Nobel laureate) Gunnar Myrdal called such spirals "vicious circles"—cumulative causation (positive feedback), but in a downward-spiraling direction.[41]

The net result of these mutually reinforcing behaviors, played out within strongly entrenched institutions that have sunk deeper and deeper roots over time, is that finance capital in the American industrial system is indeed, as the Harvard University lecturer and U.S. Secretary of Labor Robert Reich and the M.I.T. economist Lester Thurow have long asserted, systemically "impatient." One set of numbers from the Council on Competitiveness–Harvard Business School project nails the point home. In 1960, big institutional owners of stock in the United States—especially the pension and mutual funds—held on to a typical share for an average period of seven years. By the 1980s, that number had fallen to only about two years.[42]

Seeking to explain the comparatively short time horizons of American industry, Porter and his colleagues have identified many other differences between how the Germans and Japanese organize their internal and external capital markets and how Americans organize theirs. For example, hostile takeovers are—by custom or law, or as a result of the extensive cross-shareholding between customers, suppliers, and bankers—"nonexistent" in Japan and "next to nonexistent" in Germany. Moreover, the assets of German workers' pension funds are customarily dedicated to purchasing only the stock of the company employing those workers—a practice that may not be "prudent" in the eyes of American regulators, but that certainly strengthens the ties between German labor and management. Here we have the makings of a Myrdalian "virtuous circle," according to which the dedication of capital enhances the firm's prospects for survival, thereby reducing the need for heavily regulated prudence.

The capital budgets of German and Japanese companies treat expenditures on training, R&D, development of closer relations with suppliers, and the initial losses associated with entering new markets or territories as investment, on a par with spending on new plant and equipment; American companies record only the latter. In other words, even other countries' accounting practices are geared explicitly for the long run.

Finally, the quite probably growing significance of spillovers or interdependencies among investment projects, especially in high-tech sectors, means that some forms of long-term investment actually increase the payoff to other investments. This means that the serial, atomistic, "Should we do project A or project B?" calculus embodied in the discounted net earnings approach that so dominates American management practice will sometimes lead decision makers to choose less than maximally profitable packages, anyway.

A Chain Needs All Its Links

American managers' lean and mean approach to heightened global competition and greater cost consciousness in the years since 1970 continues to turn mainly on the paring down of the activities of the firm, and the greater use of outside suppliers, subcontractors, and partners. It is true that some U.S. firms *are* experimenting with strategic alliances, collaborative manufacturing with suppliers, and other forms of interorganizational networking. Some *are* investing in the training of their employees and introducing productivity-enhancing new technologies into their factories, warehouses, stores, and offices.

But operating as they do in a legal, political, and cultural environment that is so fundamentally antagonistic to cooperation in economic behavior (except for periods of shared crisis, such as wartime), American managers have been implementing these reforms piecemeal. If there is such a thing as a critical mass of collaborative activity—a threshhold above which individual decisions will, more often than not, be taken to promote networked production—that boundary has *not* yet been crossed within the United States.

In the absence of all the ingredients, of *all* the links that make up the chain—the investments in training, the long-term contracts, the greater willingness to take risks, especially when it comes to the introduction of new technology—downsizing, outsourcing, and mass layoffs are creating fear and insecurity among a large and growing fraction of the population. In the brave new world of lean production, there are winners, but there are also a growing number of losers. Such is the *dualistic* nature of networked production systems: what I call the dark side of flexibility. This aspect of the problem of global business restructuring is the focus of the next chapter.

9

THE DARK SIDE
OF FLEXIBLE PRODUCTION

It is the summer of 1983. The president, Congress, business leaders, economists, and the media are joyously celebrating the fact that, at long last, the national economy is coming out of the back-to-back recessions of the previous three years. How infelicitous that one journalist should choose that moment to cast a pall on the festivities.

According to a widely read article by Robert Kuttner, published in the *Atlantic*, the share of Americans earning middle-class wages was declining over time and would likely continue to do so. The reasons included the growing vulnerability of the economy to foreign competition, stagnating productivity growth, and the long-term shift of the economy's center of gravity from manufacturing to services.[1] To a weary establishment, both inside and outside of government, Kuttner's challenge was simultaneously too well argued to dismiss out of hand, and too scary to take seriously.

The idea of a declining middle America circulated within Washington throughout the remainder of the decade. But as the economy continued to grow after the trough of the 1982 recession, thanks to a combination of Reagan-era military spending and unprecedented borrowing by consumers and businesses, it became more and more difficult, and seemingly unnecessary, to consider the idea that America was becoming an increasingly polarized society. After all, inequality historically had always *fallen* during periods of economic growth, as the unemployed returned to work and selective labor shortages began to appear in one or another corner of the country.

As it turned out, Kuttner proved to be right on the money. From the vantage point of the 1990s, it is hard to find any serious observer who does *not* agree that inequality is on the rise.[2] The polarization of the jobs that employers are making available to people searching for work is cleaving the whole population, white and black, Anglo, and Latino, into highly paid haves and more poorly paid, increasingly insecure have-nots.

In this chapter, I argue that what the M.I.T. economist Lester Thurow has called the "surge in inequality"[3] may, at least in part, be connected with the very industrial restructuring and business reorganization discussed thus far in this book. Lean production, downsizing, outsourcing, and the growing importance of spatially extensive production networks governed by powerful core firms and their strategic allies, here and abroad, are all part of businesses' search for "flexibility," in order to better cope with heightened global competition. But this very search for flexibility is also aggravating an old American problem—economic and social *dualism*. This is an institution that was widely thought to be disappearing along with those dinosaurs of the industrial past, the vertically integrated giant corporations.

As I have documented in earlier chapters, the popular pronouncement of the imminent demise of the big firms was premature. It is they and their partner companies, *not* small business, per se, that account for most of the jobs, sales, and output in American industry, year in and year out, in both mature and high-tech sectors. But the ways in which big business has been reorganizing itself to become more competitive are proliferating low-wage, insecure employment. The trend toward our becoming an hourglass economy proceeds. This is the dark side of flexible production.

THE GROWING POLARIZATION OF WAGES

In a paper originally commissioned by the U.S. Congressional Joint Economic Committee and published in late 1986, Barry Bluestone and I presented the results of an analysis of official Census Bureau data on the earnings of individual American workers.[4] We showed that, since at least the late 1970s, it was the lowest-wage and the highest-wage groups of Americans that were increasing in number most rapidly, while the proportion earning middle-level wages was falling. This message, in the form of a qualification to the Reagan administration's bullish self-promotional claims for the wonders of its track record in creating jobs, was subsequently published as an opinion editorial in the *New York Times* in February 1987.[5]

The result was a firestorm of public and academic debate. With the passage of time, the clumsy attempts by the conservatives then running the government, and by journalists sympathetic to their cause, to dismiss these findings as either "absurd" or (worse) "cooked up" in order to suit the political needs of the Democratic party opposition now seem fairly amusing (one newspaper columnist actually called the whole thesis a "big lie").[6] What was more important was the debate that this work triggered among academic economists. Over the next four years, literally dozens of books and papers were published, conferences held, interviews granted to the media, and careers made over the question of whether, during the sustained macroeconomic expansion after 1982, inequality among wage earners was in fact continuing to grow—let alone to polarize.[7]

As active participants in the debate, Bluestone, I, and our students continued to conduct research on the subject, partly in response to methodological criticisms that had been sensibly and usefully (if not always warmly) leveled at us by critics during the exchange. Over the next four years, one or the other or both of us would publish papers in the proceedings from meetings of the American Economic Association and the Industrial Relations Research Association, in a leading British economics journal, in a report issued by the Aspen Institute (sponsored by the Ford Foundation), as a chapter in a book on job creation, and in a monograph on how the long post–World War II trend toward gradually declining inequality was reversed in the 1970s, leading to increasing polarization of the workforce that lasted throughout the economic expansion of the 1980s.[8]

Economists and statisticians justifiably worry whether, even using the same original data, different researchers might come to different conclusions about some phenomenon because of, say, how they account for the effects of inflation, or how they choose to define the variable they are measuring (income? wages? compensation, which includes both wages and benefits?). The quality and content of the four-year debate about income inequality improved as the protagonists increasingly focused on these technical questions and stopped attacking one another on ideological grounds. Gradually, the facts did come to speak for themselves, more or less, and nearly all economists looking at the problem came to agree.

They agreed that average post–World War II earnings (or at least the

rate of growth of earnings) peaked back in the early 1970s. Their computer analyses told them that inequality *was* growing procyclically—that
is, even as the economy expanded, in contradiction to all previous evidence and to the predictions of standard theory. They concurred that
the proportion of Americans earning poverty-level wages was increasing
among men and, by some accounts, among women, as well, but that
women were finally catching up to men, both because of the changing
structure of the economy (more service, office, and white-collar jobs,
and fewer factory jobs) and because men's condition had worsened so
badly. Finally, the economists agreed that the dimensions along which
American workers were drifting farther and farther apart from one
another were *work experience* (how many weeks and hours of employment a person had) and, most of all, *hourly wage rates*. This growing
polarization was observable even among workers with comparable personal traits, education, and years of experience.

The appearance in 1992 of three widely read technical documents
published in prominent places, inside and outside the government,
marked the convergence of assessments on the extent, if not the causes,
of the phenomenon of secularly widening earnings inequality. One document was a report published by the Census Bureau on workers with
low earnings. Looking only at the roughly half of all people employed
year-round and full-time (as Bluestone and I had done in most of our
own earlier writing), the government found

> a sharp increase over the past decade in the likelihood that a year-
> round, full-time worker (or a worker with a year-round, full-time
> attachment to the labor force) will have . . . annual earnings [below
> the poverty level for a four-person family]. In 1979, 7.8 million or
> 12.1 percent of all year-round, full-time workers had low annual
> earnings. By 1990, the number of year-round, full-time workers
> with low annual earnings was 14.4 million and the proportion was
> 18.0 percent.
>
> Young workers have the highest likelihood of receiving low earn-
> ings . . . but the rate has increased since 1979 for all age groups
> below 65 years of age. . . . The rate for persons with 1 or more years
> of college [also] rose from 6.2 percent to 10.5 percent.[9]

At the Economic Policy Institute in Washington, D.C., the research
director, Lawrence Mishel, and a staff economist, Jared Bernstein,

looked at *all* workers, not just those employed full-time and year-round. Mishel and Bernstein used the same data as had the government, and virtually identical methods of analysis. They concluded that, between 1979 and 1989, from one business cycle peak to another, the proportion of American workers earning hourly wages below the poverty line—about $6.50 in 1991 purchasing power—rose from 25.4 percent to 28.0 percent. Over the same decade, the fraction of the workforce earning three times the poverty line and above—about $19.50 per hour—rose from 7.3 percent to 8.7 percent.[10]

The third major milestone of 1992 was an exceptionally exhaustive review of the debate, published in the American Economic Association's official review medium, the *Journal of Economic Literature*, written by M.I.T.'s Frank Levy and Harvard's Richard Murnane, two well-known economists specializing in the study of work, education, and income.[11] Levy had himself been a major participant in the debate, so people were especially eager to read his updated views. That various drafts of this paper circulated among economists all over the country, with correspondence flying back and forth, reveals much about how seriously the research and policy communities had come to take a subject that, only half a decade earlier, had been almost completely dismissed.

Drawing heavily on a doctoral dissertation originally written at Yale by the economist Lynn Karoly and completed by her at the RAND Corporation,[12] Levy and Murnane concluded in their review that:

Nineteen-hundred-seventy-three marked the end of rapid real earnings growth and the beginning of slower growth bordering on stagnation. Nineteen-hundred-seventy-nine marked the beginning of a sharp acceleration in the growth of earnings inequality, particularly among men.

[Between] 1979 and 1987... the proportion of men earning more than $40,000 (in 1988 dollars) increased, while the proportion of men earning less than $20,000 increased as well. The combination of increased earnings inequality around a slow-growing average means that significant numbers of workers—particularly younger, less educated men—now earn less than their counterparts of the mid-1960s.[13]

... [In other words, at least] the male annual earnings distribution has "hollowed out," leaving larger percentages of workers at the top and bottom of the distribution, and a smaller percentage in the middle. At least for men, it is now clear that there were fewer middle class jobs in the mid-1980s than a decade earlier.[14]

This was, in all important essentials, the central proposition of the original magazine article by Kuttner, and of that report to Congress written by Bluestone and myself six years earlier, which had triggered the debates and research projects over income inequality to begin with. Subsequent popular books by the political commentator Kevin Phillips and others only repeated the bad news, to the point where the polarization of earnings and the crisis of the middle class became the most salient issues of the 1992 presidential campaign.[15]

WHY CONVENTIONAL EXPLANATIONS DO NOT FILL THE BILL

One possible explanation for the growth in earnings inequality, at least within the context of American institutions, dominates the thinking of mainstream economists. This is the straightfoward, plausible idea that the market value of a college education rose during the 1980s. This could have occurred because, even in the older manufacturing industries, employers were requiring more and more technical skill from their workers, whatever the color of their collars. If so, then people with advanced degrees would be expected to be in a position to command a higher rate of pay per hour than others. And indeed, several economists initially thought their data did show evidence of such a shift in the structure of managers' demand for labor.

There surely *was* a widening wage gap during the first half of the 1980s between college-educated workers and those who never progressed beyond high school. But Mishel and Bernstein demonstrate that this growing gap resulted primarily from a "precipitous decline of wages among the non–college educated work force and not [from] any strong growth of the college wage." And in any case, by 1987—well before the onset of the 1990 recession—the absolute inflation-adjusted wages of even college graduates began to decline.[16] Moreover, as Harvard's Larry Katz, currently the chief economist of the U.S. Department of Labor, has himself reported, inequality has grown among even people with the *same* levels of education. Katz and the University of Chicago economist Kevin Murphy estimate that, since 1970, there has been a 30 percent increase in such within-group inequality—a rise of stunning proportions.[17]

For Levy and Murnane, as for other economists who study income distribution and the structure of labor markets, "the most important unresolved puzzle concerns the reasons for the almost 20 year trend toward increased within-group earnings inequality."[18] After all, standard economic theory predicts that, after accounting for differences in such ascriptive traits as race, gender, and age, workers with the same education and work experience should receive about the same rate of pay. (This expected leveling is driven—at least according to the theory textbooks—by the mobility of labor. Thus, anyone with skill will move from a "bad" job to one with more attractive prospects and pay. And firms will gladly pay to attract "good" employees.) To add even further to the conundrum, wage rates in the 1980s varied significantly even among factories or offices within the same industries, and even among ostensibly identical jobs within the same business establishment. How could this be so?

Harvard's Richard Freeman is the director of labor studies at the National Bureau of Economic Research and the dean of empirically oriented American scholars on labor. Freeman opens the explanatory door one step beyond the conventional education-skill story (to which he otherwise subscribes). He finds that the long-run decline of union density in the United States—the proportion of workers belonging to unions or covered by a collective bargaining agreement—has contributed to the growth in earnings inequality, both among and within industries. Unions, he writes, have reduced wage dispersion in every country and time period for which data are available. Thus, we should not be surprised that steadily declining union membership in this country contributes to growing inequality. But, he hastens to add, declining unionization is only one of many developments promoting inequality, "a supporting player in the story . . . not the main character: Rosencrantz or Guildenstern, not Hamlet."[19]

HOW LEAN PRODUCTION AND NETWORKED FIRMS PROMOTE THE POLARIZATION OF WAGES: THE NEW DUALISM

The decade between 1965 and 1975 was a golden age for mavericks in economics, as it was for the rest of American society. Out of that period emerged a school of political economists who devoted themselves to examining the evolution of the organization of work over the course of the twentieth century. Trained at the leading graduate schools, yet grounded in practical work in the civil rights movement and in urban community development, these young economists eventually produced a body of ideas about "dual," or segmented, labor markets.[20] It turns out that these old ideas have much to say about the new forms of flexible production, in general, and about the surge in inequality, in particular.

In brief, a "primary labor market" was said to be dominated by large, vertically integrated companies that, over the course of the century, had acquired some degree of oligopolistic power and operated in markets that were relatively sheltered from foreign competition. These corporations earned above-average profits, and their employees—especially if they were organized into labor unions—earned above-average wages. In the interest of achieving stable labor relations and promoting on-the-job learning, and in interaction with unions seeking job security for their members, managers created within their firms vertical career ladders, so-called internal labor markets (ILMs). Workers in this segment of the economy could progress up through the rungs of these ladders, learning by doing, under the watchful eyes of supervisors who would periodically evaluate their progress as part of a complex administrative process of "bureaucratic control."

By contrast, outside the boundaries of these generally big firms lay a world of mostly smaller enterprises. Often, these firms acted as subcontractors to the big firms in the core of the economy. Unsheltered from intense price competition with one another, these firms formed the principal actors within the so-called secondary labor market. Here, production tended to be more standardized, wages were lower, new employees could be more easily recruited off the street, employee benefits were limited or nonexistent, and opportunities for much valuable on-the-job learning were generally hard to find.

That was then; this is now. Today, business is operating in a world in which best practice increasingly entails vertical disintegration, downsizing, outsourcing, and the formation of networks of companies in order to operate across national borders and sectoral boundaries. It would be ludicrous to suggest that *all* of the features of post–World War II labor markets have somehow miraculously survived intact.

And yet, when we characterize the prototypical business organization of the new era as a lean and mean flexible firm, embedded within networks made up of partners and dependent suppliers and subcontractors, we are implicitly recognizing that the workforce in the new economy is, arguably by managerial intent, being systematically divided into insiders and outsiders. Some people are employed on full-time, year-round schedules, receive health insurance and paid vacations, and experience continuous formal and informal job training. A few can even still look forward to having access to structured, more or less predictable opportunities for upward mobility, although in an era of flexible production networks, moving "up" more often means changing jobs by moving from one company to another.

But in this evolving system, at least as many people now confront job "opportunities" that, increasingly, are limited to involuntary part-time or part-year work, low wages, few benefits, and frequent job changing that fails to provide a rising standard of living. This is life on a treadmill. Smaller firms that act as suppliers to the big firms are being similarly stratified, according to the degree to which they collaborate with their large industrial customers and are technically or financially supported by those.

To be sure, some part-time workers *choose* that form of employment, and some fraction of those who do are well paid for it. This is especially the case for professional white women who work, for example, as freelance editors or computer programmers. But one of the most knowledgeable experts on these changes in labor market institutions, the economist Chris Tilly, of the University of Massachusetts at Lowell, concludes from his extensive econometric *and* ethnographic research that "secondary, 'bad' part-time jobs greatly outnumber ['good'] part-time jobs."[21] How does he know? Because "of the 2.9 percentage point climb in the rate of part-time employment between 1969 and 1988, 2.1 points were due to increasing *involuntary* part-time employment."[22] I return to this distinction between voluntary and involuntary contingent working later in the chapter.

The increasing polarization of wages is clearly consistent with the reinstitutionalization of labor market dualism. Indeed, several European scholars have already spied this connection. More than a decade ago, the Oxford University sociologist John Goldthorpe suggested that corporations' search for flexibility could actually heighten labor market and social duality.[23] Another British commentator who recognized early the reinstitutionalization of dualism inherent in the core-ring or network transformation of business organization was John Atkinson. Working out of the corporate-sponsored British Institute for Manpower Studies, Atkinson predicted that the largest multinational firms would attempt to retain their control over global economic affairs by transforming themselves in such a way as to create a new and far more sophisticated set of hierarchies and labor market segments, both within the core company itself and outside the firm, along its network of suppliers of products and labor. Indeed, as a management consultant, Atkinson was *advocating* the reproduction of such dualities, in the interest of promoting his clients' very survival.[24]

Several years later, building explicitly on Goldthorpe's reassessment but coming at the problem from the Left of the labor movement, the University of Milan sociologist Marino Regini conceived what he termed the "new dualism." Across the world, argued Regini, companies were combining "informalization" (mainly by relying increasingly on outsourcing) with what he called "Japanization," the introduction of teamwork and just-in-time management within the core labor processes of the leading firms.[25] Italian scholars have been especially thoughtful about all of this; recall from chapter 4 how, in her elaborately detailed case study of Benetton, the Italian clothing firm, the trade union researcher Fiorenza Belussi articulated a theory of what she named "decentralized Fordism." She developed this construct in order to account for Benetton's ways of managing to combine high-level design and development work in northern Italy with outsourced production in tiers of small suppliers, some of them sweatshops scattered around southern Europe.[26]

Meanwhile, back in the United States, the Stanford University sociologists Jeffrey Pfeffer and James Baron may have been the first Americans to state that

> the increasing separation of the work force into [core] and [peripheral] workers may increase wage inequality, as workers who are considered to be more essential to the firm will tend to be paid more, while those used as buffers will have less market power and obtain lower wages. . . . We suspect [this] inequality may be based increasingly on hours worked and the stability of employment, the dimensions which [most] distinguish core and buffer employees.[27]

Thus, it seems, while the economists who had first invented dual, or segmented, labor theory were occupied elsewhere, the sociologists and planners were telling us that the institution was alive and well, embedded within the very flexible production systems that were supposed to be replacing the old order.

THE DEVOLUTION OF INTERNAL LABOR MARKETS
AND THE EROSION OF EMPLOYMENT SECURITY

The new dualism differs from the original version in at least one important way: even high-level jobs are no longer secure. In the age of flexibility, even the most profitable big firms are inclined to shed even white-collar employees. The shrinkage in the number of safe, stable, secure occupations seems to be keeping pace with the big firms' downsizing.

One consequence of these practices is that, in sharp contrast to their Japanese and German competitors, American managers are coming to devalue internal labor markets (ILMs) as organizational mechanisms for developing loyalties and productive skills. The decline of ILMs simultaneously marks an important change in the particular character of labor market dualism and constitutes one of the sources of that growing polarization of earnings about which so many raised so much fuss during the 1980s.[28]

For a number of years, the urban planner Thierry J. Noyelle, of Columbia University, has been studying what he explicitly refers to as the "dismantling" of ILMs, especially by companies in the retailing, public utilities, and financial sectors. His argument, in a nutshell, is that the rapid growth of a highly educated urban workforce, together with successful equal employment opportunity laws and programs, has made it possible for managers to "externalize" much of the skill training they used to conduct, mostly informally, in-house. The advent of information technologies, argues Noyelle, complements this trend by leading to

a kind of universalization or homogenization of skills demanded across a wide range of industries, allowing for [even] greater externalization of training for many middle-level workers. Many occupations have now become more generic and less firm specific than they once were . . . [and] computer-oriented algorithmic logic has replaced many firm-specific idiosyncratic practices. [All of this amounts to a pronounced shift] away from on the job training [that has] further undermined the raison d'être of the old internal labor market.[29]

I would suggest that the growth of an educated, more diverse urban labor force and the standardization of so much office and other white-collar work through what Noyelle calls "algorithmic logic" are *permissive* developments, while the corporate search for flexibility and cost savings has been the driver. In any case, these are complementary forces, and their net result is the decline of the ILM as an organizational solution to the problem of training.

One serious methodological problem that has plagued policy research on labor market structure is that what is being observed in the data are *people* (workers, families), when what is often wanted are observations on *jobs*. The economists Maury Gittleman of New York University and David Howell of the New School for Social Research have made a considerable contribution in this respect. In their operationalization of these institutional constructs, they also provide strong support for the hypothesis that it is indeed those middle-level jobs—and, with them, middle-class incomes—that have been disproportionately shrinking in recent years.[30]

Using data originally constructed at the NBER by Katz and his fellow Harvard economist (and Clinton administration official) Lawrence Summers, Gittleman and Howell have studied a matrix of 621 different combinations of industries and occupations, drawn from the 1980 Census. In this way, they have been able to distinguish, for example, between custodians working for the automobile industry, where they tend to be well paid, and custodians employed in fast-food restaurants, where they are not. It is these interactions that have enabled Gittleman and Howell to inform us that about half of all high-wage blue-collar workers in 1980 did *not* work in manufacturing, and that more than two-thirds of all low-wage blue-collar workers were employed in services. This ability to finally break the false (but popular) identity between manufacturing and blue-collar work is a major achievement.

These 621 combinations accounted for fully 94 percent of total non-agricultural employment in the United States in 1980. Each of these "jobs" is then characterized along seventeen different dimensions, from average hourly wage rates to the extent of involuntary part-time employment to employer-assessed physical strength requirements for performing the job. The application of a standard statistical technique, cluster analysis, to this 621-cell matrix has led the researchers to identify in the data a latent structure according to which jobs fall into one or another of six categories.

The two categories whose jobs generally pay the highest wages, offer the greatest opportunities for on-the-job training, and so on are "professional, managerial, and sales," and "public sector." The two categories that generally pay the lowest wages, and offer the fewest employee benefits are "low-wage blue-collar" and "contingent" (part-time, casual, part-year employment). Sandwiched in between are two categories consisting of "routine white-collar" and "high-wage blue-collar" jobs. This empirically derived (rather than assumed, or imposed) taxonomy extracted from 1980 data looks remarkably like the labor market structure discovered by the economists David M. Gordon, Richard Edwards, and Michael Reich from historical data on the post–World War II era, which only reinforces the impression of continuity with the past.[31]

Comparing the distributions of actual employees in 1979, 1983, and 1988 against these benchmarks, Gittleman and Howell conclude:

> The job structure has become more bifurcated in the 1980s, as "middle class" jobs [the share of the work force employed in those middle two categories] declined sharply and the work force was increasingly employed in either the best . . . or the worst . . . jobs. White women became much more concentrated at the top, while white men and black and Hispanic women were redistributed to both ends of the job structure. Black and Hispanic men, however, increased their presence only in the two [lowest] job clusters. At the same time, the quality of [these lowest category] jobs worsened considerably, at least as measured by earnings, benefits, union coverage, and involuntary part-time employment. As these results would suggest, we found that earnings differentials by cluster, controlling for education and experience, increased in the 1980s. The male–female racial wage gap also increased.[32]

THE GROWTH IN CONTINGENT WORK
So far, I have been largely differentiating jobs by the wages they pay. What about growing variations in work experience? Richard Belous of the private corporate-sponsored National Planning Association (NPA) estimates that, in 1988, the number of workers employed in part-time, temporary, contract, and other forms of what has come to be called "contingent" labor in the United States was between 30 and 37 million—roughly a quarter to a third of the civilian labor force.[33] Research on the changing incidence of contingent work was, for many years, the province of a small group of scholars and labor unions (one of those scholars, the economist Katherine Abraham, is now the commissioner

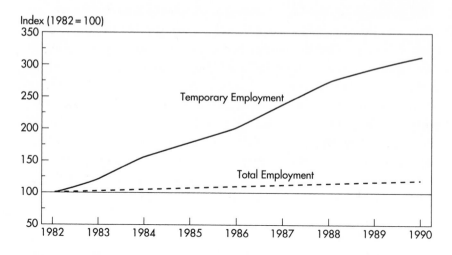

Index (1982 = 100)

FIGURE 9.1   GROWTH OF TEMPORARY EMPLOYMENT,
1982–1990

Source: Polly Callaghan and Heidi Hartmann, *Contingent Work: A Chart Book on Part-Time and Temporary Employment* (Washington, D.C.: Economic Policy Institute, for the Institute for Women's Policy Research, 1991), p. 6.

of the U.S. Bureau of Labor Statistics [BLS]).[34] At the Washington-based Institute for Women's Policy Research, Heidi Hartmann and Polly Callaghan have documented that, since 1982, temporary employment in the United States has grown three times faster than employment as a whole (figure 9.1). Between 1976 and 1990, the number of part-timers (defined as those averaging fewer than 35 hours of waged employment per week when they did work) increased by 7 percent, compared to a 2 percent growth of full-time workers—those usually employed for 35 hours a week or more.[35]

One study that has been enormously influential in shaping the public debate about contingent work was conducted at Cornell University's School of Industrial and Labor Relations by the labor economist Ronald Ehrenberg and his students. Using econometric methods, Ehrenberg's team was able to show, as (later) did Tilly, that

there has been a tendency towards increased employment of part-time workers in the United States in recent years, a trend that is observed *after* one controls for cyclical factors. Moreover, this trend

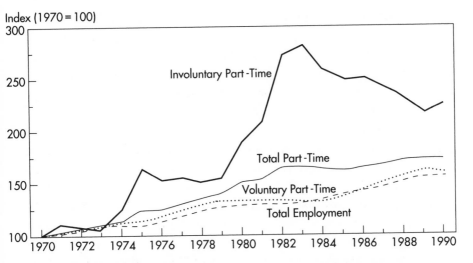

FIGURE 9.2   GROWTH OF PART-TIME EMPLOYMENT,
1970–1990

Source: Polly Callaghan and Heidi Hartmann, *Contingent Work: A Chart Book on Part-Time and Temporary Employment* (Washington, D.C.: Economic Policy Institute, for the Institute for Women's Policy Research, 1991), p. 4.

has come from an increase in "involuntary" part-time employment, not from an increase in voluntary part-time employment. Searches for explanations for the recent growth of part-time employment in the U.S. should therefore focus on the demand side of the labor market.[36]

"Demand side of the market" means the decisions of *managers*. The explosive growth of involuntary part-time work, compared with full-time employment, is graphed in figure 9.2.

These developments have serious implications for personal well-being. For example, between 1979 and 1989, the share of the private sector workforce covered by pension plans fell from 50 percent to 43 percent. The incidence of employee coverage by health insurance at least partly provided by employers declined from 69 percent to 61 percent. These are *very* substantial declines for such a short period of time.[37]

Figure 9.3 shows just how much the average decline in benefit coverage is wrapped up with the shift toward lean production and contingent work. As depicted in that figure, the proportion of voluntary part-timers without health insurance has actually been declining since 1983, while a

steadily rising share of *involuntary* part-timers have no health insurance.
This growing division of the American workforce into those who are
entitled to employee benefits and those who are not is one of the more
unfortunate aspects of the growth of insider-outsider employment arrangements: truly a dark side of flexible production.

In the late 1980s, government agencies and business research institutes began to take the phenomenon seriously. Thus, in 1989, the NPA
published a collection of case studies authored by Belous. In the same
year, two economists at the Bureau of Labor Statistics announced a
semiofficial definition of contingent work as including "any job in
which an individual does not have an explicit or implicit contract for
long term employment and one in which the minimum hours worked
can vary in a non-systematic manner."[38] We can expect more such
research and standard setting from the Clinton administration, with
Abraham now at the helm of the BLS, and with Karen Nussbaum
directing the Labor Department's Women's Bureau. (Nussbaum was
the cofounder of 9-to-5, the clerical workers' association that was eventually absorbed into the Service Employees International Union
[SEIU]. More than any other organization, SEIU has spoken up on the
subject of contingent work for more than a decade.)

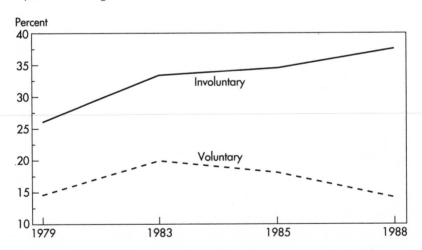

FIGURE 9.3 VOLUNTARY AND INVOLUNTARY PART-TIME
WORKERS WITHOUT HEALTH INSURANCE, 1979–1988

Source: Polly Callaghan and Heidi Hartmann, *Contingent Work: A Chart Book
on Part-Time and Temporary Employment* (Washington, D.C.: Economic Policy Institute, for the Institute for Women's Policy Research, 1991), p. 14.

Other government agencies have been looking more closely at still other dimensions of contingency. In one of the more gruesome, the General Accounting Office (the investigative arm of Congress) reported in June 1992 that the number of illegally employed minors—children under the age of fourteen—had almost tripled since 1983. William Halperin, then the associate director for surveillance at the National Institute for Occupational Safety and Health, was quoted as characterizing these findings as "astounding," yet probably only "the tip of an iceberg."[39]

The real question remains whether managers are consciously creating (or re-creating) dual labor markets by deliberately externalizing and otherwise transforming work. At least in the United States, the answer is yes. American companies *are* deliberately creating both peripheral and internal low-wage labor markets for employees on contingent work schedules.

This conclusion comes from a 1988 survey of American managers conducted by the Conference Board, the New York–based think tank for the Fortune 500.[40] The survey was conducted with representatives of 521 of the country's largest manufacturing, financial, and nonfinancial services corporations. Respondents were asked to describe their use during the previous year (1987) of internal pools of temporary workers, of outside agencies that provide "temps," of independent contractors (for example, editors or computer programmers), of so-called flextime arrangements, of regular part-time employees, of job-sharing arrangements among employees, of compressed work weeks, of phased retirement, and of home-based work.[41]

From the Conference Board data, Kathleen Christensen discovered that a significant fraction of the firms sampled *did* use part-time labor as a deliberate "contingent staffing alternative," along with outside temps, independent contractors, and internal temp pools. In some companies, "firms hire as independent contractors former employees who have been laid off, taken early retirement, or left the firm to go out on their own"—a practice that (as I know from living in Pittsburgh) is also increasingly common in such jobs as equipment maintenance in steel mills. This sort of outside contracting is common both in manufacturing and elsewhere in the economy. According to Christensen, nearly 100 percent of the finance and insurance companies surveyed reported using part-time workers as a regular activity. Managers also told the interviewers that they expected to increase their use of contingent work schedules in the future.

DUALISM IN THE NIKE PRODUCTION SYSTEM: A CASE STUDY

As discussed earlier in the case of Benetton, dual labor markets—the dark side of flexibility—can exist even within the boundaries of a single firm, especially when that firm is organized as a geographically extensive network with its core operations at one location and tiers of parallel or lower-level activities at other sites. For a U.S.-based example, take Nike.

No product sold in America (or Europe, or the Far East) today is more well known that Nike's running shoes. Like cars, hamburgers, and furniture, Nike shoes (the name *Nike* stands for the goddess of victory in Greek mythology) come in literally dozens of models, from the relatively cheap to the very expensive, from the (literally) pedestrian to the stylish. Although Nike is legally registered as an American corporation, not one of the 40 million pairs of running shoes that Nike produces annually is manufactured within the United States: *everything* is subcontracted from elsewhere.

As reconstructed by Michael T. Donaghu and Richard Barff, two Dartmouth College geographers (one of whom worked for the company for a time), Nike's global division of labor is an exemplar of the principle of what, throughout this book, I have been calling concentration without centralization. That principle is intended to capture the continuing dispersal of production, but ultimately under the technical and financial control of managers in a relatively small number of big multiregional, multisectoral, multinational corporations and their strategic allies. Nike is such a network firm, whose management has found ways to connect the lowest-paying unskilled jobs with the highest-paying skilled R&D jobs, classic mass production with flexibly automated technology, and the First World with the Third.[42]

Bill Bowerman and a partner founded Blue Ribbon Sports (BRS)—the company that would later become Nike—in 1964. Initially, BRS was purely a U.S. distributor of athletic shoes made in Japan by Onitsuka Tiger. Bowerman was a good designer, and the product began to capture the growing domestic American market as the public's interest in stylish running shoes grew. In 1971, BRS's relationship with its first Japanese partner ended, and the Americans entered into an agreement with Nisso-Iwai, a large trading company with access to its own manufacturing contractors—again, inside Japan. Research and development took place in BRS's facility in Beaverton, Oregon, under Bowerman's supervision. Production remained in Japan, under the management of the Nippon Rubber Company.

In the wake of the first oil shock of the 1970s and a major revaluation of the yen, Nippon Rubber decided to relocate much of its production operations from Japan to Taiwan and South Korea (just as the high exchange value of the yen of the early 1990s is now driving many Japanese companies to move production offshore).[43] BRS itself opened an assembly facility in Exeter, New Hampshire, sourcing components from U.S. and German suppliers. Within a short time, this network of factories was turning out a wide range of products, from inexpensive mass-market footwear to stylish running shoes. By the late 1970s, BRS was marketing its products in Europe, South America, Southeast Asia, and the United States. BRS factories in Britain and Ireland came next. In 1978, BRS legally changed its name to Nike.

The Nike production system is organized into two broad tiers. In the first of these tiers, "developed partners" located mainly in Taiwan and South Korea work closely with the R&D personnel in Oregon to make the firm's most expensive, high-end footwear (by *partners*, Nike managers mean contractors and suppliers who share some joint responsibility with the core firm, for design or for evaluation of production methods). The Asian partners contract out most of the work to local low-wage subcontractors.

"Volume producers" are considerably larger, more vertically integrated companies with their own leather tanneries and rubber factories that manufacture more standardized products and sell to several buyers, of whom Nike is only one. Production and sales are highly variable from one month to the next.

Finally within this tier, Nike has created what it calls "developing sources"—producers located in Thailand, Indonesia, Malaysia, and China. These are the lowest-wage, low- and semiskilled operations that Nike is gradually upgrading ("bringing along"). Technicians from the United States, Taiwan, and South Korea are assigned to work in these "developing" facilities, often on a rotating basis (Nike calls this its "expatriate program"). Such sources often take the form of joint ventures between the Taiwanese or South Korean "big brothers" and the less industrialized Southeast Asian "little brothers."

Nike managers explicitly acknowledge the advantages of this spatial division of labor within the first tier. Apart from "hedg[ing] against currency fluctuations, tariffs and duties [and] political climate change . . . in the long run [this arrangement] keeps pressure on the first tier producers to keep production costs low as developing sources mature into full-blown developed partners."[44]

The second tier of the Nike production network consists of the many material, component, and subassembly sources. Predictably, the least complicated elements may be produced at any of the network's locations in the United States, Europe, or Southeast Asia. But some elements, such as the air cushions that pad all modern athletic shoes, either require skilled labor to turn them out, or make use of proprietary technology or designs. These tend to be located in the vicinity of the Oregon headquarters, where engineers and others can be exchanged easily and frequently—an example of the "specialization subcontracting" that I first presented in chapter 6. As the physical infrastructure and human capital of the "little brothers" is upgraded over time, Nike anticipates that a growing fraction of its second-tier suppliers will be located in Asia.

Clearly, the Nike production system has undergone a substantial evolution since the 1970s. Yet it remains a dualistic system, combining high-wage and low-wage, specialized and standardized production, core and periphery. As Donaghu and Barff put it:

> When NIKE's shoes were first produced in Asia, Japan represented the core and other nations constituted the periphery. Today, NIKE's intraregional division of labor in South East Asia simply based on labor costs has S. Korea and Taiwan as the core and China, Thailand, Malaysia, and Indonesia as the periphery. Japan, with average manufacturing hourly wages very close to those of the United States, is still a supplier of materials, but is no longer a site of production for NIKE athletic shoes. . . . [However,] there is no evidence to suggest that production has shifted as markets have developed. In fact, there is evidence to the contrary. The English, Irish and American plants all ended production of NIKE athletic shoes in the mid 1980s, even as demand for Nike running shoes was accelerating in all three countries.

Interviews with the American managers revealed that they clearly perceived the flexibility of this production system as residing in a combination of low cost, spread risk, and—most of all—"the speed by which a design for a new model of shoe is transformed into a product at the market." This turnaround time is partly a function of using highly disciplined labor and partly of deploying "mass production techniques by simplifying work stations and updating existing machinery." For example:

a new machine was introduced that could automatically position the needle and trim excess thread for the operator. This kind of labor saving change is hardly revolutionary and can be specifically associated with the continuation of Fordist means of production. The only elements of NIKE's system that could be classified as flexible machinery [are] the computer-aided design and computer-aided engineering used in the Beaverton [U.S.] R & D facility, and some numerically-controlled molding machines used by one or two South Korean subcontractors.

The other aspect of flexibility in this system is the core firm's adeptness at shifting productive capital from one place to another. Nike "has opened plants and begun contracts only to end them within only a matter of a year or two. It also utilizes capacity subcontracting methods [see chapter 6] to meet variable market demand."

Throughout the first tier, as much as 80 percent of sales to the final assemblers of the footwear are "prepurchased," meaning that Nike effectively advances capital to the members of what, after all, clearly has features that resemble a far-flung putting-out system. And while Nike's subcontractors themselves tend to be vertically disintegrated and may in some cases be locally agglomerated, and while customer-supplier relations may sometimes be described as reciprocal, or at least not confrontational, "by any definition, [Nike's] method of physically producing athletic shoes is mass production." At the level of the global network, "the company still relies on large volume production by [mostly] semi- and unskilled labor," linked to high-tech R&D and sophisticated financial management situated in the United States.

Like Benetton, but with an even more geographically dispersed division of labor, Nike is a neo-Fordist firm, whose flexibility derives mainly from its managers' ability to construct and govern a dualistic system characterized by concentration without centralization. Its profitability derives directly from its managers being so cleverly able to manage the dark side.

PLAYING OFF INSIDERS AGAINST OUTSIDERS
To recapitulate: Collaboration among producers within networks may well help business to cope with the uncertainties and the heightened competition that are part and parcel of the post-1970s global economy. Yet at the same time, vertical disintegration of the older big firms, the

devolution of sheltered internal labor markets with their opportunities for lifetime employment, and the increasing use of outside subcontractors standing in varying degrees of dependence and independence vis-à-vis the core firms in the network all sharpen the divisions between insiders and outsiders. This in turn reinforces the long-run trend toward the polarization of American earnings.

The danger is that, instead of promoting collaboration among the more and less powerful organizational actors within a network—and, for that matter, among managers and workers within individual companies and establishments—the managers governing the system may yield to the temptation to use their power to exploit these differences in relative power to play one group within the network off against another. Thus, for example, in acknowledging the recent growth within General Motors of various forms of teamwork, the Harvard Business School's Joseph Badaracco reminds us that

> to stress only the new and important participatory elements of [General Motors'] relationship with the [United Auto Workers union] is an oversimplification. Greater cooperation has arisen against the background of a shift in power away from the UAW and toward GM. Indeed, the movement towards greater participation for workers and unions over the last ten years has been paralleled by another, perhaps even more widespread effort to limit their power through overseas sourcing, givebacks, and manufacturing strategies that expand facilities in nonunion states.[45]

Another of the leading theorists of the flexible network firm, the University of Arizona sociologist Walter Powell, is equally cautionary. "Practices such as subcontracting," he notes, "have a double edge to them; they may represent a move toward relational contracting . . . with greater emphasis on security and quality; or they could be a return to earlier times, a part of a campaign to slash labor costs, reduce employment levels, and limit the power of unions even further." Moreover, "some firms are seeking new collaborative alliances with parts suppliers while at the same time they are trying to stimulate competition among various corporate divisions and between corporate units and outside suppliers. . . . Are companies really as confused as it seems?"[46]

Powell's answer is, Perhaps not. Perhaps these contradictory combinations of integration and disintegration, of collaboration and competition, of coherence and duality are built into the very nature of how net-work forms of organization grow and develop. If so, he acknowledges, this poses an extraordinary challenge for those who must live with such a system.

Certainly, the revival of labor market segmentation further weakens the bargaining power of labor unions, making it more difficult for them to organize new workers and to pressure companies to innovate continu-ally in order to generate the additional productivity out of which to meet a rising wage bill. This is the "high road" to economic growth. From the perspective of the national—perhaps even the international—economy as a whole, growing income stratification between capital and labor, and the growth of a pool of low-wage workers act as a drag on sys-temwide technological progress and, therefore, on long-run economic growth. The reason is that dualism encourages all too many firms to build their activities on a foundation of cheap labor, thereby taking the "low road" to company profitability. This may ultimately be the most serious consequence of the dark side of flexible production and business reorganization.

THE HIGH ROAD OR THE LOW ROAD
TO LONG-RUN ECONOMIC GROWTH?

In the United States of the 1990s, the average worker brings home a paycheck that, depending on how you measure "average" and how you account for the ravages of inflation, is anywhere from 7 percent to 12 percent lower than what it was at the end of the Vietnam War, a genera-tion ago. The rich have been getting richer and the poor, poorer. And the great middle class is at best treading water, if not actually sinking below the surface of the decent life that our leaders and teachers had led them to expect as the reward for hard work. Once, these facts about the secular stagnation and polarization of the American economy were hotly contested. Today, hardly any reputable scholar would disagree. In popular journalism, stories of the two-tiered society and the hourglass economy now appear regularly on the nightly television news and in the newspaper Sunday supplements.

Secularly increasing inequality raises obvious normative concerns: Who wins? Who loses? Who is most in need? But there are also macroeconomic considerations that motivate the earnings dispersion debate. The French economist Alain Lipietz is but one among many to suggest that the impaired capacity of a growing fraction of the workforce to consume out of current income *could* create problems for effective aggregate demand, were the trend sustained for a sufficient period of time and were it not offset by government deficits or household dissaving.[47] Indeed, in the United States during the 1980s, short-term aggregate demand *was* sustained despite growing inequality, partly through the accumulation of more than $2 trillion in added federal government debt and close to $500 billion in additional consumer credit.[48]

But as I have emphasized here, and in my earlier work with Bluestone, beyond the macroeconomic impact, there are negative *microeconomic* consequences flowing from the revival of inequality in general, and of labor market dualism in particular. A growing pool of low-wage labor sends precisely the wrong signal to firms, encouraging them to compete on the principle of cheap labor rather than on the basis of technological improvement and the upgrading of their employees' skills. As the British economists Deakin and Wilkinson explain: "Dependence upon undervalued labor provides a way by which inefficient producers and obsolete technologies can survive and compete. Firms become caught in low productivity traps from which they have little incentive to escape."[49] Moreover, the process may become viciously circular, since, having deprived themselves of a technically competent workforce to begin with, these firms are subsequently ill-equipped to innovate. Hence, "when these firms are subjected to competition from more efficient [companies], improved technology and products, their only hope of survival is further to reduce wages."[50]

New Research from Freeman and Katz suggests that inequality and the erosion of wage standards may now be occurring in countries besides the United States, although to very different degrees and with less pernicious implications, given their generally stronger social safety nets.[51] Nevertheless, in both Japan and Europe, over the course of the post–World War II era, growth-oriented companies with long-term planning horizons worked together with governments committed to the practice of industrial policy. Especially in Europe, organized labor's demands for higher wages and better working conditions, and its progrowth stance in the political arena, contributed to pressing big business and the government to take the high road to economic growth and development. Along this path, companies invest continuously in their employees' skills and in manufacturing, office, warehouse, highway, railroad, and aircraft equipment, embodying new technologies. The bigger firms help to upgrade the technical capabilities of their generally smaller suppliers.

Through this combination of *technology, training,* and *technical assistance,* the productivity of the national economy increases—and, with it, the standard of living of the mass of the population.

But there is also that other, less admirable path: the low road to company profitability.[52] Along this path, managers try to beat out the competition by cheapening labor costs. They move whatever operations they can to low-wage rural areas or to Third World countries. They scrimp on training. They routinely outsource work that used to be performed in-house to independent subcontractors who will not (usually because they themselves cannot afford to) pay decent wages, let alone provide even the most basic benefits such as health insurance premiums or paid sick leave. Low-road companies try to squeeze the last ounce out of older capital equipment, rather than steadily retooling and upgrading their technical capabilities. They play off their suppliers against one another to get the cheapest price today, with no thought to the negative impact this can have on the quality of tomorrow's deliveries. At the last extreme, a company that once made its own products, using domestic workers and paying them a living wage, now hollows itself out, abandoning manufacturing altogether to become more or less strictly an importer of things made by foreign companies—or by their own overseas subsidiaries.

Throughout the twentieth century, American business has been at war with itself over whether to travel the high road or the low one. Since the 1970s, the tension between these alternatives has become even more pronounced. The consequence of a generation of managers taking the low road to a restoration of profits is the cultivation of the habit of competing mainly on the basis of cheapening labor power, rather than upgrading technology and skills. As discussed in chapter 8, not all firms in all American industries are pursuing a low-road strategy. But the evidence that a large number *are* doing so seems so compelling that we should worry about the future prospects for a restoration of the historic American economic pattern of growth at high wages with declining inequality.

Why have so many American (and, increasingly, foreign) managers elected the low road to resolving the profit squeeze? Surely the current weakness of the American labor movement is part of the answer. Drawing on data from the BLS, Freeman and the University of Wisconsin political scientist, Joel Rogers, observe that the fraction of private sector workers in America who belong to unions has fallen to a pre-1935 low of 12 percent, and that a rate of just 5 percent is a reasonable forecast for the year 2000.[53] The problem is compounded by the absence of any other structures of worker representation that could compensate for the decline in union density. A strong (or, as in the 1930s, re-emerging) trade union movement effectively forces employers to make decisions that enhance productivity, in order to contain unit labor costs.

But what explains the weakness of American labor unions and the upsurge in corporate restructuring strategies? Here, Bluestone and I have suggested that the answer lies in the sheer *suddenness* with which the American economy found itself inserted into the international trading system in the 1970s. In the brief period between 1969 and 1979, the share of U.S. gross national product (GNP) accounted for by imports *doubled*, and merchandise imports as a share of total GNP originating in the manufacturing sector nearly *tripled*. The old oligopolistic mechanisms available to the leading American corporations for absorbing cost increases by raising their price markups were thus seriously undermined by the threat (and the reality!) of foreign competition.

Add to this the continually fluctuating exchange rates, which created chronic uncertainty and wreaked havoc with investment decisions, and a portrait emerges of an environment in which it was simply easier for many firms to attempt to contain their own labor costs than to seek enhanced profits through investments in expensive new plant and equipment. As Bluestone likes to frame it, too many American firms abandoned revenue-enhancing strategies to boost profits and turned sharply toward tactics that emphasized cost reduction, instead.

The growing dependence of productive enterprise on equity financing, managed by Wall Street intermediaries, contributed to the already developing tendency in American industry for decisions to be made with an eye toward short-run profit (recall the discussion of the financial side, at the end of chapter 8). In contrast to industries in Japan, Sweden, and Germany—where the merchant banking systems have been so fundamental to the financing of long-term capital projects—U.S. industry has become more and more the prisoner of impatient capital. The surge in technologically or synergistically unsupportable mergers and acquisitions, along with rank speculation in land, currencies, and futures markets, was made that much more extreme in the country whose institutions were least well suited to providing a counterweight.

Finally, reinforcing all of these developments have been three decades of high real interest rates, with relief having come only in the 1990s. The Vietnam era spawned ever more powerful pressures for future inflation, as the Federal Reserve eased the money supply to validate competing claims on real resources in an effort to head off further political turmoil at home. The attempts by unions and other organized groups during the 1970s to maintain the living standards of their members in the face of this inflationary bias eventually convinced the government, even before the election in 1980 of Ronald Reagan to the presidency, that a tight monetary policy was essential to containing inflation and to weakening the claims of labor and of the social movements.[54]

But the monetarist experiments, first in the United States and then in the United Kingdom, only depressed first national, then international economic growth. That, in turn, suppressed whatever private investment in new plant and equipment might otherwise have been forthcoming. Now, in the 1990s, synchronized recessions in Japan and Europe have only made matters worse, by greatly weakening foreign demand for American exports.

Consequently, influenced by slow growth of demand and high real long-term interest rates throughout most of the period, the annual rate of growth of industrial capacity has fallen steadily in this country, from 3.5 percent per year in the early 1980s to about 1.5 percent per year in 1992.[55] About this trend, the chief economist of the Wall Street investment banking firm of Morgan Stanley and Company, Stephen S. Roach, has expressed the concern that "smokestack America may have gone too far in hollowing out its industrial base in order to achieve short-term efficiency gains."[56]

This is a particularly vicious circle, since slowly growing industrial capacity means that even a modest spurt of economic growth will quickly push up rates of capacity utilization. To the extent that this gets read by the always jittery Federal Reserve as a signal of impending inflation, the central bankers will be inclined to step on the monetary brakes, slowing the economy further. In this way, excessively cautious macroeconomic policy reinforces the signals to business to opt for the short run—to go for the low road.

WHITHER THE DARK SIDE?
This is a conundrum—what economists call a low-level equilibrium trap—that the search for more "flexible" production and working arrangements on the part of individual companies, their suppliers, and their strategic partners seems only to be making worse. What is needed is a combined effort by a critical mass of farsighted private businesses, a

government able and willing to overcome its own equally shortsighted fear of budget deficits and to deliver serious domestic macroeconomic stimulus, and a much greater degree of international cooordination of fiscal and monetary policies among the G-7 nations (that is, the United States, the United Kingdom, Japan, Canada, Germany, France, and Italy).

Without such interventions, we run the very grave danger of finding ourselves trapped on the dark side for years and years to come.

PART IV

RETHINKING ECONOMIC

DEVELOPMENT POLICY

10

ECONOMIC DEVELOPMENT POLICY IN A WORLD OF LEAN AND MEAN PRODUCTION

The third shift [in capitalist development] marks the rise of densely networked economies composed of both large and small . . . firms, foreign-owned subsidiaries, and joint ventures. This organizational structure has grown out of the spreading and deepening of the international market economy in a period of declining U.S. hegemony [and with the collapse of the Soviet Union]. . . .

[However] political deregulation has not created a free market of truckers, traders and petty producers. Instead, new and often more subtle forms of economic control have rushed in to fill the vacuum created by national deregulation. . . . These [network] forms of economic organization have expanded their scope and their ability to control their own spheres of economic activity.

—GARY GEREFFI AND GARY G. HAMILTON,
"MODES OF INCORPORATION IN AN INDUSTRIAL WORLD,"
AUGUST, 1990.

We are in the auditorium of a leading hotel in Washington, D.C. Two impressive sociologists, Gary Gereffi of Duke University and Gary Hamilton of the University of California at Davis, are telling their colleagues at an international conference that the capitalist system has passed through two distinct stages in its 400-year history and is now, in their judgment, entering a third.[1]

As Gereffi and Hamilton put it, during the initial period of the industrialization of Europe and subsequently of the United States, production was based primarily in small, family-owned factories that turned out commodities that were then sold mostly in local or regional markets. Then, from the late nineteenth century until roughly the end of the Vietnam War, production became dominated by the by-now familiar large, vertically integrated, multidivisional, often multinational corporation.[2]

But since the 1970s, a new organizational form of business enterprise has become increasingly common. In this book, I have explained and documented my own understanding of this still-evolving third shift in the shape of business organization, in this, the age of the corporate search for flexibility. In my formulation, the business system is increasingly taking the form of lean and mean core firms, connected by contract and by handshake to networks of other large and small organizations, including firms, governments, and communities.

Moreover, I have suggested that the seemingly benign concept of flexibility has its own contradictory aspects. In particular, we may admire the capacity of modern business managers in many places and firms to chart new ways of overcoming the barriers of time, space, and national culture, in order to coordinate spatially and sectorally decentralized and dispersed economic activities, and to join with one another and with governments to forge new institutions for promoting technological innovation. But networked forms of industrial organization also exhibit a tendency to reinforce, and perhaps to worsen, the historic stratification of jobs and earnings. This only reinforces economic and social inequality—the institutionalization of dual labor markets—within and among countries and regions, and even among the employees of the same firms.

These networks take on particular forms in different places and moments, but all the special cases display at least some underlying commonalities. The hardiest, most innovative networks typically have large corporations at their centers (or nodes), but the special cases do include such small firm–led industrial districts as those of north-central Italy—although, as discussed earlier, these are not necessarily the most important or stable examples to which one may point. High-tech regions such as California's Silicon Valley owe their durability not to their close resemblance to the Italian districts so much as to the varied ways in which the region has become inserted into the global economy. Silicon Valley has more faces—and legs on which to stand—than does Emilia-Romagna, and the implications for long-run competitiveness are already showing.

Whatever the particular forms they exhibit, one thing that the recognition of the growing importance of networking as a principle of industrial organization does is to undermine the prevailing romantic belief in the significance of atomistic, small enterprise as the dominant organizational actor in a modern industrial economy. Small firms have a role to play, especially in helping to build more economically coherent and perhaps more stable environments in low-income and rural communities, provided they are well connected to the outside and do not promote the further isolation of the community. But looking at the industrial system as a whole, as it is *really* evolving, "small" by itself turns out to be neither as bountiful nor nearly as beautiful as we have been led to believe.

Individual sites of production, distribution, and business services provision are appearing—or rather, as Michael Storper and Richard Walker have taught us, are being *created*—all over the world.[3] But it would be a grave error to see those sites as independent centers of economic activity, competing with one another on the basis of their comparative physical and human resource endowments, at arm's length, through the sole intermediation of price signals. Classical models of supply and demand cannot explain why certain insurance companies and banks continue to sponsor certain manufacturers, year after year, regardless of the availability of more immediately profitable alternative customers and clients. High-tech start-up firms go to great lengths not only to compete with foreign companies but to attract foreign investments from those same competitors. They do so because they want access to the latter's markets and because a systematic shortage of patient finance capital exists at home.

Some large customer firms in the aerospace, automobile, and electronics industries deliberately award long-term contracts to an upper tier of preferred suppliers, in order to build trust and to promote collaboration, even though, according to conventional theory, this constitutes an irrational delegation of market power by the customers. Again, these relationships should be seen for what they are: aspects of far-flung, spatially decentralized networks, governed to a significant degree by large organizations possessing concentrated economic, financial, and political power.

If this third shift is really as significant as many of us claim, then surely this must have implications for public policy. That it is vital to private corporate strategic thinking must be apparent from even the most cursory scanning of both the scholarly and the popular business magazines, which seem these days to write of almost nothing else. What does it mean to talk of—let alone to want to design—a "national" industrial or technology policy in an era in which the goals and behaviors of individual companies and establishments are so often subordinated to, or at least constrained by, the exigencies of the far-flung networks of which they are a part? How do localities within any particular country negotiate economic development incentives and responsibilities with the plant or office managers of firms whose strategic decisions are increasingly being made for them a continent or an ocean away?

On the one hand, the increasing capability to span boundaries and borders that networking affords to business would seem to have tilted the playing field decisively against locally elected and appointed economic development planners, vis-à-vis the plant location managers of the multilocational companies at the hubs or apexes of the networks.[4] Yet at the same time, precisely because the networking principle allows concentrated business organizations to coordinate operations across an ever more dispersed field of play, more decentralized production becomes increasingly feasible. But then it follows that, paradoxically, the comparative attractions of different locales actually take on an *enhanced* significance for industrial location.[5]

There are other pressing public policy questions that the advent of network forms of business has brought to the fore. Policy makers need to rethink what kinds of firms, and whose, are to be eligible for participation in national industrial and technology partnerships in the years ahead. They will have to fashion new kinds of local and regional economic development policies in a world of lean production and business networks. The network form is wreaking havoc with traditional approaches to the regulation of business; what comes next? Finally, it is going to be necessary to deal with the dark side of flexible production and to find ways of containing the tendency toward greater polarization of income and power based on whether one is inside or outside the core jobs in the hub firms.

WHO IS "US"?
Several years ago, Robert B. Reich, a Harvard University lecturer and now the secretary of labor in the Clinton administration, proposed a

challenging thesis to those thinking about the changes taking place in America's role in the world. While not writing specifically about networks as such (although he often invokes the image of "webs" of globe-spanning operations), Reich proposed a thesis about the evolving relationship between multinational corporations and individual national governments that set off a debate directly related to the questions I have been asking.

In a widely cited paper in the *Harvard Business Review* and in a subsequent monograph, Reich argued that multinational corporations have become so profoundly footloose in their ability to choose between producing and trading, and in their choice of production sites, that for all practical purposes, they no longer exhibit any important loyalty to any particular national government (let alone any particular site or region) within their range.[6] In short, they neither want nor need a home base. Instead, they have become, or are in the process of becoming, entities unto themselves, dependent only on being assured of a continual supply of highly skilled, technically well trained professionals and technicians, a group Reich calls "symbolic analysts."

From these arguments, Reich concludes that national policy ought to focus on creating more symbolic analysts—more highly trained technicians and managers—in order to make the United States a more attractive location for companies of *any* nationality. Such human capital and other policies ought *not* to favor American companies over (say) the Japanese or others. From an author who helped, over a decade of writing, to popularize the idea of public-private partnerships to promote and implement national industrial policies, this recommendation came as a bombshell to the public policy community.

Reich does marshal diverse evidence in support of his thesis. As for the attractiveness to American companies of doing more and more of their production abroad, he points to official U.S. government statistics showing that, during the late 1980s, the value of products sold from the overseas affiliates of American corporations was almost four times greater than the value of the goods exported from the United States itself. Overseas profits rose during these years, even as the profits of domestic operations declined. And the new capital expenditures invested by American firms in their offshore operations rose between two and three times faster than did investments inside the United States.[7]

For evidence in support of the growing importance of the symbolic analysts, Reich points to the extraordinary growth during the 1980s of the upper tier of the American income distribution—the top of the hourglass economy that was discussed in the previous chapter. He attributes this growth to the tremendously increased value placed by international companies on these highly skilled employees, thereby echoing the explanation of growing inequality offered by the main-stream economists.

In a debate held in 1991 at Harvard's Kennedy School of Government and elaborated in the pages of *American Prospect*, Reich and the Univer-sity of California at Berkeley economist Laura D'Andrea Tyson, who now chairs President Clinton's Council of Economic Advisers, argued the question of whether a corporation's nationality is still meaningful.[8] There, Tyson suggests that Reich's description of the dominant impor-tance of overseas production sites to American companies is "prema-ture," at best. This might well be the direction in which U.S. firms are moving, she writes, but the lion's share of production today remains sit-uated within U.S. borders. Reich responded by saying that, in the sec-tors on which America's trade deficit and future technological leader-ship depend the most, the offshore location of American firms and the foreign takeover of domestic assets have progressed the furthest.

Tyson is one of the pioneers of so-called new trade theory, which rec-ognizes that there are cases in which the firm (or region) that achieves a certain critical minimum market share or a monopoly over a new tech-nology may, by attaining economies of scale, name recognition, or a head start in learning by doing, sustain a competitive advantage over a long period of time.[9] In economists' jargon, there are first-mover advan-tages in the real world. More than any other of the new trade theorists, Tyson has been willing to take this proposition to its logical conclusion: that the competitive advantages of different companies, nations, and regions can to some extent be *created* through deliberate policies. The Japanese post–World War II strategy of export-led development is only the most obvious example.

WHY A HOME BASE REMAINS IMPORTANT TO GLOBAL COMPANIES

The Harvard Business School's Michael Porter practically invented the scholarly field known as business strategy. A consultant to many domes-tic and foreign corporations, he has written numerous technical and popular papers and monographs. While not directly confronting Reich, Porter has for some years been developing the most detailed, closely rea-soned case for why a home base is, perhaps paradoxically, *more* important to companies in the era of heightened global competition than ever before.[10]

For Porter, global competitive advantage resides in the interaction among four major aspects of economic activity. In earlier chapters I described a veritable menagerie of geometric metaphors, including "webs," "nets," "pyramids," and "hourglasses." The metaphor that ties together the four sets of institutions in Porter's model is the "diamond."[11]

The first element in Porter's model consists of what he calls "factor conditions." By this, he is referring to how businesses and government in a nation jointly find ways to continually reproduce the highest-quality workers (Reich's central theme) and other key inputs into the production process, such as infrastructure and finance. "Demand conditions" constitute a second aspect crucial to a nation's competitiveness. A sizable and growing home market enables at least some firms based there to exploit economies of scale, giving them a cost advantage that may aid them in meeting both foreign and domestic competition. "Sophisticated and demanding buyers" at home are able to force closer attention to quality, foster the adoption of more productive technologies, and offer greater effort at training throughout the supplier chain, thereby acting as a vehicle for diffusing innovation throughout at least some segments of the economy.[12]

For a domestic industry to be globally competitive, it must have globally competitive suppliers—the third aspect in Porter's diamond. The basic proposition is that better suppliers make for more efficient and higher-quality final products. While not all of these suppliers need be domestic (let alone local), such relative proximity may promote the exchange of information, the building of trust, and the pooling of the engineering resources of customer and supplier to jointly design new specialized equipment or final commodities. These mutual advantages extend to complementary industries, which, while not necessarily directly engaged in transactions with one another, may nevertheless gain from proximity, by sharing common inputs.

To the extent that the presence of a critical mass of complementary industries attracts (brings into existence) a larger and more diversified pool of workers and specialized suppliers and subcontractors than would have been available had the customers been more dispersed, there is the possibility of *all* the members of the community enjoying external economies of agglomeration—the greater-than-normal cost savings that, once again, can confer advantage on a country's (or a region's) firms in their global competition. Together, the global companies, their key domestic suppliers, and their most important domestic complementary industries constitute what Porter refers to as "industry clusters."

Porter then draws on the work of Alfred Chandler, the Harvard Business School professor who, together with John Kenneth Galbraith, is probably the most important theorist of the competitive advantages accruing to the large, powerful corporation. Following Chandler, the fourth element in Porter's diamond is "firm strategy, structure, and rivalry." The essential idea here is *rivalry*. Porter sees sharp and pervasive rivalry—among competitors, among suppliers, and even between customers and suppliers—as essential to competitiveness. It is rivalry, he believes, that makes the other three elements of the diamond work. Rivalry is mainly responsible for pressuring firms to invest and innovate, and to promote what he calls "factor creation": the continual upgrading of the quality of labor and other inputs into the production process. Without domestic rivalry of the sort that characterizes relationships among the Japanese *keiretsu* in the automobile and consumer electronics industries (to name but two of Porter's many examples), global companies risk becoming soft, comfortable with negotiated relationships, protective rather than innovative, and therefore incapable of maintaining their competitive edge.

For these reasons, Porter concludes:

> The role of the home nation seems to be as strong or stronger than ever. While globalization of competition might appear to make the nation less important, instead it seems to make it more so. . . . The home base is the nation in which the essential competitive advantages of the enterprise are created and sustained.[13]

Porter then combines this conviction with his other signature recommendation: that firms should focus their energies on their so-called core competencies, while eschewing conglomerate diversification. Taken together with the Weberian-Chandlerian assumption that large, complex business organizations are necessarily organized to at least some extent on the principle of hierarchy, Porter's logic leads him finally to conclude:

> A firm can only have one true home base for each distinct business or segment. If it attempts to have several, it will divide strategic

authority, fragment technology development, and forego the synergistic benefits of concentrating the critical skills. Most important, it will sacrifice the dynamism that arises from true integration in a national "diamond."[14]

In a close, careful, and ultimately critical but not unfriendly reading of Porter's thesis, the economic historian William Lazonick, of the University of Massachusetts at Lowell, argues that Porter has played up the relative importance of rivalry at the expense of drawing evidence from Porter's own case studies on the relative importance of cooperation within the clusters. To be sure, says Lazonick,

> domestic rivalry is an important determinant of enterprise strategies. But the substance of these competitive strategies—specifically whether they entail continuous innovation or cut-throat price-cutting—depends on how and to what extent the enterprises in an industry cooperate with one another. . . . Again and again . . . Porter articulates the centrality of domestic cooperation.[15]

Lazonick then goes on to point to many examples of institutionalized cooperation, all drawn from Porter's own book, ranging from the creation of "government supported technical institutes" to "ties through the scientific community or professional associations." At one point, Porter even mentions "family ties" as part of the glue that holds together some industrial diamonds.

Porter's theory of the diamond is then employed by Lazonick to fashion a critique of Reich's thesis on the declining importance of corporate nationality and, by extension, of a home base whose economic development ought to be an explicit objective of public, private, and public-private policy. If the offshore divisions of American firms have recently been more profitable than their U.S.-located parents, as Reich correctly reports, then perhaps the explanation is that those overseas divisions are embedded within competitively efficient *foreign* diamonds. Certainly, those companies that have most completely acquired a foreign character, such as IBM of Japan or Ford of Europe, are deeply involved in collaborative manufacturing with local suppliers, customers, and competitors in the Far East and Europe, respectively. They may be doing so much better than their headquarters operations within the United States because of the weak to nonexistent development of the clusters at *home*. If that is true, then an appropriate policy response is to start rebuilding those domestic clusters—the home bases of America's global companies.

Lazonick points to still another possible error of interpretation in Reich's thesis. As we saw a moment ago, Reich claims that the disproportionate growth of a high-wage tier of professional-technical workers during the 1980s—his symbolic analysts—proves that global companies are increasingly competing on the basis of their ability to recruit highly skilled labor, from whatever country or region. It is, says Reich, precisely that competition for the professional-technicals that drove up their relative wages, vis-à-vis workers with fewer skills or less formal education.

But as Bluestone and I demonstrated in *The Great U-Turn*, and as Lazonick points out in his critique of Reich, the high incomes of professional and technical employees in the 1980s resulted *not* from their adding value to the productive capabilities of the companies that employed them, so much as from their ability to use their skills to *extract value* from the preexisting productive base of the economy. They did so by informing and managing an unprecedented degree of speculation in real estate, and by realizing capital gains from the buying and selling of existing businesses. All too often, this merger mania was financed by the subsequent stripping of productive assets or by the termination of whole departments of the companies, including those devoted to research and development. Where hostile takeovers were avoided, it was often because one group of symbolic analysts had succeeded in blackmailing (or rather "green-mailing") another group.

All in all, Lazonick concludes,

> the emergence of a powerful market for corporate control, beginning with the conglomeration movement of the 1960s and culminating (for the time being at least) with the gargantuan leveraged buyouts of the mid-1980s, encouraged and enabled top managers to use their existing positions of corporate control to turn from creating to extracting value . . . top managers began to identify with a financial community that sought to live off the rents of previous value creation.[16]

The implication of all this for the "who is 'us'?" debate is that, again according to Lazonick, contrary to Reich, nationality *does* matter in the ownership of enterprises because norms concerning, and constraints on, the relation between ownership and control vary markedly across the advanced capitalist economies. For example, a market for corporate control has arisen in the United States and Britain, but not in Germany and Japan. And if this applies to the transnational corporations, then it surely applies as well to the networks in which these companies constitute the main players.

REGULATING THE GLOBAL COMPANIES AND THEIR NETWORKS

The countries of the world, led by the United States, Japan, and America's principal European allies, have been engaged since the closing days of World War II in negotiating multilateral rules and norms for governing the trade and foreign investment practices of all these nations' companies. General rules for all firms, aimed at the long-run goal of fewer (and, eventually, no) tariff, quota, and other legislated obstacles to "free trade" is the objective of the recently completed General Agreement on Tariffs and Trade (GATT).

Regulating the nontrade behavior of multinational corporations (MNCs) has been far more difficult and erratic a process. Prior to the 1980s, nation-states tended to see cross-border companies as threatening their national sovereignty, by virtue of their ability to move operations beyond the reach of particular governments and to engage in such practices as transfer pricing: shifting recorded transactions among locations so as to minimize exposure to high-tax places, in favor of locales with low tax rates on business.[17] Precisely because the network firms are far more disintegrated than the classic MNCs of the post–World War II era, they have an even greater, albeit imperfect, capability of evading the regulatory reach of individual governments.

Prior to the 1980s, especially in the developing world, governments often tried to reign in the power of the MNCs through licensing or local content requirements, enforced "partnerships" with the public sector, and even outright nationalization. These efforts were generally unsuccessful. Moreover, the ideological pendulum was swinging toward "liberalization," so that, by the 1980s, government attitudes toward the MNCs and the new global production networks organized around them became considerably more accommodating. Certainly as regards the U.S. government's approach to regulating its own MNCs, the Harvard University economist Raymond Vernon sees at most a "very light hand, almost to the point of invisibility."[18] Nevertheless, the ambiguities about public and private power remain, as otherwise sovereign states are forced to recognize that, in Vernon's words, "so many of their enterprises are conduits through which other sovereigns exert their influence."[19]

To be sure, efforts have been mounted by the United Nations and other bodies at the multinational level to develop codes of conduct and, in the case of the European Community, even binding rules with respect to the labor relations, environmental, health, and safety practices of the MNCs and their production networks. These attempts have met with uneven results, at best, and have often either not been enforced or have proved unenforceable.[20] The flexible boundaries of the production networks I have been examining make them that much more difficult to subject to orderly multinational regulation.

The challenge is to design new approaches to regulation of the alliances and the other network forms that do not inadvertently throw out the baby with the bathwater. As University of California–Berkeley management strategist David J. Teece argues, alliances are now quite essential to facilitating innovation. And yet, especially in the United States, there is a deeply ingrained public prejudice against all forms of interfirm cooperation, going back to the origins of the antitrust movement in the last quarter of the nineteenth century. Says Teece:

> This has manifested itself in the absence of inter-agency coordination in the federal government, with science and technology policy appearing to be weak and uncoordinated, and a reluctance in some quarters to admit and encourage the private sector to forge the inter-firm agreements, alliances and consortia necessary to develop and commercialize new technologies like high-[definition] television or superconductors. A key reason for this is the shadow that [standard economic] thinking casts over antitrust policy. It renders antitrust policy hostile to many forms of beneficial collaboration, because of fear that such arrangements are a subterfuge for cartelization and other forms of anticompetitive behavior. Until a greater understanding emerges as to the organizational requirements of the innovation process and antitrust uncertainties are removed, antitrust policy in the United States is likely to remain a barrier to innovation because it has the capacity to stifle beneficial forms of interfirm cooperation. Antitrust policy, which has always been rather hostile to horizontal agreements, must accommodate this feature of capitalist economies if it is to promote enterprise performance and economic welfare.[21]

Teece goes on to recommend a series of modifications to U.S. antitrust law, all of which add up to encouraging the courts and administrative agencies to err on the side of facilitating innovation by permitting the formation of strategic alliances and other types of network organization.[22]

GOVERNMENT INDUSTRIAL/TECHNOLOGY POLICY AND INTERFIRM ALLIANCES

The influential concept of a "postindustrial society" was always misleading. Like its foreign competitors, the United States is becoming an increasingly complex *industrial* society. The linkages between the manufacturing and nonmanufacturing sectors, between high-tech and low-tech, and between what is domestic and what is foreign are becoming more and more intricate. Recognition of these complexities—and of the possibility that they are simply overwhelming the capacity of the private market to anticipate and incorporate entirely within the framework of decentralized price-mediated supply and demand—is what has brought the debate over industrial and technology policy (ITP) back onto the tables of so many corporate board and congressional hearing rooms. Since the presidential election of 1992, the federal government has actively joined in the rethinking of industrial policy.

That government has in the past—and must in the future—play a major role in promoting technological change and, with it, the possibility of productivity growth, is argued with special force by Charles Ferguson, a former M.I.T. researcher and now a prominent management consultant. In a recent book, Ferguson and his coauthor Charles Morris write:

> Only government can create an environment that makes it possible for businesses to invest and succeed over the long term in high technology. Only government can remove the barriers to technological competitiveness that government has raised. Only government can support the precompetitive basic research capabilities that have fueled America's technological successes. And only government can act on the scale that may occasionally be necessary to counter national cartels controlling critical technology components.[23]

But presented with the fact of the proliferation of strategic alliances and the many other forms of interfirm networking within the private sector, especially when these cross national borders, advocates of specifically *national* technology and industrial policies find themselves confronting a whole new class of vexing problems. Consider, for example, what happened several years ago when Advanced Micro Devices of Silicon Valley and Japan's Fujitsu announced the formation of a strategic alliance to jointly develop flash memories (chips that do not lose the information stored on them when the power is shut off). Such devices could someday replace the hard disks upon which all of us now rely for data and program storage. The initial development work will take place in Japan. On the same day in July 1992, IBM announced that it was forming a new alliance with Japan's Toshiba and German's Siemens to develop a new generation of memory chips. At least for starters, the research will be conducted in New York State.

Both announcements reopened an old policy question, which also intersects the "who is 'us'?" debate: Why should the U.S. government continue to support such consortia of big private microelectronics companies as Sematech, the semiconductor R&D collaborative located in Austin, Texas, when so many of the member companies belonging to these consortia are independently forming alliances with foreign firms? Do not these cross-border alliances now being created in biotech, auto parts, and even software prove that support from the U.S. government is not really needed? And besides, what does the American *public* get out of it?

The existence of these technology alliances struck by individual corporations and the need for continuing national ITPs are perfectly compatible. Let us assume that a strong home base remains an essential precondition even for multinational corporations, in order (following Porter) to secure domestic markets, to develop strong relations with dependable suppliers, and to move down the learning curve—to become more familiar with the operating characteristics of new technologies. Stripped of the rhetoric, shoring up these home bases is what technology and industrial policy is all about. From training and technical assistance to the provision of expensive infrastructure and the encouragement of collaboration among rivals, ITP aims at increasing the technical and organizational capabilities of domestic firms, in order to enhance long-run productivity, profitability, and job creation. To the extent that all of this makes the Americans more attractive potential suppliers to, customers of, or technology partners with foreign firms, then domestic ITP may actually *promote* cross-border alliances.

What about letting the foreign partners of American companies have access to domestically subsidized new technology? First, it is incorrect to assume simply that an American member of one or another U.S. (or subfederal) government-sponsored consortium necessarily transfers whatever it learns from its domestic collaborations to its foreign allies. Big companies—especially those with a long history of contracting to the Defense Department—are well practiced at compartmentalizing proprietary information.

But even if the collaboratively generated new knowledge leaks like a sieve from the domestic consortium into the cross-border alliances— something that most economists would *wish* to occur—the public purpose that justifies the domestic policy is not necessarily compromised. "Precompetitive" R&D collaborations are justified to begin with on the theory that the cost of and risk to any individual firm of investing in innovative lines of research on new products or production processes have become so great, in so many areas, as to slow the overall rate of market-driven technological progress. What collaboratives such as Sematech do is to socialize part of those costs and risks. So long as the principal beneficiaries of the public investment in any particular nation's (or region's) technology consortia are domestic firms and workers who live here— whatever the color of their passports—why worry about the indirect involvement of such technically prodigious companies as a Siemens or a Fujitsu?

Publicly supported high-tech consortia located elsewhere, such as the European Community's ESPRIT and JESSI, already permit U.S. corporations to participate in their activities, either directly (as with IBM of Europe), or indirectly, via the foreigners' alliances with EC member firms. The quid pro quo is that the foreign (in this case, the American) companies actively involve themselves in research, development, or production *within Europe*. We could do the same in this country, extending to the overseas partners of the U.S. members of Sematech, or of domestic companies working with one or another of the national laboratories, the privilege of associated membership in these government-funded consortia, so long as the foreigners conduct some negotiated share of their own project-related work inside *this* country.

PLANNING FOR CONVERSION FROM MILITARY PRODUCTION

What to do with a large defense industrial base in the post–cold war era has become a major concern for economic development planners everywhere. Production networks whose central actors are the major prime contractors to the Defense Department encompass a large proportion of the country's most technically adept companies.[24] Effective planning for conversion is inconceivable unless public policy can find ways to engage these networks in the process of change.

About two million people served in the armed forces in 1990, a million civilians worked for the Pentagon, and at least four million workers were employed by defense contractors and subcontractors.[25] Moreover, unlike the situation following World War II, there is no huge pent-up consumer demand for goods and services, waiting to be filled by converting (say) tank producers back to being auto makers. And to add insult to injury, American companies now face a degree of foreign competition in every field, from clothing to aircraft to computers, that they could never have imagined in their wildest dreams back in 1946.

Yet conversion will surely proceed in the years ahead. There remain a host of national needs, from replacement of last generation's telecommunications equipment and links, and the redesign of the outdated air traffic control systems in many of the nation's airports, to the design and manufacture of high-speed rail systems. And the companies that have been the most dependent on military contracts desperately need to find new markets. Identifying new potential markets, especially those in which the federal government might play an important role as lead customer, is only one of the challenges facing the National Institute of Standards and Technology in the U.S. Department of Commerce. The Defense Department's Advanced Research Projects Agency (DARPA, recently renamed ARPA to accentuate its broadened mission) is also playing this role, liberated by the Clinton administration from its exclusively military orientation.

The industrial engineer John Ullmann identifies five categories of organizations for which new civilian (or, at any rate, nonmilitary) markets must be sought by conversion planners. The least problematic are the many companies that never sold more than a small fraction of their output to the Pentagon or its prime contractors, as just another part of their general business activities. A second category consists of the more specialized subcontractors. The conventional wisdom has it that these have over the years become so overwhelmingly dependent on Defense Department procurement that they now face a mammoth crisis. But Maryellen Kelley and Todd Watkins, the first researchers to obtain detailed data on the smallest of subcontractors as well as the big firms that anchor these defense production networks, show that *most* defense suppliers are already "dual use," selling to private markets as well as to the Pentagon or its prime contractors.[26]

Ullmann's third category includes those major prime contractors, such as General Electric, General Motors, and IBM, that have long operated parallel military and civilian divisions. He expresses grave concern that such dual use producers will not easily be able to tear down the walls they have (he believes) erected over the past four decades, separating these activities. Kelley and Watkins think that the extent of such separation has been exaggerated; if fire walls existed in the past, most of the companies they have studied spent much of the 1980s forging closer links across these boundaries. They are (at least on this score) more optimistic than Ullman about the chances that this group of companies can be successfully weaned off the Defense Department.

The fourth category in Ullmann's typology is a more serious concern. These are the big engineering firms that produce almost entirely for the military: Grumman, General Dynamics, Lockheed, and the like. What will they do? What will they make, and for which markets? What will happen to the other, often smaller firms with which they are allied? There is much concern that these are the companies that may most aggressively seek foreign markets for armaments, rather than confront the wrenching difficulties associated with converting to civilian production.

Finally, Ullmann lists military bases and their personnel. Here and there, conversion projects for particular bases have been proposed (for example, Massachusetts will build a new prison on the site of a former army base). But for the most part, the Office of Economic Adjustment in the Defense Department is concerned only with the orderly mothballing of these facilities, or the conducting of land sales to local governments and private developers.

Ullmann might have added a sixth category to his list, one that is now occupying much of the energy of the Clinton administration. I refer to the network of national laboratories, whose most famous examples are Oak Ridge in Tennessee, Los Alamos and Sandia in New Mexico, and Lawrence Livermore in California. The national labs, which constitute a collection of the nation's leading scientists and engineers, have long been dedicated to military-related R&D. There is currently a mad scramble under way, both within the individual labs and in the federal government, to find new missions for those labs—or departments within them—capable of collaborating with industry on new civilian projects, such as the electric car, new materials, and more advanced methods of computation. Already, several of the labs have entered into such cooperative planning agreements with private companies; more are expected in the future.[27]

The M.I.T. engineering professor Leon Trilling would like to see the Air Force laboratories, such as the one at Wright Patterson base in Dayton, Ohio, convert to the research and development of civilian aircraft technologies. His hope is that this might relieve the necessity for even such giant companies as Boeing and GE to look abroad for technology partners. All the evidence I have amassed in this book suggests that the practice of cross-border technology partnerships is already too widespread to reverse. But the exhortation to convert the labs in order to help in converting the industry is right on target.[28]

MANUFACTURING MODERNIZATION

Much is at stake in the effort to upgrade the technical capabilities of individual small and medium-sized enterprises (SMEs), if only so that they might become qualified suppliers to the hub firms at the center of the global networks. As noted in chapter 3, it is now well documented that, everywhere in the industrialized world, the largest companies and plants are far more likely to have adopted and deployed computer-controlled factory automation and other new technologies than have the SMEs. That is why, over the past decade, so many countries—with the United States lagging behind in both the level and quality of effort—have launched or expanded well-funded government industrial extension programs, designed to bring *appropriate*, rather than necessarily state-of-the-art, technology to the attention of those smaller private companies that have been lagging behind.

That the SMEs in the United States have been such laggards is a serious matter for American industrial competitiveness. These are the companies that make up the vast majority of suppliers and subcontractors to the multinational corporations doing battle in an increasingly demanding global marketplace. American SMEs do not install as many high-tech tools or systems as their generally larger customers (whether American or foreign owned). They have also shown themselves to be more reluctant to introduce "soft" technologies—methods of reorganizing the workplace and of linking different establishments in the supplier chain to one another. These organizational techniques include statistical process control, just-in-time movement of parts and supplies, and design for manufacturability.[29]

Moreover, this is the segment of American industry that either cannot afford to do much training or cannot accept the risk of investing in the training of employees who then jump ship. According to the nonprofit National Center on Education and the Economy, about 90 percent of all private sector training expenditures in the United States are made by one-half of one percent of U.S. companies—and two-thirds of *that* goes to college-educated employees, who arguably need it the least.[30]

To be sure, a number of federal and state programs have been launched over the last decade to promote research and development. From Sematech and MCC, the Advanced Manufacturing Research Facility, the National Center for Manufacturing Sciences, and the Department of Defense's ManTech and ARPA programs—the grand public-private partnerships between the semiconductor, microelectronics, machining, and materials corporations in the private sector and the U.S. government—to such pioneering state-level technology development efforts as Ohio's Thomas Edison Program and Pennsylvania's Ben Franklin Partnership, we seem to be making some headway on getting new manufacturing technologies from the laboratory onto the shelf. But what about getting them off the shelf and onto the shop floor?

Until recently, this was probably the greatest single shortcoming in America's industrial policy arsenal. True, some states had started programs of their own that were explicitly addressed to the challenge of diffusing appropriate technology to the SMEs. Experts in the field have given high marks to Massachusetts' Industrial Services Program, Georgia's industrial extension activities, Maryland's Technology Extension Service, the Pennsylvania Industrial Resource Centers, and (until it was abolished by a newly elected penny-wise, pound-foolish governor) the Michigan Modernization Service. But it has also been clear that these programs are drastically underfunded and inadequately coordinated, and that the variations in quality from one site to another *within* each state program leave much to be desired.

In 1988, Congress sought to remedy the situation. Included within the Omnibus Trade and Competitiveness Act was the redesignation of the old National Bureau of Standards as the new National Institute of

Standards and Technology (NIST). Even before the ascendancy of Clinton to the presidency in 1992, NIST had created seven regional Manufacturing Technology Centers (MTCs) around the country, in New York, Ohio, South Carolina, Michigan, Minnesota, California, and Kansas. More are now under construction. These demonstration and information centers provide information and education explicitly targeted to SMEs. They maintain operating tools and equipment systems that visiting owners and managers can inspect. They help firms evaluate their needs. And they encourage their regional clients to put more effort into training. Moreover, NIST's centers have a small budget for subsidizing state programs, through the federal State Technology Extension Program, or STEP.[31]

So far, so good. But compared to the competition, there are still miles to go. Shapira's field visits to Japan taught him that America's most significant competitor—approximately the size of California—now has at least 169 regional centers (*kohsetsushi*) engaged in testing, technical assistance, training, and technology diffusion aimed at SMEs. While the centers are managed by the governments of their particular prefectures, Tokyo establishes guidelines, provides some funding, and operates a national system for qualifying and registering the consultants on whom the centers draw in working with their client small firms. According to the U.S. Congressional Office of Technology Assessment, the national government of Japan spends "at least 20 times" as much on assistance to SMEs as does the U.S. government.[32]

In 1914, the U.S. government created the agricultural extension service—arguably the single most successful (and widely copied) example of government-business planning in American history. Over the course of the century, literally millions of small farmers were taught how to apply new tools, seeds, and pesticides to improve their yields and, in the process, to make a decent living for themselves and their families. The United States established an elaborate network of research stations, outreach agents, land-grant university laboratories, and public-private partnerships between local, county, state, and federal agencies and private companies.

Certainly, the challenges facing those committed to spreading the gospel of appropriate manufacturing technology are far more complex than anything envisioned by those legions of "ag extension" agents. For one thing, the small farmers tended to operate in splendid isolation from one another. But these days, as noted throughout this book, no modern manufacturers who want to be competitive can ignore the pressures to link their technical and personnel management systems to those of their principal customers and suppliers.

REGIONAL, STATE, AND LOCAL ECONOMIC DEVELOPMENT IN A WORLD OF NETWORKED PRODUCTION

In recent years, within the field of regional economic development, new initiatives are emerging in many countries—including the United States. Policy researchers, planners, and managers are becoming engaged in building local business networks, instead of focusing exclusively on trying to attract existing plants, stores, or shops away from a competing jurisdiction. Under the umbrella of the manufacturing modernization movement, policy makers are constructing public-private partnerships designed to transfer modern technology from the laboratory—or from the generally more technically adept large firms in a region—into the hands of the SMEs, which, thus equipped, might become more attractive suppliers to the big firms and business partners to one another.

The central theme of the first third of this book was that claims about SMEs as the most important job generators and technology leaders in modern industrialized economies have been exaggerated, at best, as have assertions about the competitive superiority of small firm–led industrial districts. Instead, I have argued that such formations must be seen as either special cases of, or as segments within, larger, more spatially extensive production networks, governed by the general principle of *concentration without centralization.* Seen either as stand-alone entrepreneurial enterprises or as collections of locally oriented, socially embedded "competitive cooperators," SMEs are an important component of any region's economic base. It is just that we should not mistake them for being the *drivers* of the economic development process.

At the beginning of this chapter, I quoted Gereffi and Hamilton on their characterization of networked production as constituting a third shift in the historical development of business organization in industrial capitalism. Recently, a band of energetic theorists of, and public policy advocates for, the deliberate care and feeding of localized networks of SMEs has proposed an analogous concept. Doug Ross, currently the United States assistant secretary of labor for employment and training of the Washington-based Corporation for Enterprise Development, formerly executive director (and, before that, the director of economic development for the state of Michigan), refers to the advocacy of networking among SMEs as a "third wave" of local and regional development policy thinking, succeeding the earlier preferences for recruitment—some would say "pirating"—of existing facilities from some other jurisdiction and building up a locality's human capital and physical infrastructure.[33] Obviously, these approaches are not mutually exclusive; indeed, they are surely interdependent.

Many of the principles of the small firm–led industrial district—which we now understand to be a highly localized special case of the network—are set out in a series of books written or coauthored by Stuart Rosenfeld.[34] Rosenfeld is an experienced consultant to state and local governments and to private foundations who, during the 1980s, organized a number of southern governors (including President Clinton, a former governor of Arkansas) to create the influential Southern Technology Council. From this wealth of experience, Rosenfeld has come to understand that individual small businesses lack the scale and resources for supporting a full-time staff of technical specialists and administrators. Moreover, most existing government skill-training programs do not serve small firms (especially small *manufacturing* firms) very well, and private vendors of training services charge too much money. SMEs have insufficient knowledge of conditions in all but the most local markets and are generally quite ignorant about long-term trends with respect to markets, technology, and government regulations.

To overcome these structural obstacles to their more effective performance, SMEs in several states have begun to form localized production networks of their own, influenced by the European models, especially those of the Third Italy and Denmark. Consistent with the theory of the production network, the basic unit of analysis is not a particular *industry* (for example, footwear) but rather a *filiere*, an input-output chain, consisting of (say) footwear plus its backward-linked (upstream) suppliers and its principal forward-linked (downstream) customers. Thus, for example:

> rural manufacturers in states as diverse as Oregon, Arkansas, North Carolina and Florida are creating new alliances and tighter business relationships for a variety of purposes: process development, marketing, training and equipment purchases . . . to create incentives for collaboration, usually as challenge grants for group activities [and] supporting individuals or organizations in a community to act as network facilitators and help organize collaborative efforts.[35]

Some states are developing industrial extension services to assist these local networks, while others are trying to link groups of SMEs with trade associations, technical schools, community colleges, and even on occasion with a big regional university.[36] Even union activists have been forming collaborative networks, within and across regions of the United States.[37]

In all such ventures, a key process is the development of *trust* among the generally smaller companies in the network. As a breed, small business owners are notoriously uncooperative, fiercely independent souls. Thus, in his work with local manufacturing networks in Massachusetts, Pennsylvania, and elsewhere (as well as abroad), the M.I.T. social scientist Charles Sabel coaxes groups of owners to "study trust" by collectively examining the structure of and problems within each other's industries, "to connive in a form of self-distraction which would allow them to catch sight of new possibilities," and, in so doing, to collectively identify or redefine a set of common interests. In other words, what had been atomistic competitors transform themselves (with the help of a facilitator) into a group of *stakeholders*.

C. Richard Hatch, the American architect and long-time advocate of such SME-led production networks, refers to the Emilia-Romagna region of north-central Italy as the "epicenter of the network system."[38] As discussed in earlier chapters, Italian scholars of the small firm districts themselves emphasize the crucial role of the government—that is, of public policy—in reinforcing (but not creating) the conditions that facilitate interfirm cooperation within these regions. The Bolognese industrial economist Patrizio Bianchi writes about (and actively helps to design) what he calls "collective agents," or "service centers."[39] In both the "red" (Communist) region of Emilia and the "white" (Christian Democratic, conservative) region of the Veneto, incentives and support for interfirm collaboration in production were already in place in the 1960s and underwent continued development for several decades thereafter. Local governments provided physical spaces for new firm start-ups ("incubators"), complete with shared administrative and computing services, information about technology and market opportunities, and high-quality technical and managerial assistance—what the Italian writers, following Sabastiano Brusco, call "real" services (to distinguish them from financial aid, per se).[40] A good example is CITER, the Emilian consortium including the regional government and the trade association of textile producers centered in Modena.

Nor have the regional service centers had to depend entirely on local public and private resources. In the late 1970s, the European Community began financially subsidizing such regional activities, all over the continent, but especially in southern Europe and Ireland, in connection with the EC Social Fund, the Regional Fund, and other programs earmarked for regional social and economic development. The objectives were to reduce the incentives to business to "shop around" the continent for the least costly industrial sites and to encourage the emergence of innovative clusters of SMEs. These, and many other continentwide programs, were substantially beefed up in 1989.

While the EC subsidies are still forthcoming today, the general fiscal crisis that Europe has suffered since the late 1980s has all but decimated many of these government initiatives. The Italian social planners are properly worried that, without this public sector infrastructure and facilitation, the capacity of the small private firms to continue to negotiate the precarious knife edge between cooperation and competition will deteriorate.

COORDINATING FEDERAL AND STATE/LOCAL APPROACHES TO WORKING WITH PRODUCTION NETWORKS

David Osborne is a freelance journalist whose book *Laboratories of Democracy*[41] had an enormous influence on Bill Clinton, long before the latter campaigned for the American presidency. Osborne's collection of case studies of interesting, creative, state-level economic development programs plays on a remark made early in the twentieth century by Supreme Court Justice Louis Brandeis: "It is one of the happy incidents of the federal system that a single courageous State may, if its citizens choose, serve as a laboratory; and try novel social and economic experiments without risk to the rest of the country."[42] It is certainly true that, historically, many social welfare and other programs were first introduced at the state level and only gradually worked their way up to the Congress, often because the private sector eventually found the patchwork of different laws and regulations among the states unacceptably onerous to navigate.

Nevertheless, argues the Pennsylvania State University economist Irwin Feller, the proliferation of state (and of federal-state cooperative) programs in technology policy—especially as these attempt to promote interfirm production or innovation networks—has now far outpaced our critical knowledge of what works and what does not.[43] Feller considers most state programs, including the famous ones such as Pennsylvania's Ben Franklin Partnership and its Industrial Resource Centers, Ohio's Thomas Edison Program, and New Jersey's Advanced Technology Centers to be insufficiently evaluated, not because evaluation is impossible but because the theoretical conceptualization of most of these programs is so imprecise. Different programs are implicitly addressed to very different objectives: promoting business start-ups, diffusing knowledge, encouraging collaboration, or rescuing mature industries in depressed communities. Conversion from military to civilian production is the most recent addition to the list. It is, argues Feller, premature for the federal government to begin "picking winners and losers" or to be looking for "models" among the many state and local initiatives until much more study and evaluation of this potpourri of programs has been completed.

A good example of a popular local economic development program that has only recently been adequately evaluated is the creation of so-called science and technology "parks." These are the high-tech equivalent of the standard industrial parks (or "estates") that governments all over the world have, since the 1950s, constructed to attract and house new business. The underlying principle is clear enough: the developer, typically a government agency or a nonprofit one or an agency's private sector contractor constructs a facility for housing a number of individual laboratories, plants, offices, or warehouses. The developer provides power, sewer, and water services, and perhaps additional infrastructure. R&D operations also require support services with respect to computing, sales, and distribution. Economies of scale in service provision, combined with land rent subsidies ("write-downs") and the attraction of having complementary companies nearby will, it is hoped, make such locations more desirable for companies. Once built, in the immortal words of the romantic hero in the popular American film *Field of Dreams*, "they will come."

Only all too often, they do not. Detailed evaluations of the ability of the planners of such science parks to attract new business, to create jobs in a region, and to promote technological learning (or sharing) have been conducted in the United States and the United Kingdom.[44] The researchers have come to very much the same conclusions on both sides of the Atlantic. The success rate among such parks is, on average, very low. The oldest of the American high-tech parks—Research Triangle in North Carolina, Stanford Industrial Park in California, and the University of Utah Research Park—have retained their advantages, but few newcomers have been able to attract a critical mass of tenant firms.

It may simply be too late for newer areas to enter the field. There are only so many high-tech agglomerations that can sustain a critical mass of activity. The expensive efforts of Cambridge University in England to build and maintain a high-tech park linked to that university's prestigious science and engineering departments have been especially disappointing. Yet development of new parks continues, as state and local governments try to build their own fields of dreams. Feller points to these as especially egregious examples of uncoordinated, socially wasteful local public expenditures, part of the modern equivalent of old-fashioned "smokestack chasing," and—as these recent evaluations show—likely to be just as fruitless.

Feller's critique is brusque—and long overdue. It is one thing to make lists of "innovative" programs and quite another to reach conclusions about what works, under what conditions. If only to impose some degree of coherence on a federal-state "system" that threatens to get totally out of control, Shapira, Kelley, and others advocate making NIST the centerpiece of both program design and evaluation research, although Shapira continues to urge that the actual delivery of services be as localized as possible, while Kelley (consistent with Feller's concerns about the need for focus and well-defined goals) would give considerably more power to NIST, itself. Clearly, much additional thinking is needed about the problem of how to coordinate national and local programs and policies affecting the *location* and the *habitat* of the firms that make up local, regional, and global production networks.

THE DARK SIDE, REDUX

In this concluding chapter to a long inquiry, I have been examining questions that will have to be considered and reconsidered by governments seeking to promote and manage national, regional, and local economic development in a world likely to be increasingly populated by network forms of business organization. I have presented a thicket of theoretical arguments, case studies, and statistics showing that the network form—what the University of California–Berkeley management professor Michael Gerlach calls "alliance capitalism"[45]—is likely to be with us for many years to come. It is a social arrangement with which we are going to have to live. But can we live with the dark side of flexible, networked production? Should we?

In the short run, the challenge is to find ways to ensure equal hourly pay and access to full benefits for the growing fraction of the workforce treated by managers within both the large and the small firms as contingent labor. This includes part-time, seasonal, and temporary employees.[46] This may be achieved partly through stronger enforcement of existing labor laws but will surely also require new legislation, not easily won in the present conservative political climate. Ultimately, I think, more comprehensive coverage of wages, hours, and benefits protection will only come when, and as, workers in the periphery of the economy become better and more vocally organized, whether into trade unions as we have known them, or into other forms.[47]

In support of such efforts, better intelligence is needed, through revisions to the national and international statistics-gathering programs that report on the magnitude and composition of employment. Compiling numbers on the extent and the sectoral and regional composition of contingent work arrangements should not have to depend so heavily on the heroic efforts of private researchers, as is now the case.

In the long run, what is at issue is nothing less than the matter of whether we are committed to pursuing what, in chapter 9, following Chris Freeman, I called the high road to economic growth and development. As I argued there, the key to attaining the high road is for the bigger firms to help upgrade the technical capabilities of their generally smaller suppliers. The possibilities for prosperity lie in a joining of the public and private sectors to provide what I call the "three Ts": *technology*, *training*, and *technical assistance*. As for the new network firms, the objective is to make lean organizations rather less mean.

Too many firms have been following a low-road path, especially in the United States. Even such generally progressive firms as the Ford Motor Company can make us wonder, when we learn that the corporation has been sending executives "to watch workers at McDonald's flip hamburgers and cycle customers past drive-up service windows."[48] On the other hand, Ford invested heavily in new plants, equipment, and training throughout the 1980s—the exception in American industry. As noted in chapter 9, annual growth in industrial capacity declined steadily between 1981 and 1993. And the Census Bureau reports that average real wages continue to stagnate, while earnings inequality continues its long, slow drift upward. Yet Congress continues to oppose the pursuit of a long-term growth strategy led by public investment, of the sort proposed by Clinton in the earliest days of his administration.

But when public investment strikes out, that does not mean that the underlying problems that led to advocating it magically disappear. However it is paid for, whatever the public-private mix, you cannot grow a company or an economy without the three Ts. Even—perhaps especially—in a world of network forms of industrial organization.

11

POSTSCRIPT:

REASSESSING *LEAN AND MEAN* ON

THE EVE OF THE NEW MILLENIUM

I think it fair to say that when the first edition of *Lean and Mean* appeared in 1994, most of its major and minor themes were well received, or at least widely acknowledged, in North America, Europe, and the Far East. Some of the claims I put forth in the book were novel, others were derivative, but my analysis was intentionally revisionist.

For example, and in retrospect most famously (which is why the editors at The Guilford Press have chosen to rework the book's original subtitle), I challenged the conventional wisdom about small firms being the driving force behind job creation and innovation. I found most mainstream statements on this topic to be imprecise, based on bad data, and all too often full of ideologically based wishful thinking. The undoubted trend toward decentralization of production—across regions and countries, and even expressed in the vertical disintegration within companies popularly known as the "flattening of hierarchies"—was being governed, managed, coordinated (I argued) by continued, indeed increasingly, powerful concentrated business organizations. I coined the rather infelicitous principle of "concentration without centralization" in an attempt to capture this seeming paradox.

I also discovered and reported on examples of local small-firm industrial districts in which closeness was sometimes suffocating rather than stimulating innovativeness, and others in which large transnational corporations seemed to be absorbing the best of the small local firms into their own cross-border orbits. This last trend appeared to threaten the local orientation of those small firms, even as it more tightly exposed them to the contradictory forces of national and international competition. In view of this evolution, I raised the possibility that incorporation into the global economy in this way could eventually transform the nature of the districts themselves, even as they remained vital regional economies.

In the first edition, I also elaborated upon the mounting, but then not yet ubiquitous, expressions of concern emanating from scholars around the world about a dark side to the emerging regime of what has (following the French theorists of *regulation*) come to be called "flexible accumulation." The dark side refers to the growing gap between the highest and the lowest paid workers and their families; the spread of more or less chronic contingent (part-time, temporary contract, increasingly volatile) employment arrangements; and a general decline in income and job security confronting all but the most highly educated (or otherwise sheltered) employees.

All of these observations were set within the context of an analysis of what I (at least) always thought of as the *main* theme of the book: the remarkably rapid and globally pervasive growth of network forms of interfirm (and other interorganizational) relationships. To me, this was the most dramatic and consequential aspect of "the changing landscape of corporate power in the age of flexibility"—this book's original subtitle. We were, I argued, witnessing a revolution in the spread of business-to-business and business-to-government webs and alliances, crossing old political borders and sectoral boundaries, and incorporating growing numbers of previously stand-alone institutions, from hospitals to colleges to research laboratories. Other scholars, especially those from the domains of organizational sociology and business strategy, had said some of this before I did (and I willingly became, and have tried to honestly acknowledge my status as, an apprentice to them). But perhaps I might claim to have taken the theory farther than most in proffering these transformations as constituting more than simply an alteration of form, but rather marking a true sea change in the processes of capitalist world development.[1]

It was gratifying, if sometimes sobering, to see my book, warts and all, become the object of close scrutiny in special symposia held at the annual professional conventions of Anglo-American sociologists and geographers. Two different collections of business economists and strategy theorists took the trouble to devote sections of their journals and magazines to inviting

sometimes heated, always lively debates about *Lean and Mean*.[2] There
were lots of reviews, in both academic and mainstream commercial media, generally rather favorable. So far, the original edition has been translated into four languages.

Almost five years have passed since the bulk of my research was completed. Recently, I have been engaged in extending the analysis of network
forms, and what they enable people and organizations to do differently, to
a variety of political-economic scales, especially that at which many
African-American and Latino neighborhood and community-based socioe-
conomic development groups operate.[3] This has challenged me to critically rethink the value of this whole approach. In retrospect, how well do I
think the original analyses in *Lean and Mean* have held up? Where did
my aim go most off target? What did I miss? What light do recent developments, and newly published research, shed on my two organically related
central theses: that the shift toward flexible production systems within and
across companies, countries, and sectors is increasingly characterized by
networks connecting small and big firms, substantially anchored by alliances among powerful, concentrated producers and financiers; but with
associated labor processes in which jobs are increasingly fragmented, work
more insecure, and pay more variable and uncertain?

Then consider the most salient policy developments of this last decade
of the twentieth century: continued devolution of public authority from
the federal government to state governments, the decision to effectively
force millions of poor welfare recipients into the labor markets of their
cities, the further privatization of the provision of infrastructure and formerly publicly provided services, a substantial retreat from an activist national technology policy, and a general nervousness at the highest levels of
government and on Wall Street about the dangers of permitting (let alone
deliberately stimulating) more rapid economic growth. How is this so-
called neoliberal policy constellation, now dominant to one degree or another across the industrialized world, likely to affect the growing inequality, insecurity, volatility, and social polarization that nearly everyone now
agrees are accompanying global economic integration on the eve of the
new millennium? Can I now say something more about what an enlight-
ened, progressive policy stance in this context might be, compared to what
I addressed in chapter 10 of the original edition?

These are the matters I have been invited to re-engage, in this postscript
to the new Guilford edition of *Lean and Mean*. To explore them in depth
would require writing another whole book.[4] What follows, then—in no

special order—are some ruminations, a bit of additional evidence, and, as before, many more questions than answers.

THE STUBBORN MYTH ABOUT SMALL BUSINESS AS THE ENGINE OF ECONOMIC DEVELOPMENT

During the policy debates about economic development and the rise of the theory of flexible specialization during the 1980s, I became discontented with the extent to which the proposition was being repeated that new developments in competition and technology were systematically privileging smaller forms of business enterprise. But these were seriously theorized arguments, so in my book I tried to accord them the respect they were due, in the course of offering my critique.

Less deserving of respectful treatment were the seemingly endless repetitions of the line, originating with David Birch, that "nearly x percent of all new jobs are created by [small, medium-sized, 'gazelle'-like—take your pick] firms," a statement in which x was sometimes said to be "50," sometimes "75," sometimes "90," sometimes even "virtually 100." Policymakers, including governors and mayors, repeated these assertions, and drew upon them as "evidence" that policies should systematically favor these small-business "job generators." Much of this was, of course, traditional rank ideology. But much of it really did take the "research" as providing an objective rationale or basis for such policy designs as across-the-board tax breaks or environmental pollution variances for small businesses. This, even though a number of first-rate technical critiques of the Small Business Administration–based data and Birch-like analyses have been published, of late, that come to the same conclusions that I did.[5]

Even confronted with the revived merger and acquisition wave of the 1990s, the believers that small is both inherently beautiful and on the rise have stuck to their guns. (Item: with the recent merger of the Boeing Company and McDonnell Douglas, there are now effectively only two producers of commercial aircraft in the entire world, Boeing-McDonnell and the European Airbus Industrie Consortium. Another recent aerospace merger, between Lockheed and Martin Marietta, is devoting itself entirely to manufacturing planes for the military and to managing a variety of what used to be publicly provided services, as recipients of government privatization contracts). Concentration is even on the rise in the traditionally fragmented retail sector, both among discount chains like K-Mart and among the specialty retailers.[6]

Many policymakers who casually use these numbers and repeat the conventional wisdom like a mantra fail to distinguish between individual sites of business—that is, establishments—and whole companies, or enterprises. A majority of all businesses have only one site of operation; they are truly small. But the minority that do have multiple facilities in different locations account for a disproportionate share of all jobs, sales, and profits, in both manufacturing and the rest of the economy. I presented a detailed statistical accounting of this distinction in table 3.1 of chapter 3. When this book was first written, Census Bureau data were available only through 1987. From less complete but more recent official data for 1992, I have prepared an abbreviated version of that same table, for entire enterprises—whole companies—to make this point yet again. The findings are contained here in table 11.1.

For the economy as a whole, firms with fewer than 100 employees (including the many sole proprietors who have no one on their payrolls at all) accounted for fully 98.4 percent of all companies in the United States in 1992. But that mass of businesses was responsible for only 38.6 percent of all jobs, and received just 34.3 percent of all sales revenues. At the other end of the size distribution, companies that were so large as to employ 500 or more workers made up only three-tenths of 1 percent of all businesses, but that tiny fraction of the business community employed 47 percent of all workers and raked in 53 percent of all sales dollars. As table 11.1 shows, the gaps were even more vast in the manufacturing and financial sectors.

Let me spell this out. In the finance sector (which, in the official definition of the Census Bureau, includes insurance and real estate, two indus-

	Firms With 100 Workers			Firms With < 500 Workers			Firms With 500+ Workers		
	Share of All Firms (%)	Share of Employment (%)	Share of Sales (%)	Share of All Firms (%)	Share of Employment (%)	Share of Sales (%)	Share of All Firms (%)	Share of Employment (%)	Share of Sales (%)
All private industry	98.4	38.6	34.3	99.7	53.0	47.0	0.3	47.0	53.0
Manu- facturing	93.7	22.0	14.6	98.6	38.2	27.9	1.3	61.8	72.1
Finance	97.8	30.8	17.9	99.3	42.9	27.7	0.7	57.1	72.3

Source: Burea of the Census, U.S. Department of Commerce, *Statistics of U.S. Businesses: 1992* (Washington, DC: U.S. Government Printing Office, 1994).

TABLE 11.1　ENTERPRISE EMPLOYMENT AND SALES SIZE DISTRIBUTIONS IN U.S. PRIVATE SECTOR: 1992

tries that are popularly conceived to be dominated by the little guys), small firms accounted for almost 98 percent of the total number of businesses, but for less than one of every three jobs and fewer than a fifth of all sales revenues. By contrast, finance, insurance, and real estate companies employing 500 or more employees in 1992 made up only seven-tenths of 1 percent of all such businesses, but this tiny group accounted for 57 percent of all jobs in the sector and 72.3 percent of all sales. To repeat: this overwhelming dominance of the big firms is not new news in modern American economic life—and that is precisely my point. This, at least, has not changed.

Now consider the situation in a particular city: New York. If there is any place where we might expect small operations to have a dominant presence, it would be in a crowded metropolis that is also the world's financial capital. And there are plenty of policymakers who believe this to be the case, most recently the comptroller of New York City's government. In November 1996 the comptroller's chief economist issued a widely quoted report, presenting yet another set of numbers that purported to show that

Pct. of estabs. with fewer than	1988	1993
10 employees	79.7	80.5
100 employees	98.0	98.2
500 employees	99.7	99.7
Pct. of total employment in estabs. with fewer than		
10 employees	13.7	14.9
100 employees	45.5	48.2
500 employees	67.5	69.8

So in 1993, only three-tenths of 1 percent of all private-sector business establishments in NYC were larger than 500 employees, but that tiny number of establishments employed 30.2 percent of all workers! That's the SBA concept of "small firm" as being anything under 500 employees. If instead we use the more sensible OECD definition of "small" being a business with fewer than 100 employees, then in NYC in 1993 only 1.8 percent of all establishments employed more than 100 people, yet that tiny number of businesses accounted for fully 51.8 percent of all workers—more than half!

Moreover, these numbers are all for businesses, including the smallest retailers. If we could break out manufacturers, or financial firms, or wholesalers, the disproportionate influence of the biggest firms as job creators would be even greater.

Source: Division of Research and Statistics, New York State Department of Labor, *Tomorrow's Jobs, Tomorrow's Workers: New York City* (Albany, N.Y.: New York State Department of Labor, 1994), p. 7.

TABLE 11.2   DISTRIBUTIONS OF EMPLOYMENT IN NYC
PRIVATE-SECTOR BUSINESS ESTABLISHMENTS, 1988
AND 1993

small business is now "the engine of job growth."[7] The office openly adopted the ridiculous definition created by the U.S. Small Business Administration (SBA) of a "small" business as one that employs fewer than 500 persons—a ridiculous definition (as I noted back in chapters 1 and 2) precisely because it includes nearly 100 percent of all companies! The comptroller's calculations also suffer from many of the technical problems that I exposed in those earlier chapters, and that so many other economists and thoughtful journalists have openly criticized too.

What is the reality about the economy of New York City? Reliable time series data on whole enterprises are simply not available for an area as small as New York. But for individual establishments—that is, plants, offices, warehouses, stores, and shops—the true patterns are revealed in table 11.2. In 1988, and again in 1993, according to data from the New York State Department of Labor, we learn that, for the private sector as a whole, the big establishments (those with 500 or more employees) made up only three-tenths of 1 percent of all such operations, but accounted for two-thirds of all the private sector jobs in the city. Indeed, if anything, the job share of these biggest establishments actually rose a bit over this period.

Finally, again using data from the U.S. Census Bureau and now examining only manufacturing, if we look all the way back to 1980 (see figure 11.1) we can see that, over the last fifteen years, the share of such jobs created in smaller establishments has—depending on your definition of what constitutes "small"—either been absolutely unchanging over all that time, or actually *falling*! This is decidedly not the conventional wisdom.

To the extent that the New York City comptroller's department was motivated in part by a desire to slow down the rush of almost unbridled tax breaks that the mayor and governor have been lavishing on very large corporations in the office and financial sectors to entice them to remain located in the city, I am sympathetic. Nevertheless, the comptroller's report is factually misleading, commits every methodological error I exposed in the first edition of this book, and lends itself to the no-more-sensible (or warranted) opposite extreme policy of subsidizing small businesses just because they are small.

A richer perspective on all this comes from, of all places, a magazine advertisement taken out last spring by the New York City Partnership and the city's Department of Business Services.[8] New York University urban planner Mitchell Moss was commissioned to research and write the article, which is of course unapologetically boosterish concerning the new oppor-

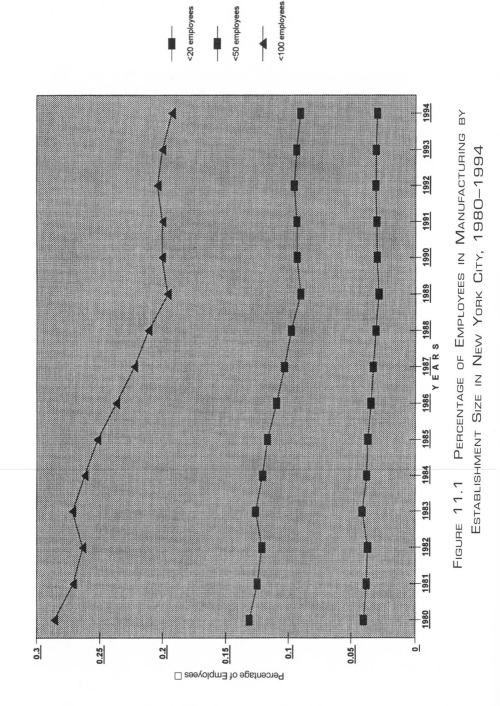

FIGURE 11.1   PERCENTAGE OF EMPLOYEES IN MANUFACTURING BY
ESTABLISHMENT SIZE IN NEW YORK CITY, 1980–1994

tunities for private investment in the city's reviving economy. The emphasis is predictably on small business (and here, Moss, too, begins with that nonsensical SBA definition of "small" as any firm employing fewer than 500 people).

He apparently did have access to data on whole companies, because he reports that firms with fewer than 500 employees now account for some 70 percent of all private-sector jobs in the city. If such firms make up 99 percent of the total (the national average), that means that less than 1 percent of the city's companies employ 30 percent of its workers. The disproportionate influence of big firms is less in the big city than in the national economy, which makes sense because cities have always been incubators for and homes to small firms. But the disproportion is still there. Moss then proceeds to name the sectors in which smaller companies have been thriving, or at least struggling to restructure: jewelry design and production, multimedia and financial services, recycling, furniture manufacture. The advertisement presents a rich mix of vignettes about these sectors, featuring stories about dynamic entrepreneurs in each.

These numbers, like mine, show clearly that the economy of the city of New York—like that of the national economy as a whole—depends on both small *and* large firms. And the sponsors of the ad know this. That is why the Partnership's subsidiary, the New York City Investment Fund, created in 1996, has made one of its prime objectives the strengthening of relationships among large and small firms. This—not some mindless attack on small business, per se—was both the substance of my analysis in the earlier chapters of *Lean and Mean* and a key element in my policy recommendations. If, as Steve Davis, John Haltiwanger, Jonathan Leonard, James Medoff, and others I cite in chapter 2 have shown, small firms are the most vulnerable to cyclical instability, have the shortest expected survival rates, and tend to pay the lowest wages and the fewest benefits to their workers, then policy needs to address these structural problems directly. The solution does not involve tax breaks (which startups, especially, can't use, since they aren't yet bringing in enough revenue to have to pay income taxes). The solution is with public (or public–private) investments in infrastructure, technical assistance, targeted venture capital, and, most of all, closer working relationships with their city's or region's big companies, for example, through supplier chains. That the New York City Partnership and the Investment Fund recognize this nuance bodes well for the course of economic development policymaking in the city in the years ahead.

I should add a word about the allegation that smaller firms have at least become (if they were not already) the engines of technological progress. Again without in any way denying the importance (only the dominance) of smaller entrepreneurs, after the publication of the first edition of *Lean and Mean* I encountered a marvelous study of the U.S. semiconductor industry, written by economic geographer David Angel. He provides further evidence for what was the main theme of my original chapter 3: that even innovative activity needs now to be understood in the context of interfirm networks and alliances governed by the big firms:

> It is large and medium-sized semiconductor firms—most notably IBM, Intel, Motorola, and Texas Instruments—that have been the leading agents of change in manufacturing practice within the semiconductor industry. To a significant degree, it is the ability of these large producers to improve yields and accelerate the pace of product development that has allowed the United States to increase its [worldwide] share of high-value product markets. . . . [S]mall start-up firms continue to play a crucial role. . . . [H]owever, few small firms stand alone; the majority are now bound into a complex array of cooperative alliances with large established producers. [We are witnessing the birth of] a new organizational model in which large firms serve as anchors for a shifting network of alliances and manufacturing agreements.[9]

FURTHER EVIDENCE OF A DARK SIDE TO FLEXIBILITY

Now, well into the seventh year of an economic upturn, it is not surprising that declining unemployment, tightening labor markets, and expanding consumption and investment demand are finally conjoining to begin to raise average real wages after years of stagnation. That Europe and Japan may finally be recovering from their own (very different) experiences of stagnation only adds to the much awaited improvement in domestic earnings. After adjusting for inflation using the new chain-weighted price index for consumer spending, the U.S. Department of Labor estimates that real average wages, which were falling at an annual rate of 1 percent in the last business cycle peak year of 1989, are now rising at an annual rate of a bit over 1 percent.[10] This is not much growth, but it is growth nonetheless.

Equally important, it now appears that this tightening of labor markets in the aggregate and in particular regions of the country is finally beginning to bring up the wage rate that low- skilled (or, at any rate, poorly paid) workers are able to command. With this development, the gap between high-paid and low-paid workers—that much-cited worsening earnings inequality—appears to be narrowing at last. Economists Rebecca Blank and David Card estimate that, all other things being equal, a 1 percent drop in the jobless rate produces a 3.7 percent rise in annual earnings (mainly through increased hours of employment) for the lowest fifth of the work force, whereas the top fifth's gain is just 1.4 percent.[11]

Welcome as they are, these are macroeconomic developments, subject to the usual ups and downs of the business cycle, with the associated tightening and loosening of labor markets. What about long-run trends? What about structural developments that go beyond (and across) cycles? Here, as I argued in chapter 9, far-ranging changes are underway in how companies organize work: how they produce, where they locate the work to be done, and whom they hire to do the work. A central feature of these changes is what now appears to be a long-run ("secular" as opposed to periodic or "cyclical") move away from what, at least for the most profitable operations of the biggest, most visible, and most influential firms and agencies, constituted the dominant (if never universal) employment system of the post–World War II era.

To recapitulate: in the quarter century after 1945, with job tasks substantially standardized and broken down into a finely grained division of labor, managers commonly hired in new young workers at the bottom of job ladders, trained them on the job, promoted people within the organization, and paid wages more in accord with seniority and experience than with individual productivity or current firm performance. While few companies ever formally committed themselves to literally "lifetime" employment, there was widespread expectation that (at least for white men, but gradually for others as well) there would be a high degree of long- term job security, with occasional disruptions in work triggering the receipt of unemployment insurance and other forms of temporary income maintenance. Seniority systems were promoted by unions, and within the civil service, as a means of achieving fairness. Business came to support them because, by according greater job security to more experienced employees, business could expect in return that senior employees would more willingly train and otherwise teach skills and know-how to younger workers within the organization.

Whether there is a well-defined, coherent new "system" of work coming into existence—one accommodating to globalization, deregulation, the shortened shelf lives of products, the enhanced capability of more and more competitors to quickly erode the advantages of (play catchup with) the innovators through "reverse engineering," and generally heightened competition—is still an open question. But there is a growing consensus among scholars that the old system is coming apart. This is not the place to examine all of the many aspects of this "devolution" of internal labor markets, the growth of "flexible" forms of work organization, and the implications for training and economic development.[12] But I can at least sketch the main features of what a number of researchers believe is happening to how work and labor markets are coming to be reorganized.[13]

The changes are motivated by a complex of reasons: deregulation, greater actual or potential competition from abroad, growing numbers of corporate hostile takeover attempts and other signals from stockholders that put a premium on short-term firm performance. All of this has made managers increasingly conscious of short-term fixed (or, as economists like to describe the wages of workers on long-term implicit contracts, "quasifixed") costs, and committed to reducing them whenever and as much as possible. First with IBM, then with Xerox, most recently with AT&T and the big banks, the stock market has instantly rewarded companies that cut costs through consolidations, mass layoffs, and wage and benefit rollbacks by bidding up share values, which only further encourages this sort of management behavior. Even as progressive voices—from the former U.S. secretaries of labor Ray Marshall and Robert Reich, to labor economist Richard Freeman, to political theorist and activist Joel Rogers—were advocating that more companies should take (and stick to) the "high road" in labor policy, the business community has been organizing itself (e.g., through the Labor Policy Association) to advocate for, and legitimate, greater "flexibility" and management discretion in work arrangements.

What this amounts to, in practice, is a proliferation of different forms of work organization, blurring the traditional distinctions between "core" and "periphery," "permanent" and "contingent," "inside" and "outside" employees and between "primary" and "secondary" labor markets. Thus managers employ some workers in more or less routine wage and salary positions inside the firm or agency. But then they also hire temporary help agencies, management consulting firms, and other contractors to provide employees (ranging from specialized computer programmers to janitors and clerical personnel) to work alongside the "regular" people but on short-term assignments and under the management of the contractor. Companies, colleges, and hospitals outsource work formerly performed in-house to outside suppliers, here and abroad. They also shift some work from full-time to part-time schedules, in part to avoid federal labor regulations covering wages and benefits that have been interpreted by the courts as not covering "leased" workers. The "temp" agencies and other contractors are being used increasingly by managers in the "focal" firm as a mechanism for screening potential regular employees, with candidates serving their "probation" on the payrolls of the contractor before moving into the focal company or agency. This creates the possibility for further inequities, since two persons working side by side for an outside contractor may be equally competent, yet one will eventually be absorbed into a full-time job while the other remains bouncing from one temporary assignment to another.

Two concrete expressions of this growing heterogeneity in work organization and practices, both among and within particular employers, are declining employment security and more uncertain wage and salary prospects over time. Surveys conducted by the American Management Association show that managers increasingly regard layoffs as "strategic or structural in nature," rather than as a response to short-term, temporary business conditions.[14] The fraction of laid-off workers who can expect recall into their old jobs has been lower in the post-1991 recovery than in the previous four national business cycles, while those who do get re-employed somewhere are more likely than before to land in part-time jobs.[15]

Flatter organizational structures—yet another workplace "innovation" whose introduction in recent years has been responsible for so many of the layoffs of middle-level managers—contribute to reduced promotion opportunities within the surviving internal labor markets of companies and agencies. More generally, the payoff to seniority, as measured by age-earnings profiles, is shrinking over time, both across the workforce and, more precisely, over the careers of specific individuals, including those who stay with the same employer. The flattening of average age-earnings profiles is suggestive, but longitudinal studies on specific workers or on cohorts are more definitive. There is now considerable evidence that younger workers cannot expect the same rate of pay increases over time that their parents received.[16] For example, a new study from the Employee Benefit Research Institute reports that between 1991 and 1996 median job tenure for men aged 25 to 64 fell by nearly 20 percent, which means that people don't remain in a particular job long enough to move up to the same extent as did earlier cohorts.[17]

Two researchers affiliated with the National Bureau of Economic Research have measured declining upward mobility more directly. Studying the U.S. Department of Labor's National Longitudinal Surveys for Youth, they found that, among younger men and women since 1979, there has been a significant increase in immobility over time *within* each quintile of the earnings distribution, with the degree of growth in immobility being greatest for the bottom quintiles. More generally, what these researchers have found is a convergence over time in interquintile transition probabilities toward the diagonal. That is "econospeak" for saying that people are increasingly likely to stay from one period to the next within their original position in the earnings distribution than to move up or down in position. This is a major discovery. Moreover, these estimates control for personal differences in schooling and other factors relating to skill. The overall conclusion, therefore, is that there has been, in the authors own words, a sharp decrease in mobility over time, across all skill groups.[18]

And there is more. Compared to expectations formed in earlier periods (such as the early 1970s) or to the average earnings profile for their age group, workers in the 1980s and 1990s may now be facing increasingly *fluctuating* earnings from one year to the next. The jury is still out on the relative magnitudes of the transitory (unexpectedly fluctuating) versus permanent (long-term, predictable) components of annual increases in wage inequality. Economists Peter Gottschalk and Robert Moffit were the first to report a large transitory component.[19] Annette Bernhardt and her colleagues are finding the permanent component to be relatively larger, which, while not evidence for "fluctuation," certainly confirms the growing structural trend toward inequality. In their words, we are witnessing

> the very real possibility that the structure of mobility opportunity for young workers has changed, and that this change has been for the worse. . . . The recent cohort has substantially more workers who experience both low and high wage [permanent, as distinct from "transitory"] growth. In short, . . . what we are potentially looking at is a rise in the inequality of opportunity.[20]

Researchers are vigorously comparing notes and methodologies on the question of the relative importance of increasingly fluctuating earnings. Stay tuned; this is important. Quite apart from questions of "fairness," Cappelli and others are questioning the long-run implications of any growing instability in earnings streams for the continued smooth functioning of a consumer economy built on households borrowing against future income and having to make small regular payments over time for everything from housing and cars to their children's education.

In sum, the mix of types of jobs and work schedules is becoming more diverse. According to Cappelli's careful and sober formulation, "[T]here is a continuum between 'pure' internalized arrangements and complete market determination along which employers are moving, and the argument here is that, on average, practices are shifting along that continuum," from the former in the direction of the latter.[21] If this were mostly a voluntary development, serving the needs of a greater share of the population, that would be one thing. But all serious researchers agree that these changes are being initiated mainly on the "demand side" of the labor market, with managers seeking to reduce their exposure to long-term fixed obligations. It appears that a growing fraction of pay is becoming "contingent"—on individual job performance, on the fortunes of the employer, on the current state of the "animal spirits" in the stock market, or on what the company thinks it can get for its money by turning to suppliers in other, lower cost locations. Whatever the causes, the results are stagnating long-run average wages, declining personal mobility over time, and (possibly) increasing uncertainty in what a person's earnings next year will be.

This is all part of what I have called the dark side of the era of flexible accumulation. It would appear that many of the concerns I raised in the first edition of *Lean and Mean* have been bourne out.

AS FOR AGGLOMERATION AND INDUSTRIAL DISTRICTS, IT DEPENDS

If the original edition of *Lean and Mean* conveyed the impression that I did not "believe" in the existence, let alone the importance, of forces of spatial agglomeration or geographic "clustering" of economic activity, that impression only reflects inadequacies in my own writing. I have always understood agglomeration to be important—but contingent. There are historical moments in which centrifugal forces dominate centripetal forces, and vice versa. In particular, as I insisted originally, and continue to insist, what is an undeniably rising tendency—toward greater interfirm collaboration, for example, in the form of just-in-time inventory management and somewhat reduced dependence on arms-length relations between customer firms and their most important suppliers—does not, in and of itself, *require* spatial contiguity. It depends.

Maryellen Kelley's research on agglomerative tendencies in what she calls the "machining-intensive durable goods sector" (which accounts for a quarter of all U.S. manufacturing jobs) reflects these contingencies. In a series of papers,[22] she and her colleagues have reported that certain sources of geographic clustering are more likely to affect how (and even to what extent) nonadopters of various information technologies learn from spatially proximate prior adopters than are other forms of agglomeration. For relatively inexpensive stand-alone technologies, being located within a dense complex of similar firms—"localization," the essential property of, industrial districts—matters. So does proximity to institutes and colleges granting engineering degrees and to more diverse pools of labor—thereby conferring the benefits of what Jane Jacobs and other writers and re-searchers have long referred to as "urbanization economies." For other technologies, urbanization matters, whereas localization (specialization in the sector) does not. Moreover, firm size and business organization matter, with the smallest firms tending to be the most likely to learn from their neighbors.

In other words, clusters may exist as a historical leftover—as I strongly suspect is the case for many of the small-firm industrial districts celebrated by the "post-Fordists"—or they may actually contribute to tomorrow's tech-nological trajectories. They may be relevant to explaining why small firms survive within difficult environments, or they may be old news. What mat-ters may be co-location—or it may be embedment within an interfirm sys-tem of relational subcontracting that, while it originated in a particular place and time, has become geographically dispersed since then. As Hugh Whittaker has shown in the case of small machine shops in Tokyo, the "re-production" of the industrial district may turn on how the big-firm cus-tomers treat their key (first-tier) suppliers during recessions; in the 1990s, "orders from 'new key factories are less stable, and work tends to be taken in-house during a downturn. . . . [Flexibly specialized small firm] net-works, therefore, have disadvantages of their own, and are not a panacea for small subcontractor problems."[23] Moreover, survival of the district may also partly depend (as I suggested for the Italian district of Prato) on na-tional economic and political decisions having nothing to do with the lo-cality, for example, decisions about whether to permit the importation of competing parts and other products into the country.[24]

The interrelationship between macro- and local forces has, I think, been insufficiently appreciated in most studies of industrial districts. To take another example, an important contingency that can affect the ag-glomeration/deagglomeration axis in an increasingly interdependent glob-al economy is the movement of exchange rates and corporate expectations of further movements. Even the Japanese *keiretsu* are trying to cope with the sharply worsening conditions for exporting from production complexes located within the home islands. Consider the recent strategic accommodation of the Sumitomo Electric Corporation to the rising value of the yen since the mid-1980s. As table 11.3 shows, in an astonishingly short time, between 1985 and 1993, the share of offshore production of consumer durables by Sumitomo factories in overall company output *tripled*. According to the deputy CEO of the company, this hollowing of domestic operations (almost entirely to the Asian mainland) was triggered mostly by the expected continuing high value of the yen in international trade, a trend that was pricing Japanese products out of world markets. As I noted earlier in regard to southwestern Germany, so too in Japan local clusters have in some instances been coming apart.

Ultimately, the depiction by Nigel Thrift, Allen Scott, Michael Storper, Ash Amin, and others of the most successful industrial districts as "local nodes in global webs" may well be the best approximation of this complex reality. This formulation emphasizes extralocal context as well as properties of the locale—and, in the most creative theorizing, actually imagines that causality runs in both directions. Here's how UCLA planner-geographer Michael Storper sees it:

The formation of [a] *global context* of trade, investment, and communications and organized networks of human relations in production is perhaps the clearest dimension of globalization. The global economy is being constructed as an increasingly widely spread and

Product Line	1985	1993
Color televisions	38.8%	71.9%
VCRs	6.3	41.6
Tape recorders	26.1	55.2
Microwave ovens	22.7	63.3
Refrigerators	18.6	40.5
Auto parts	4.9	25.1

Source: Speech by Tsuneio Nakahara, vice-chairman and deputy CEO, Sumitomo Electric Corporation, to a conference at M.I.T. organized by the Industrial Performance Center, Cambridge, MA, March 30, 1995.

TABLE 11.3   PERCENTAGE OF CONTENT PRODUCED OUTSIDE OF JAPAN BY THE SUMITOMO ELECTRIC CORPORATION: 1985–1993

accepted grid of these sorts of transactions, akin to a new global lingua franca of commerce, investment, and organization, based on historical and secular advances in transportation and communication technologies, and the development and diffusion of modern organizational "science," both of these in the context of the increasingly global political order of trade. . . . But the paradox is that it is precisely this global grid or language that leads both to . . . break[ing] down barriers of taste, transport, and scale . . . [and opening] up markets to products based on superior forms of local knowledge; it consolidates markets and leads to such fantastic product differentiation possibilities that markets refragment and, with them, new specialized and localized divisions of labor reemerge.[24]

This analysis seems right, points to future directions for research, and provides a healthy antidote to the romantic inclination to unqualifiedly celebrating the privilege of the local in the face of powerful forces of globalization.

WHAT ABOUT THIS IDEA OF "GLOBALIZATION"?

When I wrote the original book, incorporating and embellishing the idea of "globalization," I could never have expected that this would become such a contested conception. I thought that the work of Castells, Saskia Sassen, Peter Dicken, Gary Gereffi, and others had by now clearly established the analytical (as distinct from the purely rhetorical) basis for the idea. Apparently, I was wrong.

The "internationalization" of economic affairs is of course not a new development. Multilocational and multinational firms have been part of world capitalism for centuries. Nor is it the increasing relative importance of world trade per se, in the post World War II era (the GATT reports that exports have grown considerably faster than the national domestic outputs of all trading nations since 1960[26]) that distinguishes globalizing tendencies.

Yet in recent years, critics—especially on the left—have raised precisely such markers of continuing internationalization to buttress the claim that, when push comes to shove, at best there is nothing especially new about developments in the world economy, certainly nothing that should replace class struggle as the explanation for wage stagnation and rising inequality. Leading left scholars and journalists, starting with the late David

Gordon and including Frances Fox Piven, Joel Rogers, Doug Henwood, and Simon Head, have all recently expressed the judgment that globalization is, after all, just an ideological construction of international business and its political supporters, one that is being used (by, I suppose, writers like myself) to weaken the resolve of domestic governments and labor organizers as big capital flexes its collective muscle, employing the threat to geographically reconfigure production and trade to whipsaw one group of workers, cities, and countries against the other in order to reduce labor costs and, more generally, encourage a more favorable business climate.[27]

Having written many previous books and papers over the years about just such uses of corporate power against labor and local government—the working title of my most well-known book, *The Deindustrialization of America*, was, after all, "Capital vs. Community"—I don t feel particularly guilty about my efforts to call a pitchfork a pitchfork. Moreover, having been close to the labor movement all of my life, I am not now especially worried about demotivating or de-energizing union activists, especially when so much of the best analysis of the qualitatively heightened "bargaining power" of global capital vis-à-vis labor is coming from union intellectuals and activists themselves! Nevertheless, it appears useful to restate, and perhaps amplify, what it is that is so structurally new about globalization.

As Hector Figueroa, assistant research director of the Service Employees International Union, puts it, in contrast to internationalization, "Globalization occurs when, say, IBM integrates the financing, design, production and marketing of its Thinkpad [laptop] computers across several countries," much as Ford has for years been doing in the production of its cars and American Airlines has been doing in the management of its customer information systems. According to Figueroa, an insurance company like Axa of France also dances to the globalization tune when it spins off a division of its North American subsidiary, Equitable, and uses part of the proceeds to finance the acquisition of an Australian firm.[28]

Analytically speaking, there are two key processes here. The pace and extent of what Barry Bluestone and I originally termed the "hypermobility" of financial capital is unprecedented. In his textbook/primer on the subject, geographer Peter Dicken (citing Peter Drucker) reports:

[T]he London Eurodollar market, in which the world's financial institutions borrow from and lend to each other, turns over $300 billion each working day, or $75 trillion a year, a volume at least twenty-five

times that of world trade. . . . [F]oreign exchange transactions [now]
run around $150 billion a day or about $35 trillion a year—twelve
times the world-wide trade in goods and services.[29]

Social critics such as Will Hutton, editor of the *Observer*, and even the international financier George Soros, have written that unless the destabiliz-
ing effects of these hypermobile international financial transactions are
subjected to some new form of social regulation, we will have no chance
of permanently improving wages and providing workers with a measure of
income security within any of our countries.

The second element in Figueroa's analysis of what distinguishes globalization from mere internationalization is that, even within particular firms
(and, I have added, within their interorganizational networks), the opportunities for cross-border coordination and the proliferation of competitive,
accommodating operating sites now make possible a qualitative transformation in the nature and locus of *actual production*. As Dicken puts it, "In
effect, then, the traditional international economy of traders is giving way
to a world economy of international producers."

There are severe measurement problems in documenting these
changes, especially when it comes to trying to capture the organizational
complexities of distributed (networked) production. Part of the debate
within the Left seems to be over these measurement issues. For example,
in an influential paper published in *New Left Review* back in the late
1980s, David Gordon suggested that, since most U.S. trade was conducted
with a handful of other highly industrialized, high-average-wage countries—the same countries to which most U.S. direct foreign investment
(DFI) was being directed—such transactions could not possibly explain
domestic wage stagnation or the decline in union membership at home.[30]
Interestingly, an analogous argument is made by the mainstream economists, such as Paul Krugman and Robert Z. Lawrence, who believe that
U.S. trade with all countries amounts to too small a fraction of GDP to be
able to account for changes in the material conditions of American workers, especially those at the low end of the earnings distribution.

There is of course no way to know just exactly how much actual trade or
relocation of capital it takes to constitute a credible threat by business
when it demands tax breaks from some local or national government, or
wage and benefit givebacks by unions, or intimidates labor organizers.

But how can anyone deny the sometimes quite explicit willingness of
business to use its enhanced ability to relocate production out of the coun-

try in order to—there is no other word for it—blackmail American workers? The AFL-CIO's Thea Lee summarizes much of the recent evidence:

Businesses have not hesitated to use the existence of mobility-enhancing trade rules to whipsaw workers at the bargaining table. As early as 1992, 40 percent of corporate executives polled by the *Wall Street Journal* (September 24, 1992, p. R7) admitted that it was likely or somewhat likely that their company would shift some production to Mexico within a few years. Twenty-four percent of the executives polled said their companies were likely to "use NAFTA [the North American Free Trade Agreement] as a bargaining chip to keep wages down in the U.S." Since the implementation of NAFTA in January 1994, these expectations have been fulfilled. In a report prepared for the NAFTA Labor Secretariat's Commission on Labor Cooperation, Cornell University researcher Kate Bronfenbrenner documents numerous examples of employers using the possibility of relocating production to Mexico under NAFTA as a threat during wage negotiations or union-organizing campaigns. Some employers posted a large map of North America with arrows pointing from the current plant location to Mexico. Others provided statistics to workers detailing "the average wage of a Mexican auto worker, the average wage of their U.S. counterparts," and how much the company stood to gain from moving to Mexico. ITT Automotive in Michigan parked tractor trailers loaded with production equipment labeled "Mexico Transfer Job" in front of a plant where a union campaign was under way. Clearly, labor–management relations and the balance of bargaining power between workers and employers are affected by trade flows and trade rules.[31]

Economist Lee's long-standing skepticism about NAFTA received powerful support last year, upon publication of numbers on the effects of the first three years of the agreement with Mexico. The Clinton administration estimated that between 90,000 and 160,000 new jobs were created inside the United States as a result of the agreement. But the U.S. Labor Department's NAFTA Trade Adjustment Assistance Program also certified that about 132,000 workers permanently lost their jobs as a result of the agreement. By one estimate, United States based multinational corporations more than doubled their shipments of American-made parts and components to Mexico after the passage of NAFTA, to be assembled there and reimported into the States. Overall, what had been a $2 billion U.S. trade surplus with Mexico just prior to NAFTA turned into a record $16 billion deficit in 1996. And what of the Mexicans? Wages along their side of the border fell by 30 percent, according to the International Monetary Fund (IMF).[32]

I certainly do not imagine trade and investment to be motivated entirely by business' search for cheap labor. Whether in the Far East, on the European periphery, or even in the developing countries, multinational capital and the markets are also attracted by rapid rates of economic growth, trained workforces, reliable infrastructure, and political stability. Still, cross-national differences in unit labor costs surely do help to drive capital mobility. In this regard, based on research conducted by Alan Taylor and Jeffrey Williamson, three progressive economists—Dean Baker, Gerald Epstein, and Robert Pollin—report that

> wages across 15 OECD countries were converging at an average annual rate of 0.35 percent in 1870–1913, but that this process of convergence ended between 1913–1950. The rate of convergence then accelerates to 0.79 percent per year in the period 1950 87. However, the rate of productivity convergence over these same periods is far more rapid. Productivity convergence was 0.36 percent per year from 1870–1913. This convergence process is then essentially reversed in the 1913–1950 period, but then accelerates tremendously to 3.14 percent from 1950–1987. . . . Over 1950–1987, wages converged at only 25 percent the rate of productivity.[33]

That productivities across the OECD are converging more rapidly than wages, which creates the possibility for companies to reduce costs without forgoing proportional output, is almost surely true across many of the non-OECD countries, as well, certainly in Asia and parts of Latin America. This is the objective, "material" basis for my own long-standing argument that globalization does indeed matter. These data are fully consistent with an older Marxist conception, too: as capitalist relations of production penetrate more and more corners of the globe, the technical division of labor converges across space more rapidly than does the cost of reproduction of labor power. That wedge offers the possibility for "arbitrage" in the form of capital flight.

Moreover, trade, per se, understates the significance of globalization.

Not even DFI captures the full extent of the growing interdependences among companies and countries, although it cannot be irrelevant that, according to the United Nations Centre on Transnational Corporations, global DFI, which grew at an average annual pace of about $10 billion between 1960 and 1973, has been expanding nearly four times as fast since then.[34] Or that a growing fraction of world DFI is in services rather than in manufacturing. Or that a still small, but growing, share emanates from the transnational corporations of developing countries.

Most serious, official trade and DFI statistics almost completely ignore the extent of strategic alliances and other forms of cross-border networking. One country's companies building branch plants abroad to increase market share, or to escape market saturation at home, may not be a new thing. But entering foreign locations through alliances with their (and with third-party) companies and governments to defend existing market positions and create new offensives against other networks and alliances surely is. This is occurring in any number of industries, as exemplified by steel minimill Nucor's recent alliance with Brazil's Companhia Siderurgica Nacional to build a new steel mill in that country, or in Birmingham Steel Corporation's new joint venture with Mitsui to export 700,000 tons of scrap a year from California to Asia.[35]

Despite the anxieties on the left and the denials from the mainstream, I maintain that the proliferation of viable production sites around the globe, and of competitive private and quasi-public producers—in manufacturing, software, retail, finance, construction, and infrastructure provision (i.e., engineering)—is continuing, and that the networked reorganization of relations of production does indeed constitute something new to world capitalism. Whether this amounts to the emergence of a coherent form of post-Fordism I do not know. Whether such technological developments as distributed information processing and telecom are essential to this transformation in capitalist production—a central premise of Castells's work—or are essentially permissive of global coordination, as Bluestone and I have long argued, globalization is, I think, what is most new about the post-1973 world economy.

Note that none of this suggests, even for a moment, that the evolving international, multinational regulatory regime is somehow separate or independent from the politics of nation-states—yet another unfair charge that many of those nervous about "exaggerating the global" have leveled at me and at others. For example, the capital markets have shown themselves to be powerful arbiters of policy. Beginning with the international financial backlash against French socialist president Mitterand's efforts to reflate the economy after 1981, in the form of a sudden and decisive flight from the franc, the markets have shown that they can, according to Saskia Sassen, "force governments to take certain measures and not others. Investors vote with their feet, moving quickly in and out of countries, often with massive amounts of money . . . they have emerged as a sort of a global, cross-border economic electorate, where the right to vote is predicated on the possibility of registering capital."[36]

At the same time, even if paradoxically, the very deregulation of intranational financial markets, and the ceding of power over international capital mobility to private commercial arbitration and to such private organizations as Moody's Investors Service and Standard and Poor (which in only a decade have become the principal bond-rating entities that stand as gatekeepers for investment funds sought by companies and governments throughout the world), are clearly the result of political decisions taken by individual elected leaders and national governments. In Sassen's words, "The state itself has been a key agent in the implementation of global processes."

Thus, when countries seeking to pursue expansionary, job-creating policies find themselves facing challenges to their "credibility" from the financial markets, which then exercise their discipline by forcing up interest costs, which the offending governments must then pay, we are witnessing deregulation, yes, but also the emergence of a new regulatory regime, in which states are full participants. This very different interpretation of developments, so central to Sassen's brilliant analysis, implies a continuing role for national politics.

Nor does any of this negate the validity of the "local." To quote again from Storper, there

> may well be a new kind of "nexus" business organization, whose impacts on economic development processes we have barely begun to glimpse; but such a model of the global firm does not so much imply deterritorialization of the economic process as a recasting of the role of territories in complex, intraorganizationally and interorganizationally linked global business flows.[37]

In other words, local nodes within global webs.

(A Bit) More on Economic and Social Policy in the Age of Lean and Mean Business

In a recent paper, motivated by precisely those left concerns about how to defend local power and autonomy (although not some romanticized autarky) in the emerging era, Rutgers University urban planner Susan Fainstein advises:

> The reality that giant, multinational . . . corporations dominate economic transactions means that their critics must find ways of tapping into their economic power rather than dismissing them on moral grounds. Public–private partnerships under these conditions are inevitable; what needs to be done is assure that the public component is more controlling and shares more in the proceeds.[38]

In chapter 10, I argued that economic development and other policies that ignore the advent of the network principle, or that romantically and unthinkingly buy into the belief in the superiority of small business per se, are bound to fail. These conclusions led me to advocate greater public policy attention to stimulating networks and collaboratives, generally, in the interest of better connecting localities and marginalized populations to the global economy, on terms that might buy those communities the advantages that derive from networking (especially the stimulation to organizational learning), without totally eradicating the "local" in the process. It is far worse to be isolated, excluded from the evolving networks, than to have to struggle, à la Fainstein, to figure out on what terms and through what existing or newly designed institutions to engage.

Since completing the book, I have been following my own advice, devoting much of my time to studying, writing about, and selectively advocating for economic development networks. Unaware until just this year of the creative theorizing of Oxford University economic geographer Erik Swyngedouw on the simultaneous devolution of national-level politics and economics to the local *and* the further elaboration of global relations, I, too, found myself operating within what Swyngedouw calls a "politics of scales."[39]

In the United States (and not only here), working seriously at the subnational level is compelled by quite profound changes in the philosophy (if not yet in the actual legal structure) of American government. Chief among these is the fiscally and ideologically driven thrust, since the late

1970s, toward devolution of power and responsibility from the national government to the states. Since state governments differ among themselves enormously in terms of levels of spending on infrastructure, social welfare programs, schools, public health, and so forth, and since the states have in few cases been willing agents for redistribution or fair resource allocation among their own constituent jurisdictions, for example, between central cities and suburbs, this devolution is effectively unleveling the playing field and reinforcing regional differences—a property that business has always been able to use well to extract concessions from governments and unions.[40]

A particular case in point these days is the radical transformation of the national system of guranteeing minimal incomes to women and dependent children in poverty. In the summer of 1996, with a stroke of his bill-signing pen, a Democratic Party president, Bill Clinton, ended "welfare as we know it," by requiring all welfare recipients to take jobs of some sort and by devolving control over most aspects of welfare to the governors (this trend had actually been underway for years, as more and more states attained variances from federal regulations to permit local "experiments"). By almost every expert forecast, the devolution of national social welfare into workfare is expected to greatly worsen the relative income position of the poorest citizens of the country. By virtue of their generally limited skills and spotty prior work experience, these millions of new labor force entrants will for the most part be confined to the lowest paying segments of their respective local labor markets, or will be bused out of state to regions facing temporary shortages of unskilled labor. Moreover, this "crowding" effect suggests that many existing working poor are likely to be displaced, while the average earnings of the entire low-skilled labor force are likely to be depressed—perhaps by as much as 9 or 10 percent in the case of the New York City workforce.[41]

Workfare, and devolution more generally, should be seen as part and parcel of the shift toward labor market flexibility. Unless things change, the dark side is about to get very much darker. Since there no longer seems to be any political opportunity for halting the program itself, it becomes necessary to design new approaches to creating and targeting jobs for the new welfare-to-work force; to organize them so as to head off throwing these citizens into direct competition with those of their neighbors who are already working, and with the unions; and to rewrite statutes so as to make these persons eligible for wage supplements already on the books— the Earned Income Tax Credit (EITC)—but available only to "real" workers. All of these approaches are presently being hotly debated in cities across the United States.[42]

Devolution and workfare necessitate a local perspective. On the subject of the technical and organizational modernization of U.S. firms, especially those engaged in manufacturing, there is a continuing national attention to the problem, based largely in the programs of the National Institute of Standards and Technology (NIST) in the U.S. Department of Commerce. Through the Manufacturing Extension Partnership network of technical assistance agencies that NIST has been creating throughout the country, via the Advanced Technology Program that explicitly encourages interfirm alliances, and through other efforts at the state and local level, "manufacturing modernization" continues to be a fertile area for local and regional (and, for that matter, national) institution building. Chapter 10 already contains a number of examples and references to the growing literature on the subject. This work now occupies many scholars and activists, such as the Boston-based firm, Jobs for the Future.[43] An entire recent issue of the leading American scientific journal on research and development and technology policy was devoted to setting out the theory and identifying on-going projects.[44] I myself have contributed to this burgeoning literature in recent work conducted for the U.S. Economic Development Administration.[45]

The consortia, supplier relations development projects, technical assistance providers, and systems designers all trying to find both humane and effective ways to propel the creative diffusion of new technologies—especially to the least resourceful smaller firms that constitute the overwhelming majority of late adopters—are doing Fainstein's work: building public–private collaborations that acknowledge the growing complexity of global forces while being committed to rebuilding localities and communities.

Of course, manufacturing modernization hardly exhausts the agenda for local work. Another area, to which I have recently returned after many years away from the subject, concerns the sources and alleviation of urban poverty, especially in communities of color. Even as the research on what was to become *Lean and Mean* was proceeding, I was approached by the Ford Foundation to undertake an exploratory look at the employment and training activities of community development corporations (CDCs) and other community-based organizations (CBOs). Between October 1991 and June 1992, a team of economists, community development business consultants, and graduate students conducted interviews in, and studied documents pertaining to, ten cities around the country that had been selected by the Ford Foundation. Over the course of the following year, we had follow-up correspondence with more than 100 informants including researchers, public officials, congressional staffers, and—most important—directors and staff members from the CDCs and CBOs we had visited the year before.

The initial findings were published by the Ford Foundation in January 1995.[46] We found that a few CDCs in each region of the country—for example, New Community Corporation (NCC) in Newark, the Center for Employment Training (CET) based in San Jose, and Chicanos por la Causa in Phoenix, have successfully managed to mount their own job training programs. But the greatest successes were examples of citywide (and even metrowide) *networks* of community organizations—for example, COPS/METRO in San Antonio, the Pittsburgh Partnership for Neighborhood Development, WIRE NET in Cleveland, STRIVE in New York City, Focus HOPE in Detroit, Pioneer Human Services in Seattle, and the Chicago Jobs Council—geared to connecting neighborhood residents to mainstream institutions that can get them quality training and, eventually, real jobs. CET itself has since become a nationwide network of both wholly owned and jointly collaborative operations.[47]

The extent of innovative cross-boundary networking among CBOs and between certain CBOs and helpful mainstream institutions was the principal "discovery" of the book *Building Bridges*. By design, that report was exploratory, sketchy, and impressionistic. It seemed a logical next step to select a small number of dramatic, creative, interesting examples of these and other CBO-related economic development and job training networks, and then to write in-depth case studies of each. The cases would be compared and contrasted, to identify what is unique about each, but also to search for general principles that might inform the propagation of replicable models that other community development groups in other places might usefully pursue. Just such a project was sponsored by the Ford, MacArthur, and Annie E. Casey Foundations. The results of the case studies, together with a reassessment of urban labor market theory, will be published in early 1998.[48]

Again, my colleagues and I found myriad examples of the network principle in action. The best of the CBOs are already experimenting with such networking, as in the cases of San Jose's CET and Newark's NCC; the COPS/METRO Alliance in San Antonio; Cleveland's WIRE NET; the New York-New Jersey job training/recruiting/placement network formed around the Port Authority's Regional Alliance for Smaller Contractors;

and in such citywide CDC umbrella organizations as the Pittsburgh Partnership for Neighborhood Development and the Chicago Jobs Council. Many work closely with key commercial banks, initially around fair housing mortgage lending but subsequently on job creation and training, too. Many CBO networks are crafting networking relationships with their community colleges.

In a nutshell, it definitely appears that interorganizational networking is occurring, at this scale of development as well as at the interregional and international scales that made up the central focus of *Lean and Mean*. This principle has electrified the field of community development, and helped to spark a number of new foundation, nonprofit-sector, and government initiatives.[49] It is beginning to appear that a key to the long-sought-after expansion of the scale of operations of these community-based antipoverty organizations will be to join and form collaborative networks, rather than to always attempt to "build a new room on their old houses."

Yet another area in which the global meets the local is the agenda for action of the newly reviving American labor movement. As attested to by seemingly endless strategy meetings (in which many of us have been involved) and as reflected in journal issues and even entire new magazines devoted to the subject,[50] the unions recognize the reality that global capital must be engaged—negotiated with, collaborated with, and sometimes confronted—at *all* of Swyngedouw's political scales. Existing transnational institutions, such as shop stewards' networks, must be sustained and expanded. Side agreements regulating highly visible transnational trade agreements such as NAFTA and the GATT are essential, even if logic and the early evidence suggest that these are not likely to be terribly effective or easily enforced. Unions and international secretariats are mounting corporate campaigns to pressure global companies to assume greater responsibility for the labor practices of their subcontractors. Indeed, there are already a surprising number of such campaigns underway, with signs of the possibility of real progress in textiles and sporting goods.[51] The at least implicitly pro-union, explicitly higher living standard ("high road") philosophies embodied in International Labor Organization treaties and in the European Social Chapter are worth rallying around. Unions based in different countries and the international union secretariats can, and are, assisting in cross-border organizing.

At the same time, unions recognize that hypermobile capital has not magically done away with the value that workers place on locality, or their inclination to become politically engaged (when they do) at this scale.

This may be especially true for low-wage workers and at least some recent immigrants, which is why organizing union–community coalitions has been a priority for labor activists since the earliest days of the plant closure movement.[52] These sorts of coalitions and even networks can only succeed when workers and communities of color stand at the center of the action, since they constitute the fastest growing fraction of the subjects of such organizing efforts.

As for the intercorporate networks themselves, greater local, or at least national, regulation is certainly feasible. Firms and their partners depend on a host of local and nation-specific property rights and other enabling laws and entitlements. That these have tended to be so permissive of restructuring, whipsawing, and other business muscle-flexing reflects prevailing political power relations and ideology. In other words, they are to some extent subject to democratic contest.

But not entirely. It is by now well understood that supralocal governments constitute the optimal (as well as the obvious) levels for requiring and enforcing redistributive goals, since localities can too easily fall into competing with one another for mobile buisiness at the expense of their poorest, least powerful residents. Similarly, at some point, the national government must be the level at which to insist upon new quids pro quo with business to ameliorate rampant (or threatened) capital flight. There is only so much that (say) the mayor of San Juan or the governor of Rhode Island can do alone.

One area in which effective regulation is unthinkable at the local level is the development and enforcement of a coherent industrial relations regime. Such regimes evolve over long historical periods, and engage national governments and aggregated class actors, for example, whole trade union movements and corporate associations. In chapters 9 and 10, I distinguished "low road" and "high road" approaches to achieving national economic growth. Just how difficult (and rare) is the high road alternative? To what extent could we imagine implementing it at a local level? British social scientists David Ashton and Francis Green offer five "conditions" for the evolution of a consistent high road industrial relations/economic growth regime: the firm and highly publicized commitment of ruling political elites; the commitment by a critical mass of employers to a high training/high wage competitive strategy; the presence of a body of law and regulation that ensures the continued availability of high-quality workplace (or other forms of what Paul Osterman calls employer-centered) training; widespread incentives and deeply ingrained norms that encourage young people to seek to acquire new skills; and a commitment by government and the education establishment to the continued development of an integrated off-the-job and on-the-job training system.[53] As I said earlier, who could imagine creating and reproducing such conditions locally?

Which brings us finally to the literal water's edge: the need to invent new institutions for regulating global financial and other markets. As Hutton[54] has forcefully argued, the hegemony of the financial markets, with their characteristic lender's bias against reflation, is now acting as a brake on the growth and development of the entire global economic system. Certainly in Europe, economic growth has become hostage to a built-in deflationary bias, as a result of the budget deficit caps called for in the Maastricht Treaty. Absent more rapid economic growth, the three goals of contemporary social democratic policy—competitiveness, profitability, but also a rising standard of living, more equitably shared—are irreconcilable.[55] To the extent that Castells is right, that globally networked firms can (for a time) escape neo-liberal speed limits to growth within any particular country by shifting operations to the offshore components of their nets, this bias is only reinforced, since few corporate leaders see a short-run need to explicitly endorse more rapid economic growth within their home countries.[56]

The solution may ultimately lie in the creation of new truly international—that is, *global*—institutions. But no single country can afford to wait. Moreover, the trajectory of evolving multinational institutional development is itself not independent of the actions of individual nation-states. I am quite persuaded by the stance on this that is being developed by economist Robert Pollin. In the conclusion of a compelling new paper, Pollin writes:

[T]he constraints on . . . viable [domestic] short-term expansionary policies . . . are real and serious, increasingly so as the integration of the global economy within a neo-liberal policy framework proceeds apace. At the same time, it is futile for progressive governments to await the formation of a reconstituted Bretton Woods or some other supportive institutional arrangement to relieve the various external constraints before they are willing to pursue an expansionary program. [There exists a] range of policy initiatives that can be successfully deployed in different national settings, even after taking the full measure of the external constraints and other difficulties.

Of course, international cooperation in support of full employment would greatly facilitate any domestic expansionary program. But to create pressure for such forms of cooperation will require that successful domestic political movements throughout the world demand them. Their success thus creates demonstration effects. This is the path through which international cooperation for full employment and egalitarian growth is most likely to become a reality.[57]

I ended *Lean and Mean* by urging activists not to throw the baby out with the bath water. I read Fainstein and Pollin as cautioning much the same thing. Network forms of organization—and not only in business—are now a permanent part of our landscape. Without in any way abandoning a progressive domestic *and* international politics—Swengydouw's politics at all scales—we had best get on with the task of living with, and finding ways to civilize, the form as we enter the new millennium.

NOTES

CHAPTER 1: BIG FIRMS, SMALL FIRMS, NETWORK FIRMS

1. Calculated from U.S. Bureau of the Census, *1987 Enterprise Statistics, Company Summary*, document ES87-3 (Washington, D.C.: U.S. Government Printing Office, 1991), table 3.

2. "How Every Team Could Win in the HDTV Derby," *Business Week*, March 29, 1993, p. 91; Diane Duston, "HDTV Competitors Agree to Work Together," *Boston Globe*, May 25, 1993, p. 37.

3. Jane Perlez, "Toyota and Honda Create Global Production System," *New York Times*, March 26, 1993, p. A1; Eike Schamp, "Towards a Spatial Reorganisation of the German Car Industry? The Implications of New Production Concepts," in *Industrial Change and Regional Development: The Transformation of New Industrial Spaces*, ed. Georges Benko and Mick Dunford (London: Belhaven Press/Pinter, 1991), pp. 159–70.

4. David Harvey, *The Condition of Postmodernity* (Oxford: Basil Blackwell, 1989).

5. Theodore Levitt, "The Globalization of Markets," in *Strategy: Seeking and Securing Competitive Advantage*, ed. Cynthia A. Montgomery and Michael E. Porter (Boston: Harvard Business School Press, 1991), pp. 187–204.

6. Yves Doz, "International Industries: Fragmentation versus Globalization," in *Technology and Global Industry*, ed. Bruce R. Guile and Harvey Brooks (Washington, D.C.: National Academy Press, 1987), p. 98.

7. Robert B. Reich, *The Work of Nations: Preparing Ourselves for 21st Century Capitalism* (New York: Knopf, 1991).

8. Thus, I am adding the dimensions of power and inequality to the pioneering theorizing of Walter W. Powell, "Neither Market Nor Hierarchy: Network Forms of Organization," in *Research in Organizational Behavior*, ed. Barry M. Straw and Larry L. Cummings (Greenwich, Conn.: JAI Press, 1990), pp. 295–336. On the revitalization and reorganization of the big firms, see Rosabeth Moss Kanter, *When Giants Learn to Dance* (New York: Simon and Schuster, 1990).

9. "In the Labs, the Fight to Spend Less, Get More," *Business Week*, June 28, 1993, pp. 102–4.

10. James P. Womack, Daniel T. Jones, and Daniel Roos, *The Machine That Changed the World* (New York: Rawson/Macmillan, 1990).

11. C. K. Prahalad and Gary Hamel, "The Core Competence of the Corporation," *Harvard Business Review* (May–June 1990), pp. 79–91.

12. Michael L. Dertouzos, Richard K. Lester, and Robert M. Solow, *Made in America: Regaining the Competitive Edge* (Cambridge, Mass.: M.I.T. Press, 1989); Womack, Jones, and Roos, *Machine That Changed*.

13. Andrew Sayer and Richard Walker, *The New Social Economy: Reworking the Division of Labor* (Oxford: Basil Blackwell, 1992), chap. 4.

14. Joseph L. Badaracco, Jr., "Changing Forms of the Corporation," in *The U.S. Business Corporation: An Institution in Transition*, ed. John R. Meyer and James M. Gustafson (Cambridge, Mass.: Ballinger, 1988); Badaracco, *The Knowledge Link: How Firms Compete through Strategic Alliances* (Boston: Harvard Business School Press, 1991); David Mowery, ed., *International Collaborative Ventures in U.S. Manufacturing* (Cambridge, Mass.: Ballinger, 1988).

15. Ronald Dore, *Flexible Rigidities: Industrial Policy and Structural Adjustment in the Japanese Economy 1970–1980* (London: Athlone Press, 1986); Michael L. Gerlach, *Alliance Capitalism: The Social Organization of Japanese Business* (Berkeley: University of California Press, 1992); James R. Lincoln, Michael L. Gerlach, and Peggy Takahashi, "*Keiretsu* Networks in the Japanese Economy: A Dyad Analysis of Intercorporate Ties," *American Sociological Review* 57 (October 1992): 561–85.

16. Nick Oliver and Barry Wilkinson, *The Japanization of British Industry* (Oxford: Basil Blackwell, 1988); and Werner Sengenberger and Duncan Campbell, ed., *Is the Single Firm Vanishing? Inter-Enterprise Networks, Labour and Labour Institutions* (Geneva: International Institute for Labour Studies, International Labour Office, 1992).

17. Maryellen R. Kelley and Todd Watkins, "The Defense Industrial Network: A Legacy of the Cold War" (Heinz School of Public Policy and Management, Carnegie Mellon University, August 1992, unpublished MS).

18. Robert E. Cole, "Diffusion of Participatory Work Structures in Japan, Sweden, and the United States," in *Change in Organizations: New Perspectives on Theory, Research, and Practice*, ed. Paul S. Goodman (San Francisco: Jossey-Bass, 1982); Richard Freeman, "Employee Councils, Worker Participation, and Other Squishy Stuff," in *Proceedings of the Forty-second Annual Meeting, Industrial Relations Research Association* (Madison, Wisc.: IRRA, 1990), pp. 328–37.

19. Barry Bluestone and Irving Bluestone, *Negotiating the Future* (New York: Basic Books, 1992); William N. Cooke, *Labor-Management Cooperation* (Kalamazoo, Mich.: Upjohn Institute for Employment Research, 1990); Maryellen R. Kelley and Bennett Harrison, "Unions, Technology, and Labor-Management Cooperation," in *Unions and Competitiveness*, ed. Lawrence Mishel and Paula Voos (New York: Sharpe, 1992); and David I. Levine and Laura D'Andrea Tyson, "Participation, Productivity, and the Firm's Environment," in *Paying for Productivity*, ed. Alan S. Blinder (Washington, D.C.: Brookings Institution, 1990), pp. 183–236.

20. The literature on labor market segmentation is vast. For a sampling, see Robert Averitt, *The Dual Economy: The Dynamics of American Industrial Structure* (New York: Norton, 1968); Barry Bluestone and Mary Huff Stevenson, "Industrial Transformation and the Evolution of Dual Labor Markets," in *The Dynamics of Labor Market Segmentation*, ed. Frank Wilkinson (New York: Academic Press, 1981); Peter B. Doeringer and Michael J. Piore, *Internal Labor Markets and Manpower Analysis* (Lexington, Mass.: Heath, 1971; reprinted, with a new introduction by M. E. Sharpe, Armonk, N.Y.: 1985); Richard Edwards, *Contested Terrain* (New York: Basic Books, 1979); David M. Gordon, *Theories of Poverty and Underemployment* (Lexington, Mass.: Lexington Books, 1973); David M. Gordon, Richard Edwards, and Michael Reich, *Segmented Work, Divided Workers* (New York: Oxford University Press, 1982); Mark Granovetter and Charles Tilly, "Inequality and Labor Processes," in *Handbook of Sociology*, ed. Neil J. Smelser (Newbury Park, Calif.: Sage, 1988), pp. 175–222; Bennett Harrison, *Education, Training, and the Urban Ghetto* (Baltimore: Johns Hopkins University Press, 1972), chap. 5; Bennett Harrison and Andrew Sum, "The Theory of 'Dual' or Segmented Labor Markets," *Journal of Economic Issues* 13(September 1979): 687–706; Paul Osterman, ed., *Internal Labor Markets* (Cambridge, Mass.: M.I.T. Press, 1984); Michael J. Piore, "The Technological Foundations of Dualism and Discontinuity," in *Dualism and Discontinuity in Industrial Societies*, ed. Suzanne Berger and Michael J. Piore (New York: Cambridge University Press, 1980); and Chris Tilly and Charles Tilly, "Capitalist Work and Labor Markets," in *Handbook of Economic Sociology*, ed. Neil Smelser and Richard Swedberg (New York: Russell Sage, forthcoming).

Long after most writers believed that the dualism of an earlier period of industrialization was disappearing, a British sociologist, John Goldthorpe, presciently forecast an exacerbation of labor market segmentation. See Goldthorpe, "The End of Convergence: Corporatist and Dualist Tendencies in Modern Western Societies," in *Order and Conflict in Contemporary Capitalism*, ed. Goldthorpe (Oxford: Clarendon Press, 1984), pp. 315–43. This paper has been more influential than any single other work I have read in a decade in pointing me in the direction of the research that has culminated in my own new writing on this subject.

21. Susan Christopherson, "Flexibility in the U.S. Service Economy and the Emerging Spatial Division of Labour," *Transactions of the Institute of British Geographers* 14 (1989): 131–43; and Katherine Nelson, "Labor Demand, Labor Supply, and the Suburbanization of Low-Wage Office Workers," in *Production, Work, Territory*, ed. Allen J. Scott and Michael Storper (Boston: Allen & Unwin, 1986), pp. 149–71.

22. Eileen Appelbaum, "Restructuring Work: Temporary, Part Time, and At-home Employment," in *Computer Chips and Paper Clips: Technology and Women's Employment*, ed. Heidi Hartmann (Washington, D.C., National Academy Press, 1987), pp. 268–310; Virginia L. duRivage, ed., *New Policies for the Part-time and Contingent Workforce* (Armonk, N.Y.: Sharpe, for the Economic Policy Institute, 1992), containing papers by Appelbaum, Françoise Carre, Chris Tilly, and the editor; Beverly Lozano, *The Invisible Work Force: Transforming American Business with Outside and Home-Based Workers* (New York: Free Press, 1989); and Womack, Jones, and Roos, *Machine That Changed*.

23. Bennett Harrison and Barry Bluestone, *The Great U-Turn: Corporate Restructuring and the Polarizing of America* (New York: Basic Books, 1988); Frank Levy and Richard Murnane, "U.S. Earnings Levels and Earnings Inequality: A Review of Recent Trends and Proposed Explanations," *Journal of Economic Literature* 30 (September 1992): 1333–81; Tim Smeeding, Michael O'Higgins, and Lee Rainwater, eds., *Poverty, Inequality, and Income Distribution in Comparative Perspective: The Luxembourg Income Study* (London: Harvester Wheatsheaf, 1990).

24. David L. Birch, *Job Creation in America: How Our Smallest Companies Put the Most People to Work* (New York: Free Press, 1987), p.16.

25. "Small Is Beautiful Now in Manufacturing," *Business Week*, October 22, 1984, pp. 152–56.

26. "The Rise and Rise of America's Small Firms," *Economist*, January 21, 1989, pp. 73–74.

27. Tom Peters, "New Products, New Markets, New Competition, New Thinking," *Economist*, March 4, 1989, pp. 27–32.

28. George Gilder, *The Spirit of Enterprise* (New York: Basic Books, 1984) and *Microcosm: The Quantum Revolution in Economics and Technology* (New York: Simon & Schuster, 1989).

29. David Friedman, *The Misunderstood Miracle* (Ithaca, N.Y.: Cornell University Press, 1988).

30. Michael Best, "Sector Strategies and Industrial Policy: The Furniture Industry and the Greater London Enterprise Board," in *Reversing Industrial Decline?*, ed. Paul Hirst and Jonathan Zeitlin (Oxford: Berg, 1989), pp. 191–222; Best, *The New Competition* (Cambridge, Mass.: Harvard University Press, 1990); and Robin Murray, ed., *Technology Strategies and Local Economic Intervention* (Nottingham: Spokesman, 1989).

31. Michael J. Piore and Charles Sabel, *The Second Industrial Divide: Possibilities for Prosperity* (New York: Basic Books, 1984).

32. The key early Italian works are: Arnaldo Bagnasco, *Tre Italie: La Problematica Territoriale Dello Sviluppo Italiano* (Bologna: Il Mulino, 1977); Giacomo Becattini, "Sectors and/or Districts: Some Remarks on the Conceptual Foundations of Industrial Economics," in *Small Firms and Industrial Districts in Italy*, ed. Edward Goodman, Julia Bamford, and Peter Saynor (London: Routledge, 1989), pp. 123–35; Sabastiano Brusco, "Small Firms and Industrial Districts: The Experience of Italy," in *New Firms and Regional Development*, ed. David Keeble and Francis Weever (London: Croom Helm, 1986); and Brusco, "The Emilian Model: Productive Decentralization and Social Integration," *Cambridge Journal of Economics* 6 (June 1982): 167–84. The book that was most responsible for alerting a larger non-Italian public to the existence of the Italian industrial districts was by Piore and Sabel, *Second Industrial Divide*.

An entire United Nations conference in 1990 was devoted to comparing stories about industrial districts in different countries. See Frank Pyke and Werner Sengenberger, eds., *Industrial Districts and Local Economic Regeneration* (Geneva: International Institute for Labour Studies, International Labour Office, 1992).

33. Mark Granovetter, "Economic Action and Social Structure: The Problem of Embeddedness," *American Journal of Sociology* 91 (November 1985): 481–510; Edward Lorenz, "Neither Friends Nor Strangers: Informal Networks of Subcontracting in French Industry," in *Trust: Making and Breaking Cooperative Relations*, ed. Diego Gambetta (Oxford: Basil Blackwell, 1988), pp. 194–210; and Charles Sabel, "Studied Trust: Building New Forms of Cooperation in a Volatile Economy," in Pyke and Sengenberger, *Industrial Districts*.

34. Becattini, "Sectors and/or Districts"; Alfred Marshall, *Industry and Trade* (London: Macmillan, 1927, 3rd ed., originally published in 1919); Alfred Weber, *Theory of the Location of Industry*, trans. Carl Friedrich (Chicago: University of Chicago Press, 1929).

35. Allen J. Scott, *New Industrial Spaces* (London: Pion, 1988).

36. Piore and Sabel, *Second Industrial Divide.*

37. See Best, "Sector Strategies and Industrial Policy"; Best, *New Competition;* and Murray, *Technology Strategies.*

38. C. Richard Hatch, "Learning from Italy's Industrial Renaissance," *Entrepreneurial Economy* 6 (July/August 1987): 1–5.

39. Steven J. Davis, "Size Distribution Statistics from County Business Patterns Data" (Graduate School of Business, University of Chicago, September 1990, unpublished MS).

40. Bennett Harrison and Maryellen R. Kelley, "Outsourcing and the Search for 'Flexibility'," *Work, Employment and Society* 7 (June 1993): 213–35; Kelley and Harrison, "The Subcontracting Behavior of Single vs. Multiplant Enterprises in U.S. Manufacturing: Implications for Economic Development," *World Development* 18 (September 1990): 1273–94.

41. Bo Carlsson and Erol Taymaz, "Flexible Technology and Industrial Structure in the U.S.," *Small Business Economics,* forthcoming.

42. For example, on the United States, see Jonathan Leonard, "On the Size Distribution of Employment and Establishments" (Haas School of Business, University of California at Berkeley, 1985, unpublished MS); and Leonard, "In the Wrong Place at the Wrong Time: The Extent of Frictional and Structural Employment," in *Unemployment and the Structure of Labor Markets,* ed. Kevin Lang and Jonathan Leonard (Oxford: Basil Blackwell, 1987). For similar evidence on Great Britain, see David Branchflower, N. Millward, and A. Oswald, "Unionism and Employment Behaviour," *Economic Journal* 101 (July 1991): 815–34.

43. David J. Storey and Steven G. Johnson, *Job Generation and Labour Market Change* (London: Macmillan, 1987), pp. 24–25.

44. In this country, such research has been most significantly advanced by Timothy Dunne, Mark Roberts, and Larry Samuelson; see "Patterns of Firm Entry and Exit in U.S. Manufacturing Industries," *Rand Journal of Economics* 19 (Winter 1988): 495–515; "Plant Turnover and Gross Employment Flows in the U.S. Manufacturing Sector," *Journal of Labor Economics* 7 (January 1989): 48–71; and "The Growth and Failure of U.S. Manufacturing Plants," *Quarterly Journal of Economics* 104 (November 1989): 671–98. See also Steven J. Davis and John Haltiwanger, "Gross Job Creation, Gross Job Destruction, and Employment Reallocation," *Quarterly Journal of Economics* 108 (August 1992): 819–63.

For similar analyses conducted in Germany, see Josef Bruderl and Rudolf Schussler, "Organizational Mortality: The Liabilities of Newness and Adolescence," *Administrative Science Quarterly* 35 (September 1990): 530–47.

45. David Wiesel and Buck Brown, "The Hyping of Small-Firm Job Growth," *Wall Street Journal*, November 8, 1988, p. B1.

46. "Small Businesses Tend to Stay Pint-Size," *Business Week*, July 31, 1989, p. 20. The full report is by Douglas P. Handler, "Business Demographics," Economic Analysis Department, Dun & Bradstreet Corporation, New York, unpublished MS, 1989. For a more theoretically sophisticated critique, see Stephen Fothergill's review of Birch's book in *Environment and Planning* A 21 (June 1989): 842–43.

47. Storey and Johnson, *Job Generation.*

48. David J. Storey and Steven G. Johnson, *Job Creation in Small and Medium Sized Enterprises*, vol. 1 (Luxembourg: Office for Official Publications of the European Community, 1987), p. 16.

49. Gary Loveman and Werner Sengenberger, "Introduction: Economic and Social Reorganization in the Small and Medium-Sized Enterprise Sector," in *The Re-Emergence of Small Enterprises: Industrial Restructuring in Industrialized Countries*, ed. Sengenberger, Loveman, and Michael J. Piore (Geneva: International Institute for Labour Studies, International Labour Office, 1990), p. 21.

50. David M. Gordon, *The Working Poor: Towards a State Agenda* (Washington, D.C.: Council of State Planning Agencies, 1979); Doeringer and Piore, *Internal Labor Markets and Manpower Analysis.* Gordon's professors at Harvard in the late 1960s had theorized such a relationship, but Gordon was the first to actually measure it in detail.

51. Charles Brown, James Hamilton, and James Medoff, *Employers Large and Small* (Cambridge, Mass.: Harvard University Press, 1990).

52. The leading spokespersons for this view are two American economists, Zoltan Acs and David Audretsch. See *Innovation and Small Firms* (Cambridge, Mass.: M.I.T. Press, 1990); "Innovation, Market Structure, and Firm Size," *Review of Economics and Statistics* 69 (November 1987): 567–74; Acs, Audretsch, and Bo Carlsson, "Flexibility, Plant Size, and Industrial Restructuring," in *The Economics of Small Firms: A European Challenge*, ed. Acs and Audretsch (Boston: Kluwer, 1990). See also Carlsson, "The Evolution of Manufacturing Technology and Its Impact on Industrial Structure: An International Study," *Small Business Economics* 1 (Spring 1989): 21–37. This is also a central theme in Piore and Sabel, *Second Industrial Divide.*

53. Gilder, *Microcosm*.

54. Wesley M. Cohen and Richard C. Levin, "Empirical Studies of Innovation and Market Structure," in *Handbook of Industrial Organization*, vol. 2., ed. Richard Schmalense and Richard Willie (New York: North-Holland, 1989), pp. 1059–107; Cohen and Steven Klepper, "Firm Size versus Diversity in the Achievement of Technological Advance," *Small Business Economics*, forthcoming.

55. A close reading of statistics from McGraw-Hill's *American Machinist* magazine reveals that, between 1983 and 1989, the share of a plant's tool stock consisting of computer-controlled equipment—what engineers call programmable automation (PA)—grew sytematically larger, the *larger* the initial size of the plant. Moreover, in the course of the period, while the penetration rate rose in every plant size category, it rose proportionately *more*, the larger the plant. Thus, between 1983 and 1989, the PA penetration rate within the class of very small plants increased by 75 percent, but among plants with 500 or more employees, it rose by 142 percent (Bo Carlsson and Erol Taymaz, "Flexible Technology and Industrial Structure in the U.S.," *Small Business Economics*, forthcoming).

56. Quoted by Lawrence M. Fisher, "Intel Raising Capacity of Chip Factory," *New York Times*, April 2, 1993, p. C2.

57. Loveman and Sengenberger, "Economic and Social Reorganization," p. 48.

58. Ash Amin and Kevin Robins, "The Re-Emergence of Regional Economies? The Mythical Geography of Flexible Accumulation," *Environment and Planning D: Society and Space* 8 (March 1990): 7–34; Fiorenza Belussi, "Benetton Italy: Beyond Fordism and Flexible Specialization to the Evolution of the Network Firm Model," in *Information Technology and Women's Employment: The Case of the European Clothing Industry*, ed. S. Mitter (Berlin: Springer Verlag, 1989); Flavia Martinelli and Erica Schoenberger, "Oligopoly Is Alive and Well: Notes for a Broader Discussion of Flexible Accumulation," in *Industrial Change and Regional Development: The Transformation of New Industrial Spaces*, ed. Georges Benko and Mick Dunford (London: Belhaven Press/Pinter, 1991), chap. 6.

59. Gilder, *Microcosm*. See also Everett Rogers and Judith Larsen, *Silicon Valley Fever* (New York: Basic Books, 1984).

60. AnnaLee Saxenian, "Regional Networks and the Resurgence of Silicon Valley," *California Management Review* 33 (Fall 1990): 89–112.

61. Charles H. Ferguson, "From the People Who Brought You Voodoo Economics," *Harvard Business Review* 66 (May–June 1988): 55–62; Richard Florida and Martin Kenney, *The Breakthrough Illusion* (New York: Basic Books, 1990);

Richard Gordon, "Innovation, Industrial Networks, and High Technology Regions," in *Innovation Networks: Spatial Perspectives*, ed. Roberto Camagni (London: Belhaven Press, 1991), pp. 174–95; Gordon, "State, Milieu, Network: Systems of Innovation in Silicon Valley," in *Systems of Innovation*, ed. Patrizio Bianchi and M. Quere (Paris: Groupe de Recherche Europeen sue les Milieux Innovateurs, forthcoming); Ann Markusen, Peter Hall, Scott Campbell, and Sabina Deitrick, *The Rise of the Gunbelt* (New York: Oxford University Press, 1990); David Teece, "Foreign Investment in Silicon Valley," *California Management Review* 34 (Winter 1992): 88–106.

62. This description is based on an exhaustive, multiyear case study of Benetton, conducted by Fiorenza Belussi, an Italian trade union researcher and a University of Sussex scholar in technology policy. See Belussi, "Benetton: Information Technology in Production and Distribution: A Case Study of the Innovative Potential of Traditional Sectors," Science Policy Research Unit, University of Sussex, Brighton, U.K., SPRU Occasional Paper No. 25, 1987; Belussi, "Benetton Italy"; Belussi, "La Flessibilita si fa Gerarchia: la Benetton," in *Nuovi Modelli D'Impresa Gerarchie Organizzative E Imprese Rete*, ed. Belussi (Milan: Franco Angeli, 1992), pp. 287–340; and Belussi and Massimo Festa, "L'Impresa Rete Del Modello Veneto: Dal Post-Fordismo Al Toyotismo? Alcune Note Illustrative Sulle Strutture Organizzative Dell'Indotto Benetton," IRES-CGIL, Maestre, Veneto, November 1990, unpublished MS.

63. AnnaLee Saxenian, "The Urban Contradictions of Silicon Valley: Regional Growth and the Restructuring of the Semiconductor Industry," in *Sunbelt-Snowbelt: Urban Development and Regional Restructuring*, ed. Larry Sawers and William K. Tabb (New York: Oxford University Press, 1984), pp. 163–97.

64. Also see Florida and Kenney, *Breakthrough Illusion*, chap. 7; Lenny Siegel and Herb Borock, *Background Report on Silicon Valley: Report to the U.S. Civil Rights Commission* (Mountain View, Calif.: Pacific Studies Center, 1982); Siegel and John Markoff, *The High Cost of High Tech* (New York: Harper & Row, 1985).

65. Richard Gordon and Linda M. Kimball, "High Technology, Employment and the Challenges to Education," Silicon Valley Research Group, University of California at Santa Cruz, July 1985, unpublished MS; Paul M. Ong, "The Widening Divide: Income Inequality and Poverty in Los Angeles," Department of Urban Planning, School of Architecture and Urban Planning, University of California at Los Angeles, 1989, unpublished MS; Michael Storper and Allen Scott, "Work Organization and Local Labour Markets in an Era of Flexible Production," working paper no. 30, World Employment Programme Research, International Labour Office, Geneva, 1989.

66. Robert B. Reich, "Who Is Us?" *Harvard Business Review* 68 (January–February 1990): 53–64; Reich, *The Work of Nations*; Laura D'Andrea Tyson,

"They Are Not Us: Why American Ownership Still Matters," *American Prospect* (Winter 1991): 37–49; Reich, "Who Do We Think They Are?" *American Prospect* (Winter 1991): 49–53; Michael Porter, *The Competitive Advantage of Nations* (New York: Free Press, 1990); William Lazonick, "Industry Clusters versus Global Webs: Organizational Capabilities in the U.S. Economy" (Department of Economics, Barnard College, Columbia University, unpublished revised MS of March 15, 1992); D. Eleanor Westney, "The U.S. Multinational Corporation and the Globalization of Competition" (paper presented at the Thematic Session of the Annual Meeting of the American Sociological Association, Washington, D.C., August 13, 1990); and John H. Dunning, "The Global Economy and the National Governance of Economies: A Plea for a Fundamental Re-Think" (Graduate School of Management, Rutgers University, Newark, N.J., November 1992, unpublished MS).

67. Compare Paul Krugman, ed., *Strategic Trade Policy and International Economics* (Cambridge, Mass.: M.I.T. Press, 1986); Laura D'Andrea Tyson, *Who's Bashing Whom?* (Washington, D.C.: Institute for International Economics, 1992).

68. Norman J. Glickman and Douglas P. Woodward, *The New Competitors: How Foreign Investors Are Changing the U.S. Economy* (New York: Basic Books, 1989); Edward M. Graham and Paul R. Krugman, *Foreign Direct Investment in the United States* (Washington, D.C.: Institute for International Economics, 1989.

69. Barry Bluestone and Bennett Harrison, *The Deindustrialization of America* (New York: Basic Books, 1982); U.S. Department of Commerce, Bureau of the Census, "Workers with Low Earnings: 1964 to 1990," *Current Population Reports: Consumer Income*, series P-60, no. 178 (Washington, D.C.: U.S. Government Printing Office, March 1992); Richard B. Freeman and Lawrence F. Katz, "Rising Wage Inequality: The United States vs. Other Advanced Countries," in *Working Under Different Rules*, ed. Richard B. Freeman (New York: Russell Sage, 1994), pp. 29–62; Harrison and Bluestone, *Great U-Turn*; Lynn A. Karoly, "The Trend in Inequality among Families, Individuals and Workers in the United States: A Twenty-five Year Prospective," in *Uneven Tides: Rising Inequality in America*, ed. Sheldon Danziger and Peter Gottschalk (New York: Russell Sage Foundation, 1993), pp. 19–97; Robert Kuttner, "The Declining Middle," *Atlantic Monthly*, July 1983, pp. 60–69; Levy and Murnane, "U.S. Earnings Levels and Earnings Inequality." A history of the often contentious debate during the 1980s over the trend toward polarizing incomes in the United States is offered by James Lardner, "The Declining Middle," *New Yorker*, May 3, 1993, pp. 108–14.

70. See, for example, Werner Sengenberger and Duncan Campbell, eds., *Lean Production and Beyond: Labour Aspects of a New Production Concept* (Geneva: International Institute for Labour Studies, 1993).

71. Richard B. Freeman and Joel Rogers, "Who Speaks for Us? Employee Representation in a Non-Union Labor Market," in *Employee Representation: Alternatives and Future Directions*, ed. Bruce E. Kaufman and Morris M. Kleiner (Milwaukee, Wis.: Industrial Relations Research Association, 1993), pp. 13–80.

72. Robert Kuttner, *The End of Laissez-Faire* (New York: Knopf, 1991).

73. David Harvey, *The Limits to Capital* (Oxford: Basil Blackwell, 1982); Andrew Sayer and Richard Walker, *The New Social Economy: Reworking the Division of Labor* (Oxford: Basil Blackwell, 1992), p. 201; Erica Schoenberger, "Some Dilemmas of Automation," *Economic Geography* 65 (July 1989): 232–47.

74. Amy K. Glasmeier, *The High-Tech Potential: Economic Development in Rural America* (New Brunswick, N.J.: Center for Urban Policy Research, Rutgers University, 1991); Stuart A. Rosenfeld, *Competitive Manufacturing: New Strategies for Regional Development* (New Brunswick, N.J.: Center for Urban Policy Research, Rutgers University, 1992); Sabel, "Studied Trust."

75. Richard Barnet and Ronald Muller, *Global Reach* (New York: Simon & Schuster, 1973); Stephan Hymer and Robert Rowthorn, "Multinational Corporations and International Oligopoly," in *The International Corporation*, ed. Charles P. Kindleberger (Cambridge, Mass.: M.I.T. Press, 1970), pp. 57–91; Raymond Vernon, *Sovereignty at Bay* (New York: Basic Books, 1971); and Gordon L. Clark, *Unions and Communities Under Siege* (New York: Cambridge University Press, 1989).

76. Thomas M. Jorde and David J. Teece, "Innovation and Cooperation: Implications for Competition and Antitrust," *Journal of Economic Perspectives* 4 (Summer 1990): 75–95.

CHAPTER 2: THE MYTH OF SMALL FIRMS AS JOB GENERATORS

1. Gary Loveman and Werner Sengenberger, "Introduction: Economic and Social Reorganization in the Small and Medium-Sized Enterprise Sector," in *The Re-Emergence of Small Enterprises: Industrial Restructuring in Industrialized Countries*, ed. Sengenberger, Loveman, and Michael J. Piore (Geneva: International Institute for Labour Studies, International Labour Office, 1990), pp. 1–61; Steven J. Davis, "Size Distribution Statistics from County Business Patterns Data" (Graduate School of Business, University of Chicago, September 1990, unpublished MS).

2. During the last complete peak-to-peak business cycle, bounded by the years 1979 and 1989, annual productivity growth across all U.S. industries averaged only 1.09 percent per year—far behind the rates of productivity growth of France, Italy, or Japan. Even the United Kingdom managed to outperform the United States during the 1980s. See Lawrence Mishel and Jared

Bernstein, *The State of Working America: 1992–93* (New York: Sharpe, 1993), p. 422. Over the longer period from 1973 through 1989, the American experience of *less than* 1 percent annual growth in productivity was bettered by every other major industrialized country in the world except for the Netherlands (ibid.).

3. Quoted in David Wiesel and Buck Brown, "The Hyping of Small-Firm Job Growth," *Wall Street Journal*, November 8, 1988, p. B1.

4. On the United States, see Barry Bluestone and Bennett Harrison, *The Deindustrialization of America* (New York: Basic Books, 1982); and Harrison and Bluestone, *The Great U-Turn: Corporate Restructuring and the Polarizing of America* (New York: Basic Books, 1988). On Italy, see Richard M. Locke, "Local Politics and Industrial Adjustment: The Political Economy of Italy in the 1980s" (Ph.D. diss., M.I.T., February 1989); and Locke and Serafino Negrelli, "Il Case FIAT AUTO," in *Strategie di Riaggiustamento Industriale*, ed. Marino Regini and Charles Sabel (Bologna: Il Mulino, 1989).

5. Thomas Kochan, Harry C. Katz, and Robert B. McKersie, *The Transformation of American Industrial Relations* (New York: Basic Books, 1986); Anil Verma and Kochan, "The Growth and Nature of the Nonunion Sector within a Firm," in *Challenges and Choices Facing American Labor*, ed. Kochan (Cambridge, Mass.: M.I.T. Press, 1985), pp. 89–118.

6. Barry Bluestone, Peter Jourdan, and Mark Sullivan, *Aircraft Industry Dynamics* (Boston: Auburn House, 1981); Susan Christopherson and Michael Storper, "New Forms of Labor Segmentation and Industrial Politics in Flexible Production Industries," *Industrial and Labor Relations Review* 42 (1988): 331–47; and the papers in *Production, Work, Territory*, ed. Allen Scott and Michael Storper (London: Allen & Unwin, 1986).

7. Bluestone and Harrison, *Deindustrialization of America*.

8. Mishel and Bernstein, *State of Working America*, chap. 4.

9. Michael L. Dertouzos, Richard K. Lester, and Robert M. Solow, *Made in America: Regaining the Competitive Edge* (Cambridge, Mass.: M.I.T. Press, 1989); C. K. Prahalad and Gary Hamel, "The Core Competence of the Corporation," in *Strategy: Seeking and Securing Competitive Advantage*, ed. Cynthia A. Montgomery and Michael E. Porter (Boston: Harvard Business School Press, 1991), pp. 277–300; James P. Womack, Daniel T. Jones, and Daniel Roos, *The Machine That Changed the World* (New York: Rawson/Macmillan, 1990).

10. "The Job Generation Process" (Department of Urban Studies and Planning, M.I.T., 1979, unpublished MS).

11. The sole exception was a short essay, "Who Creates Jobs?" published in *Public Interest*, no. 65 (Fall 1981): 4–14.

12. George Gilder, *The Spirit of Enterprise* (New York: Basic Books, 1984).

13. Organization for European Co-operation and Development, "Employment in Small and Large Firms: Where Have the Jobs Come From?" *Employment Outlook* (Paris: OECD, September 1985), chap. 4.

14. David L. Birch, *Job Generation in America* (New York: Free Press, 1987).

15. Ibid., p. 16.

16. See Case's "The Disciples of David Birch," *Inc.*, January 1989, pp. 39–45; and Case, *From the Ground Up: The Resurgence of American Entrepreneurship* (New York: Simon & Schuster, 1992), esp. pp. 20–41. So politically charged is this subject that Case, too, seems not always to hear precisely what some of his interviewees say to him about the "Birch numbers." Thus, for example, Case reports on interviewing Sammis White, the director of the Urban Research Center and an associate professor of urban planning at the University of Wisconsin at Milwaukee. Case reports that White's studies of the sources of job creation and destruction in Milwaukee, using the U.S. Labor Department's ES-202 payroll records, supported "the phenomenon that David Birch discovered—in spades" (Case, *From the Ground Up*, p. 40).

   Yet in the conclusion to his own published professional paper on this same research, White and a colleague wrote: "The prevailing wisdom on job generation is not a universal truth, in that small establishments do not create the majority of jobs in most industries, geographic areas, and time periods" (Sammis B. White and Jeffrey D. Osterman, "Is Employment Growth Really Coming from Small Establishments?" *Economic Development Quarterly*, 5 [August 1991]: 256).

17. Catherine Armington and Marjorie Odle, "Small Business—How Many Jobs?" *Brookings Review* 1 (Winter 1982): 14–17; Douglas P. Handler, "Business Demographics" (New York: Economic Analysis Department, Dun & Bradstreet, 1989); Gene Koretz, "Small Businesses Tend to Stay Pint-Size," *Business Week*, July 31, 1989, p. 20; and Wiesel and Brown, "Hyping of Small-Firm Job Growth." For a more theoretically sophisticated critique, see Stephen Fothergill's review of Birch's book in *Environment and Planning A*, June 1989.

18. Quoted in Wiesel and Brown, "Hyping of Small-Firm Job Growth," p. B1.

19. Sengenberger, Loveman, and Piore, *Re-Emergence of Small Enterprises*.

20. Michael J. Piore and Charles Sabel, *The Second Industrial Divide: Possibilities for Prosperity* (New York: Basic Books, 1984).

21. This cyclical sensitivity of small and large firm employment shares appears in the European data, as well. See David J. Storey and Steven G. Johnson, *Job Creation in Small and Medium Sized Enterprises*, 2 vols. (Luxembourg: Office for Official Publications of the European Community, 1987). In an even more technically sophisticated analysis of the U.S. data, Davis and Haltiwanger conclude that rates of job creation and destruction show "no systematic cyclical variation" among "younger, smaller, and single-unit plants," but that "rates among older, larger, and multi-unit plants show pronounced countercyclic variation" (Steven J. Davis and John Haltiwanger, "Gross Job Creation, Gross Job Destruction, and Employment Reallocation," *Quarterly Journal of Economics* 107 [August 1992]: 861). In other words, during recessions, employment in large firms and establishments is much more volatile than in small firms and establishments.

22. Bennett Harrison, "The Myth of Small Firms as the Predominant Job Generators," *Economic Development Quarterly* 8 (February 1994): 3–18.

23. U.S. Bureau of the Census, *1987 Enterprise Statistics, Company Summary*, bulletin ES87-3 (Washington, D.C.: U.S. Government Printing Office, June 1991), table 3.

24. Michael B. Teitz, Amy Glasmeier, and Philip Shapira, "Small Business and Employment Growth in California," working paper no. 348 (Berkeley: Institute of Urban and Regional Development, University of California, March 1981).

25. Douglas P. Handler, "Business Demographics" (New York: Economic Analysis Department, Dun & Bradstreet, 1989).

26. Jonathan S. Leonard, "On the Size Distribution of Employment and Establishments," working paper no. 1951 (Cambridge, Mass.: National Bureau of Economic Research, June 1986).

27. These technical problems—the "size distribution fallacy," which results from doing the arithmetic in terms of net employment changes but fixed size categories, and the "regression fallacy" first observed by Jon Leonard—have now been definitively studied and documented by Steven J. Davis, John Haltiwanger, and Scott Schuh. Compare "Small Business and Job Creation: Dissecting the Myth and Reassessing the Facts" (Cambridge, Mass.: National Bureau of Economic Research, October 1993), Working Paper 4492; and their forthcoming book, *Job Creation and Destruction in U.S. Manufacturing.*

28. Bennett Harrison and Maryellen R. Kelley, "Outsourcing and the Search for 'Flexibility'," *Work, Employment and Society* 7 (June 1993): 213–35; Kelley and Harrison, "The Subcontracting Behavior of Single vs. Multiplant Enterprises in U.S. Manufacturing: Implications for Economic Development," *World Development* 18 (September 1990): 1273–94; Kelley and Harrison, "Unions,

Technology, and Labor-Management Cooperation," in *Unions and Competitiveness*, ed. Lawrence Mishel and Paula Voos (New York: Sharpe, 1992), pp. 247–86.

29. Timothy Dunne, Mark J. Roberts, and Larry Samuelson, "The Growth and Failure of U.S. Manufacturing Plants," *Quarterly Journal of Economics* 104 (November 1989): 671–98.

30. Josef Bruderl and Rudolf Schussler, "Organizational Mortality: The Liabilities of Newness and Adolescence," *Administrative Science Quarterly* 35 (September 1990): 530–47.

31. David M. Gordon, *The Working Poor: Towards a State Agenda* (Washington, D.C.: Council of State Planning Agencies, 1979); Charles Brown, James Hamilton, and James Medoff, *Employers Large and Small* (Cambridge: Harvard University Press, 1990).

32. René Morissette, "Canadian Jobs and Firm Size: Do Smaller Firms Pay Less?" Business and Labor Market Analysis Group, Analytical Studies Branch, Statistics Canada, Ottawa, Ontario (Research Paper Series 35, 1991).

33. Loveman and Sengenberger, "Economic and Social Reorganization," p. 38. Among all the OECD countries for which there are data, Loveman and Sengenberger found the wage gap by firm or plant size to be smallest in Germany. But even there, employees in facilities employing fewer than 100 workers earn on average as much as 20 percent less than the largest employers.

34. F. Blackaby, ed., *Deindustrialization* (London: Heinemann, 1979); Robert Martin and Robert Rowthorn, eds., *The Geography of De-industrialization* (London: Macmillan, 1986); Agit Singh, "U.K. Industry and the World Economy: A Case of De-industrialization?" *Cambridge Journal of Economics* 1 (June 1977): 113–36.

35. Werner Sengenberger and Frank Pyke, "Industrial Districts and Local Economic Regeneration: Research and Policy Issues," in *Industrial Districts and Local Economic Regeneration*, ed. Pyke and Sengenberger (Geneva: International Institute for Labour Studies, International Labour Office, 1992), p. 11.

CHAPTER 3: ARE SMALL FIRMS THE TECHNOLOGY LEADERS?
1. David L. Birch, *Job Creation in America: How Our Smallest Companies Put the Most People to Work* (New York: Free Press, 1987); George Gilder, *The Spirit of Enterprise* (New York: Basic Books, 1984); and Gilder, *Microcosm: The Quantum Revolution in Economics and Technology* (New York: Simon & Schuster, 1989).

2. Giovanni Dosi, "Sources, Procedures, and Microeconomic Effects of Innovation," *Journal of Economic Literature* 26 (September 1988): 1120–71; Michael J. Piore and Charles F. Sabel, *The Second Industrial Divide: Possibilities for Prosperity* (New York: Basic Books, 1984).

3. John Kenneth Galbraith, *American Capitalism: The Concept of Countervailing Power* (Boston: Houghton Mifflin, 1956), p. 86.

4. Joseph L. Schumpter, *The Theory of Economic Development* (Cambridge, Mass.: Harvard University Press, 1934); Schumpeter, *Capitalism, Socialism, and Democracy* (New York: Harper & Row, 1942).

5. F. M. Scherer and David Ross, *Industrial Market Structure and Economic Performance*, 3rd ed. (Boston: Houghton Mifflin, 1990).

6. Sidney G. Winter, "Schumpeterian Competition in Alternative Technological Regimes," *Journal of Economic Behavior and Organization* 5 (1984): 287–320.

7. Wesley M. Cohen and Steven Klepper, "The Tradeoff Between Firm Size and Diversity in the Pursuit of Technological Progress," *Small Business Economics* 4 (January 1992): 1–14. Their formulation builds on Richard Nelson's conviction that competition and diversity can force technological change in the face of institutional resistance, on Winter's theory that large and small firms take qualitatively different paths to innovation, and on Cohen and Levinthal's very important idea that different firms have different organizational capacities to *absorb* new information and to accommodate the new routines associated with a change of industrial technique (Wesley M. Cohen and Daniel A. Levinthal, "Absorptive Capacity: A New Perspective on Learning and Innovation," *Administrative Science Quarterly* 35 [1990]: 128–52); Richard Nelson, "Capitalism as an Engine of Progress," *Research Policy* 19 (1990): 119–32; Winter, "Schumpeterian Competition."

8. Zoltan J. Acs and David B. Audretsch, "Innovation, Market Structure, and Firm Size," *Review of Economics and Statistics* 69 (November 1987): 567–75; Acs and Audretsch, "Innovation in Large and Small Firms: An Empirical Analysis," *American Economic Review* 78 (September 1988): 678–90; Acs and Audretsch, *Innovation and Small Firms* (Cambridge, Mass.: M.I.T. Press, 1990); Acs and Audretsch, eds., *The Economics of Small Firms: A European Challenge* (Boston: Kluwer Academic, 1990).

9. Keith L. Edwards and Theodore J. Gordon, "Characterization of Innovations Introduced on the U.S. Market in 1982," The Futures Group, unpublished report prepared for the SBA under contract no. SBA-6050-OA-82, March 1984.

10. Acs and Audretsch, *Innovation in Small Firms*, chap. 2, esp. p. 16 and the appendix.

11. Ibid., p. 21.

12. Acs and Audretsch, "Small Firms in the 1990s," in Acs and Audretsch, *Economics of Small Firms*, p. 6.

13. Given these limitations, this research seems almost *too* precise. For example, Acs and Audretsch report that, in 1982, there were precisely 28 innovations in the metal office furniture industry, of which 25 were generated by large firms with 500 or more employees; and that there were 395 innovations in electronic computing equipment, of which 10 were not allocable by firm size and 227 of which were produced within small firms. Surely there must be at least some error in the underlying measures!

14. Gary Loveman and Werner Sengenberger, "Introduction: Economic and Social Reorganization in the Small and Medium-Sized Enterprise Sector," in Sengenberger, Loveman, and Michael J. Piore, eds., *The Re-Emergence of Small Enterprises: Industrial Restructuring in Industrialized Countries* (Geneva: International Institute for Labour Studies, International Labour Office, 1990), p. 47. It should be noted that, as discussed in the last two chapters, the Piore-Sabel thesis places at least equal weight on the crisis of mass production industries— a decline in the importance of standardization, per se—as on such enabling technologies as computer-controlled factory automation.

15. Dosi, "Microeconomic Effects of Innovation," p. 1155.

16. Zolton J. Acs, David B. Audretsch, and Bo Carlsson, "Flexibility, Plant Size and Industrial Restructuring," in Acs and Audretsch, *Economics of Small Firms*, p. 141.

17. Here I am working principally from Acs and Audretsch, *Innovation and Small Firms*, esp. chap. 6. But see also Bo Carlsson and Erol Taymaz, "Flexible Technology and Industrial Structure in the U.S.," *Small Business Economics*, forthcoming.

18. Acs and Audretsch estimate ten alternative statistical, that is, multiple regression interindustry models to explain changes between the mid-1970s and the mid-1980s in the small organization share of sales. "Small" units are first defined as plants (establishments) with fewer than 500 employees, then again as those with fewer than 100. The share of each industry's machine tools that are computer controlled—the variable at the center of their story— is statistically significant at a generally acceptable (95 percent) level of confidence in only one equation. Among the five equations intended to explain declining mean plant size over time as a function of the NC-CNC penetration rate, the technology effect is significantly different from zero in only two. It is difficult to see how such strong conclusions can be offered from such weak results.

19. Such discussions of the properties of the *American Machinist* sample as have been made public appear in Carlsson and Taymaz, "Flexible Technology," footnote 1; in "The 13th American Machinist Inventory of Metalworking Equipment, 1983," *American Machinist* 127 (November 1983): 113–44; and in "The 14th Inventory of Metalworking Equipment: Big Gains for Smaller Plants," *American Machinist* 133 (November 1989): 91–110.

20. Bela Gold, "Changing Perspectives on Size, Scale and Returns: An Interpretive Survey," *Journal of Economic Literature* 19 (March 1981): 31.

21. Maryellen R. Kelley, "Organizational Resources and the Industrial Environment: The Importance of Firm Size and Inter-Firm Linkages to the Adoption of Advanced Manufacturing Technology," *International Journal of Technology Management* 8 (November 1993): 36–68; Kelley, "Productivity and Information Technology: The Illusive Connection," *Management Science*, forthcoming; Kelley and Harvey Brooks, "External Learning Opportunities and the Diffusion of Process Innovations to Small Firms: The Case of Programmable Automation," *Technological Forecasting and Social Change* 39 (April 1991): 103–25.

22. Timothy Dunne, "Technology Usage in U.S. Manufacturing Industries: New Evidence from the Survey of Manufacturing Technology," paper CES 91-7 (Washington, D.C.: Center for Economic Studies, U.S. Bureau of the Census, November 1991), pp. 2–3; U.S. Bureau of the Census, *Manufacturing Technology 1988, Current Industrial Reports* (Washington, D.C.: U.S. Government Printing Office, May 1989).

23. W. W. Daniel, *Workplace Industrial Relations and Technical Change* (London: Policy Studies Institute/Pinter, 1987); Michael White, H.-J. Braczyk, A. Ghobadian, and J. Niebuhr, *Small Firms' Innovation: Why Regions Differ* (London: Policy Studies Institute/Pinter, 1988).

24. Ray P. Oakey and Paul N. O'Farrell, "The Regional Extent of Computer Numerically Controlled (CNC) Machine Tool Adoption and Post Adoption Success in Small British Mechanical Engineering Firms," *Regional Studies* 26 (1992): 163–75.

It is surprising that Acs, Audretsch, and others seem to take for granted that new technology, once acquired by a firm, is implicitly more immediately productive (if they did *not* assume this, then from whence comes the alleged competitive advantage?). There are many reasons why the additional efficiency associated with the introduction of new technology should be seen as a research question, rather than taken as fact. The British evidence I just cited demonstrates that adoption may be more and less successful. The Harvard Business School professor Ramchadran Jaikumar has shown that, compared with comparable Japanese firms, American adopters of flexible manufacturing systems are more likely to use them in a "taylorist" context, to reduce costs, rather than for broadening the range or increasing the quality of products (Jaikumar, "Postindustrial Manufacturing," *Harvard Business Review* 64 [1986]: 69–76).

For a review of the theoretical issues, and for new findings on the comparative efficiency of programmable automation among U.S. metalworking establishments, see Kelley, "Productivity and Information Technology."

25. Kevin McCormick, "Small Firms, New Technology and the Division of Labour in Japan," *New Technology, Work and Employment* 3 (Autumn 1988): 134–42.

26. The latter are reported in Congress of the United States, Office of Technology Assessment, *Making Things Better* (Washington, D.C.: U.S. Government Printing Office, 1990), chap. 6.

27. Hans-Jurgen Ewers, Carsten Becker, and Michael Fritsch, "The Effects of the Use of Computer-Aided Technology in Industrial Enterprises: It's the Context That Counts," in *Technical Change and Employment*, ed. Ronald Schettkat and Michael Wagner (New York: deGruyter, 1989), pp. 25–64.

28. Guido Rey, "Small Firms: Profile and Analysis, 1981–85," in *Small Firms and Industrial Districts in Italy*, ed. Edward Goodman, Julia Bamford, and Peter Saynor (London: Routledge, 1989).

29. Fabio Arcangeli, Giovanni Dosi, and Massimo Moggi, "Patterns of Diffusion of Electronics Technologies: An International Comparison with Special Reference to the Italian Case" (University of Rome, "La Sapienza," November 1989, unpublished MS).

30. Sergio Mariotti and Marco Mutinelli, "Diffusion of Flexible Automation in Italy," in *Computer Integrated Manufacturing, Volume III: Models, Case Studies, and Forecasts of Diffusion*, ed. Robert U. Ayres, William Haywood, and I. Tchijov (London: Chapman & Hall, 1992), p. 145.

31. Ewers, Becker, and Fritsch, "Context That Counts."

32. Kelley and Brooks, "External Learning Opportunities."

33. Birch, *Job Creation in America*, pp. 70–76.

34. Walter W. Powell and Peter Brantley, "Competitive Cooperation in Biotechnology: Learning through Networks?" in *Networks and Organizations: Structure, Form, and Action*, ed. Nitin Nohria and Robert G. Eccles (Boston: Harvard Business School Press, 1992), chap. 14.

35. Richard Florida and Martin Kenney, *The Breakthrough Illusion* (New York: Basic Books, 1990); Luigi Orsenigo, *The Emergence of Biotechnology:*

Institutions and Markets in Industrial Innovation (New York: St. Martin's Press, 1989).

36. Martin Kenney, *Biotechnology: The University-Industry Complex* (New Haven: Yale University Press, 1986).

37. James M. Utterback, "Innovation and Industrial Evolution in Manufacturing Industries," in *Technology and Global Industry: Companies and Nations in the World Economy*, ed. Bruce R. Guile and Harvey Brooks (Washington, D.C.: National Academy Press, 1988), pp. 16–48; Raymond Vernon, "International Investment and International Trade in the Product Cycle," *Quarterly Journal of Economics* 80 (May 1966): 190–207.

38. Orsenigo, *Emergence of Biotechnology*, p. 59.

39. Powell and Brantley, "Competitive Cooperation," p. 375.

40. Ibid., p. 391.

41. Cohen and Levinthal, "Absorptive Capacity." See also Ashish Arora and Alfonso Gambardella, "Division of Labor and Inventive Activity," working paper 93-3 (Heinz School of Public Policy and Management, Carnegie Mellon University, January 1993).

42. Gary Pisano, W. Shan, and David Teece, "Joint Ventures and Collaboration in the Biotechnology Industry," in *International Collaborative Ventures in U.S. Manufacturing*, ed. David C. Mowery (Cambridge, Mass.: Ballinger, 1988), pp. 183–222.

43. Arora and Gambardella, "Division of Labor."

44. Michael A. Cusamano, "Factory Concepts and Practices in Software Development," *Annals of the History of Computing* 13 (1991): 3.

45. Alfred D. Chandler, Jr., *The Visible Hand* (Cambridge, Mass.: Harvard University Press, 1977); David A. Hounschell, *From the American System to Mass Production: 1800–1932* (Baltimore: Johns Hopkins University Press, 1984); Stephen Wood, ed., *The Transformation of Work?* (London: Unwin & Hyman, 1989); Joan Woodward, *Industrial Organization: Theory and Practice* (New York: Oxford University Press, 1965).

46. Cusamano, "Factory Concepts," p. 5.

47. Michael A. Cusamano, *Japan's Software Factories: A Challenge to U.S. Management* (New York: Oxford University Press, 1991).

48. Michael A. Cusamano, *The Japanese Automobile Industry: Technology and*

Management at Nissan and Toyota (Cambridge, Mass.: Harvard University Press, 1985); Cusamano, *Japan's Software Factories.* See also "Now Software Isn't Safe from Japan" (*Business Week*, February 11, 1991, p. 84), which cites Cusamano's finding that, in a comparison of comparable projects undertaken by Japanese and American software houses and divisions, the Japanese wrote about 70 percent more lines of source code in the same period of time and had fewer than half as many measurable defects.

49. "Software Made Simple: Will Object-Oriented Programming Transform the Computer Industry?" *Business Week*, October 30, 1991, pp. 92–100.

50. John Markoff, "U.S. Lead in Software Faces a Rising Threat," *New York Times*, October 25, 1992, p. 16.

51. Quoted in ibid.

52. Richard Gordon, "Markets, Hierarchies, and Alliances: Beyond the Flexible Specialization Debate"(Silicon Valley Research Group, University of California at Santa Cruz, March 1989, unpublished MS), p. 17.

53. Yves Doz, "International Industries: Fragmentation versus Globalization," in Guile and Brooks, *Technology and Global Industry*, p. 100.

54. Keith Pavitt, "What We Know about the Strategic Management of Technology," *California Management Review* 32 (Spring 1990): 23. Emphasis in the original.
    For other critiques of product cycle theory, see Giovanni Dosi, Keith Pavitt, and Luc Soete, *The Economics of Technical Change and International Trade* (London: Harvester/Wheatsheaf, 1990), chap. 5; and Michael Storper, "Oligopoly and the Product Cycle: Essentialism in Economic Geography," *Economic Geography* 61 (July 1985): 260–82.

55. David Harvey, *The Limits to Capital* (Oxford: Basil Blackwell, 1982); Andrew Sayer and Richard Walker, *The New Social Economy: Reworking the Division of Labor* (Oxford: Basil Blackwell, 1992), p. 201; Erica Schoenberger, "Some Dilemmas of Automation," *Economic Geography* 66 (1990): 232–47.

56. Jeffrey R. Williams, "How Sustainable Is Your Competitive Advantage?" *California Management Review* 34 (Spring 1992): 51.

CHAPTER 4: THE EVOLUTION (AND DEVOLUTION?) OF THE ITALIAN INDUSTRIAL DISTRICTS
    1. The epigraph is from Philip Cooke and Kevin Morgan, "The Network Paradigm: New Departures in Corporate and Regional Development," *Society and Space* (October 1993): 553. The principal work on the "suburbanization"

of manufacturing jobs in the United States was pioneered by John F. Kain, "The Distribution and Movement of Jobs and Industry," in *The Metropolitan Enigma: Inquiries into the Nature and Dimensions of America's "Urban Crisis,"* ed. James Q. Wilson (Cambridge, Mass.: Harvard University Press, 1968). For a more recent discussion, see John D. Kasarda, "Jobs, Migration, and Emerging Urban Mismatches," in *Urban Change and Poverty*, ed. Michael G. H. McGeary and Laurence E. Lynn (Washington, D.C.: National Academy Press, 1988).

On the standard treatments of the interregional and international mobility of productive capital, including the theory of the so-called new international division of labor, see Barry Bluestone and Bennett Harrison, *The Deindustrialization of America* (New York: Basic Books, 1982); Manuel Castells, ed., *High Technology, Space and Society* (Beverly Hills, Calif.: Sage, 1985); Foulker Froebel, F. Heinrichs, and Otto Kreye, *The New International Division of Labor* (Cambridge, U.K.: Cambridge University Press, 1980); Bennett Harrison and Barry Bluestone, *The Great U-Turn: Corporate Restructuring and the Polarizing of America* (New York: Basic Books, 1988); R. D. Norton and John Rees, "The Product Cycle and the Spatial Decentralization of U.S. Manufacturing," *Regional Studies* 13 (1979): 141–51. A new view of these developments, sometimes called the theory of "commodity chains," is presented in Edna Bonacich, Lucie Cheng, Norma Chinchilla, Norma Hamilton, and Paul Ong, *The Globalization of the Garment Industry in the Pacific Rim* (Philadelphia: Temple University Press, 1993); and Gary Gereffi and Miguel Korzeniewicz, eds., *Commodity Chains and Global Capitalism* (Westport, Conn.: Greenwood Press, 1993).

The extent, significance, and character of the international mobility of capital have been questioned by David M. Gordon, "The Global Economy: New Edifice or Crumbling Foundations?" *New Left Review* (March/April 1988); and Timothy Kochlin, "The Determinants of the Location of U.S. Direct Foreign Investment," *International Review of Applied Economics* 6 (1992).

2. The literature on the industrial districts of Europe, North America, and the Far East—but especially on Italy—is by now prodigious. I review and offer critiques of this extensive body of work in Bennett Harrison, "Industrial Districts: Old Wine in New Bottles?" *Regional Studies* 26 (October 1992): 469–83; and Harrison, "The Italian Industrial Districts and the Crisis of the Cooperative Form," *European Planning Studies*, forthcoming.

On industrial districts in Germany, see Gary Herrigel, "Industrial Organization and the Politics of Industry: Centralized and Decentralized Production in Germany" (Ph.D. diss., Department of Political Science, M.I.T., 1990); Herrigel, "Large Firms, Small Firms, and the Governance of Flexible Specialization: Baden Wuerttemberg and the Socialization of Risk," in *Country Competitiveness: Technology and the Organizing of Work*, ed. Bruce Kogut (New York: Oxford University Press, 1992). On Spain, see Lauren Benton, "Industrial Subcontracting and the Informal Sector: The Politics of Restructuring in the Madrid Electronics Industry," in *The Informal Economy: Studies in Advanced* and *Less-Developed Countries*, ed. Alejandro Portes, Manuel Castells, and Lauren Benton (Baltimore: Johns Hopkins University Press, 1989).

3. Arnaldo Bagnasco, *Tre Italie: La Problematica Territoriale Dello Sviluppo Italiano* (Bologna: Il Mulino, 1977); Michael J. Piore and Charles F. Sabel, *The Second Industrial Divide: Possibilities for Prosperity* (New York: Basic Books, 1984). Two useful collections on this region are *Small Firms and Industrial Districts in Italy*, ed. Edward Goodman, Julia Bamford, and Peter Saynor (London: Routledge, 1989); and Frank Pyke, Giacomo Becattini, and Werner Sengenberger, eds., *Industrial Districts and Inter-Firm Co-operation in Italy* (Geneva: International Institute for Labour Studies, 1990).

4. Roberto Camagni, "Regional Deindustrialization and Revitalization Processes in Italy," in *Industrial Change and Regional Economic Transformation*, ed. Lloyd Rodwin and Hidehiko Sazanami (London: HarperCollins Academic, 1991); Richard M. Locke, "Local Politics and Industrial Adjustment: The Political Economy of Italy in the 1980s" (Ph.D. diss., Department of Political Science, M.I.T., February 1989); Locke and Serafino Negrelli, "Il Case FIAT AUTO," in *Strategie di Riaggiustamento Industriale*, ed. Marino Regini and Charles Sabel (Bologna: Il Mulino, 1989).

5. Ash Amin and Kevin Robins, "The Re-Emergence of Regional Economies? The Mythical Geography of Flexible Accumulation," *Environment and Planning D: Society and Space* 8 (March 1990): 16–17; and Amin and Robins, "Industrial Districts and Regional Development: Limits and Possibilities," in Pyke, Becattini, and Sengenberger, *Industrial Districts*.

6. Giacomo Becattini, "The Marshallian Industrial District as a Socio-Economic Notion," in Pyke, Becattini, and Sengenberger, *Industrial Districts*; Becattini, ed., *Mercato e Forze Locali: Il Distretto Industriale* (Bologna: Il Mulino, 1987); Becattini, "Sectors and/or Districts: Some Remarks on the Conceptual Foundations of Industrial Economics," in Goodman, Bamford, and Saynor, *Industrial Districts in Italy*; Sabastiano Brusco and Charles Sabel, "Artisan Production and Economic Growth," in *The Dynamics of Labour Market Segmentation*, ed. Frank Wilkinson (New York: Academic Press, 1981); Brusco, "Small Firms and Industrial Districts: The Experience of Italy," in *New Firms and Regional Development*, ed. David Keeble and Francis Weever (London: Croom Helm, 1986); Brusco, "The Emilian Model: Productive Decentralization and Social Integration," *Cambridge Journal of Economics* 6 (June 1982): 167–84.

7. Alfred Marshall, *Industry and Trade* (London: Macmillan, 1919); J. K. Whitaker, ed., *The Early Economic Writings of Alfred Marshall: 1860–1890* (London: Macmillan, 1975). For commentaries, see Becattini, "The Marshallian Industrial District"; and Bellandi, "The Role of Small Firms in the Development of Italian Manufacturing Industry," in Goodman, Bamford, and Saynor, *Industrial Districts in Italy*.

8. Bruce Herman, "Economic Development and Industrial Relations in a Small Firm Economy: The Experience of Metal Workers in Emilia-Romagna" (Center for Labor Management Policy Studies, Graduate Center of the City University of New York, October 1990, unpublished MS).

9. Philip Cooke and Kevin Morgan, "Growth Regions under Duress: Renewal Strategies in Baden-Württemberg and Emila-Romagna," in *Holding Down the Global: Possibilities for Local Economic Policy*, ed. Ash Amin and Nigel Thrift (New York: Oxford University Press, forthcoming).

10. The following is based on research conducted by Giuseppina Gualtieri of the Laboratory for Industrial Policy of NOMISMA, a research institute situated in Bologna, and on interviews conducted with executives of Sasib by Gualtieri, Patrizia Faraselli, and me in July 1989.

11. *Sasib: Annual Report 1988* (Bologna: Sasib, May 1989).

12. However, this inclination to a very high degree of outsourcing by the "new" Sasib—seen by Vacerri as a cornerstone of the management philosophy of "lean production"—may not differ greatly from what is already conventional practice in this sector, at least in Italy. See Andrea Lipparini and Maurizio Sobrero, "Inter-Organizational Networks and SMEs: Reshaping the Innovative Process through Architectural Innovation" (University of Bologna and the Wharton School of Finance and Commerce, November 1992, English translation of unpublished MS). I am grateful to Mark Lazerson for bringing this empirical study of the food processing machinery sector in northern Italy to my attention.

13. Similarly, Garibaldo characterizes Fiat's acquisition of Comau in 1982 as having been followed by a "neo-Tayloristic" reorganization of work in what had formerly been a craft-based, loosely knit interfirm collaborative (Francesco Garibaldo, "The Crisis of the 'Demanding Model' and the Search for an Alternative in the Experiences of the Metal Workers Union in Emilia-Romagna" [paper prepared for a meeting at Bielefeld University, March 30, 1989, available from FIOM-Bologna]).

14. Thus, in 1973 the large manufacturing firms (with 51–500 employees; Italy has very few firms larger than that) had capital equipment-to-employment ratios that averaged only 11 percent above the ratios of the firms with 11–50 employees. By 1984, that gap had doubled, to 25 percent. Similarly, in 1973 the ratio of profits to assets in the large firms was 10 percent *below* that of the smaller firms, but by 1984 it had become 3 percent larger (Marco Bellandi, "The Role of Small Firms in the Development of Italian Manufacturing Industry," table A3.4).

According to Guido Rey, then the president of ISTAT, the national social statistics agency, in 1981 the decennial census showed that only 44 percent of the country's manufacturing firms had 100 or more employees, while by 1985 that fraction had grown to two-thirds. Because the intercensal government surveys truncate at the low end of the establishment size distribution, the latter figure excludes the smallest operations (with fewer than 20 employees), so it overstates the small firm–large firm gap somewhat. Nevertheless, Rey's own interpretation of these numbers is that there has been "a clear comeback by the medium-to-large firms, which [have] gain[ed] considerable ground in relation to the small-to-medium firms" (Guido Rey, "Small Firms: Profile and Analysis, 1981–85," in Goodman, *Industrial Districts in Italy*, p. 92 and tables A4.2 and A4.4). Rey adds that since 1981, the large firm advantage has grown, especially in terms of "productivity per employee, which is linked to technological innovation and to reductions in staff."

Some of this increased concentration is surely the result of internal growth of individual firms. Indeed, the NOMISMA researchers conclude from their extensive statistical and case study research that "Italy is experiencing, in fact, a process of concentration which is taking place on the one hand through the buying of small independent firms by larger ones, and on the other through the increasingly frequent phenomenon of small firms growing rapidly and aggregating other small firms which were previously operating within a network configuration" (Bianchi and Gualtieri, "Mergers and Acquisitions in Italy and the Debate on Competition Policy" [NOMISMA, Bologna, July 1989, unpublished MS], p. 22; subsequently published as "Emilia-Romagna and Its Industrial Districts: The Evolution of a Model," in *The Regions and European Integration: The Case of Emilia-Romagna*, ed. R. Leonardi and R. Nanetti [London: Pinter, 1990], pp. 83–108).

But there has also been a sizable increase in acquisitions in Italy in the past several years. According to my analysis of statistics collected by the Laboratory of Industrial Policy at NOMISMA, since 1983 the number of Italian businesses acquired by other Italian firms has been growing by 11 percent a year. For acquisitions where the acquiring firm is foreign, the annual trend growth rate has been even higher: 14 percent. These are the results of my fitting a single-log model to a 1983–1988 time series on acquisitions compiled by NOMISMA and presented in Marco Sassatelli and Antonella Pancaldi, "Le Acquisizioni in Italia (1983–1988): Trends e Prospettive," in Laboratorio di Politica Industriale, *Acquisizioni Fusioni Concorrenza*, no. 1 (Bologna: NOMISMA, June 1989); English summary by Bianchi and Gualtieri, "Mergers and Acquisitions"; and Bianchi, Gualtieri, Antonella Pancaldi, and Marco Sassatelli, "The Determinants of Mergers and Acquisitions: Evidence from Italy"(Bologna: NOMISMA, July 1989, unpublished MS). A quadratic specification shows that the rate of acquisitions by foreign firms—a third of whom made the acquisition through a previously acquired Italian firm or financial holding—is actually accelerating. By nationality, the greatest number of foreign firms acquiring Italian companies are American, British, French, German, Swiss, and Swedish, in that order.

In preparation for the next phase of European political and economic integration, cross-border mergers and acquisitions of this sort are rising rapidly throughout that part of the world; see "Europe's Giants Are Hungrier Than Ever," *Business Week*, July 17, 1989, p. 144.

15. Cooke and Morgan, "Growth Regions under Duress."

16. "Oligopoly Is Alive and Well: Notes for a Broader Discussion of Flexible Accumulation," in *Industrial Change and Regional Development: The Transformation of New Industrial Spaces*, ed. Georges Benko and Mick Dunford (London: Belhaven Press/Pinter, 1991), chap. 6.

17. Cooke and Morgan, "Growth Regions under Duress." At least one public official has gone on record with just this concern; Carlo Tolomelli, "Policies to Support Innovation Processes: Experiences and Prospects in Emilia-Romagna" (paper presented to the Regional Science Association's European Summer Institute, Quarco, Italy, July 1988; cited in Amin and Robins, "The Re-Emergence of Regional Economies?," p. 18). Tolomelli repeated this concern to me in the summer of 1989.

18. "Benetton Targets a New Customer—Wall Street," *Business Week*, May 29, 1989, pp. 32–33.

19. This entire section on Benetton draws extensively on the pathbreaking research of, and my long conversations with, Belussi, who is now with the Faculty of Statistics at the Universtiy of Padua. See Fiorenza Belussi, "Benetton: Information Technology in Production and Distribution: A Case Study of the Innovative Potential of Traditional Sectors," occasional paper no. 25 (Science Policy Research Unit, University of Sussex, Brighton, U.K., 1987); Belussi, "Benetton Italy: Beyond Fordism and Flexible Specialization to the Evolution of the Network Firm Model," in *Information Technology and Women's Employment: The Case of the European Clothing Industry*, ed. S. Mitter (Berlin: Springer Verlag, 1989); Belussi, "La Flessibilita si fa Gerarchia: la Benetton," in *Nuovi Modelli D'Impresa Gerarchie Organizzative E Imprese Rete*, ed. Belussi (Milan: Franco Angeli, 1992); and Belussi and Massimo Festa, "L'Impresa Rete Del Modello Veneto: Dal Post-Fordismo Al Toyotismo? Alcune Note Illustrative Sulle Strutture Organizzative Dell'Indotto Benetton" (IRES-CGIL, Maestre, Veneto, November 1990, unpublished MS).

20. Patrizio Bianchi and Nicola Bellini, "Public Policies for Local Networks of Innovators," *Research Policy* 20 (October 1991): 487–97.

21. According to Italian social security records, in 1989 manufacturing firms with 500 or more employees paid their blue-collar workers about 16 percent more than did firms employing fewer than 20 workers. For white-collar jobs, the gap was even greater, averaging about 40 percent. These data, however, do not break out the industrial districts; these are national data (M. Ferrero and D. Invernizzi, "La Struttura Retribuzioni dei Lavoretori alle Dipendenze" (Turin,

December 1991, unpublished MS; communicated to me by Mario Pianta of the National Research Council in Rome).

22. Mark Lazerson, "Subcontracting in the Modena Knit Wear Industry," in Pyke, Becattini, and Sengenberger, *Inter-Firm Co-operation in Italy*, p. 130.

23. Knuth Dohse, Ulrich Jurgens, and Thomas Malsch, "From 'Fordism' to 'Toyotism'? The Social Organization of the Labor Process in the Japanese Automobile Industry," *Politics and Society* 14 (1985): 115–46.

24. Flavia Martinelli, "Productive Organization and Service Demand in Italian Textile and Clothing 'Districts': A Case Study" (United Nations Commission on Trade and Development, UNCTAD/MTN/RLA/CB.6, Rome, December 1988).

25. This section draws principally on yet another case study conducted by NOMISMA: Silvano Bertini, "Prospettive del Conto Terzismo nel Quadro Dell'Evoluzione del Modello Pratese" (Laboratory for Industrial Policy, NOMISMA, April 1989; synthesized for me in English by Bertini as "The Textile District of Prato" in April 1990, and translated for me in its entirety in July 1990 by Lisa Stanziale, in Pittsburgh).

26. Martinelli, "Productive Organization and Service Demand."

27. For the sake of completeness, the system also includes a smaller group of middlemen, the *lanifici*, who not only coordinate but also engage in some directly productive finishing work, and a few remaining big textile firms with mills and shops in several countries (Martinelli, "Productive Organization").

28. Lazerson, "Subcontrating in the Modena Knit Wear Industry."

29. Martinelli, "Productive Organization and Service Demand."

30. Capecchi, "Flexible Specialization in Emilia-Romagna."

31. The principal American theorists of the potential role of *trust* in promoting interorganizational collaboration are Mark Granovetter, Edward Lorenz, and Charles Sabel. Compare Mark Granovetter, "Economic Action and Social Structure: The Problem of Embeddedness," *American Journal of Sociology* 91 (1985): 481–510; Edward Lorenz, "Neither Friends nor Strangers: Informal Networks of Subcontracting in French Industry," in *Trust: Making and Breaking Cooperative Relations*, ed. Diego Gambetta (Oxford: Basil Blackwell, 1988), pp. 194–210; Lorenz, "The Search For Flexibility: Subcontracting Networks in French and British Engineering," in *Reversing Industrial Decline?*, ed. Paul Hirst and Jonathan Zeitlin (Oxford: Berg, 1989), pp. 122–32; Charles F. Sabel, "Studied Trust: Building New Forms of Co-operation in a Volatile Economy," in

Industrial Districts and Local Economic Regeneration, ed. Frank Pyke and Werner Sengenberger (Geneva: International Institute for Labour Studies, 1992), pp. 215–50.

32. Sabel, "Studied Trust."

33. Marshall, *Industry and Trade*, p. 287.

34. Stuart A. Rosenfeld, *Technology, Innovation, and Rural Development: Lessons from Italy and Denmark* (Washington, D.C.: Rural Economic Policy Program, Aspen Institute for Humanistic Studies and the Ford Foundation, 1990).

35. Michael L. Blim, *Made in Italy: Small-Scale Industrialization and Its Consequences* (New York: Praeger, 1990).

36. Blim, *Made in Italy*, p. 266.

37. Blim, "Flexibly Specialized Industrial Districts at Middle Age" (Department of Sociology and Anthropology, Northeastern University, Boston, October 1992, unpublished MS).

38. Camagni, "Regional Deindustrialization."

39. Amin and Robbins, "Re-Emergence of Regional Economies?"

40. Sergio Mariotti and Marco Mutinelli, "Diffusion of Flexible Automation in Italy," in *Computer Integrated Manufacturing*, vol. 3, *Models, Case Studies, and Forecasts of Diffusion*, ed. Robert U. Ayres, William Haywood, and I. Tchijov (London: Chapman & Hall, 1992), chap. 6.

41. Amin and Thrift, *Holding Down the Global*.

CHAPTER 5: IS SILICON VALLEY AN INDUSTRIAL DISTRICT?

1. John Markoff, "Not Everyone in the Valley Loves Silicon-Friendly Government," *New York Times*, March 7, 1993, p. E7.

2. There are still other perspectives on what makes Silicon Valley work, notably George Gilder's emphasis on the formative role of small, entrepreneurial high-tech companies. See Gilder, *The Spirit of Enterprise* (New York: Basic Books, 1984); and Gilder, *Microcosm: The Quantum Revolution in Economics and Technology* (New York: Simon & Schuster, 1989). In a series of widely read debates, Gilder has been challenged by the management consultant Charles Ferguson. See Gilder, "The Revitalization of Everything: The Law of the Microcosm," *Harvard Business Review* 66 (March–April 1988): 49–61; and

Ferguson, "From the People Who Brought You Voodoo Economics," *Harvard Business Review* 66 (May–June 1988): 55–62.

I review these debates, and explicitly compare their main arguments to those of the industrial district theorists, in Bennett Harrison, "Concentrated Economic Power in the Reproduction of Silicon Valley," *Environment and Planning A*, forthcoming.

3. Bennett Harrison and Jean Kluver, "Reassessing the 'Massachusetts Miracle': Reindustrialization and Balanced Growth, or Convergence to 'Manhattanization'?" *Environment and Planning A* 21 (June 1989): 771–801.

4. AnnaLee Saxenian, "The Origins and Dynamics of Production Networks in Silicon Valley," *Research Policy* 20 (October 1991): 423.

5. AnnaLee Saxenian, "Regional Networks and the Resurgence of Silicon Valley," *California Management Review* 33 (Fall 1990): 91. See also Saxenian, *Regional Networks: Industrial Adaptation in Silicon Valley and Route 128* (Cambridge, Mass.: Harvard University Press, forthcoming).

6. Saxenian, "Regional Networks," p. 96.

7. Ibid., p. 97. Richard Gordon agrees that, in Silicon Valley, "the professional network transcends corporate organization," and that, when it works well, such a craft-type labor market structure acts as a mechanism for promoting the diffusion of both tacit and codified knowledge (Gordon, "Innovation, Industrial Networks, and High Technology Regions," in *Innovation Networks: Spatial Perspectives*, ed. Roberto Camagni (London: Belhaven Press, 1991), pp. 174–95; and Gordon, "State, Milieu, Network: Systems of Innovation in Silicon Valley," in *Systems of Innovation*, ed. Patrizio Bianchi and M. Quere (Paris: Groupe de Recherche Europeen sue les Milieux Innovateurs, forthcoming).

Indeed, the *absence* of this fluid interchange in other high-tech regions arguably explains much about why so few of the "science parks" constructed around the world during the 1980s in attempts to emulate Silicon Valley have succeeded in fostering sustained local economic development. See Michael I. Luger and Harvey A. Goldstein, *Technology in the Garden: Research Parks and Regional Economic Development* (Chapel Hill: University of North Carolina Press, 1991); and Doreen Massey, Paul Quintas, and David Wield, *High Tech Fantasies: Science Parks in Society, Science and Space* (London: Routledge, 1992).

8. Richard Florida and Martin Kenney, "Why Silicon Valley and Route 128 Won't Save Us," *California Management Review* 33 (Fall 1990): 68–88; Florida and Kenney, *The Breakthrough Illusion* (New York: Basic Books, 1990).

9. In a complementary argument, the Penn State University planner-geographer Amy Glasmeier concludes, from her detailed historical studies of the

twentieth-century evolution of the Swiss watch industry, that "while networks can and do promote innovation within an existing technological framework, historical experience suggests their fragmented, atomistic structure is subject to disorganization and disintegration during periods of technological change" (Glasmeier, "Technological Discontinuities and Flexible Production Networks: The Case of Switzerland and the World Watch Industry," *Research Policy* 20 [October 1991]: 469).

The French economist Benjamin Coriat has offered a similar critique of production systems based on flexible specialization. See Coriat, "Crise et Électronisation de la Production: Robotisation d'Atélier et Modèle Fordièn d'Accumulation du Capital," *Critiques de l'Économie Politique* (January–June 1984).

10. Florida and Kenney, *Breakthrough Illusion*, p. 100.

11. Ibid., p. 188.

12. Mark Granovetter, "Economic Action and Social Structure: The Problem of Embeddedness," *American Journal of Sociology* 91 (1985): 496.

13. Richard Florida and David Browdy, "The Invention That Got Away," *Technology Review* 94 (August–September 1991): 43–44.

This could change—but only thanks to the intervention of a government-sponsored industrial policy consortium led by the Defense Department's recently renamed Advanced Research Projects Agency (ARPA). New R&D grants have recently been awarded to several American companies, to help them get "back in the race" for flat panel displays, widely seen to be a "critical technology" for "everything from fighter plane cockpits to wall-sized television screens"—not to mention ordinary notebook-sized computers ("Industrial Policy, or Industrial Folly?" *Business Week*, May 17, 1993, p. 38).

14. Richard Gordon, "Collaborative Linkages, Transnational Networks, and New Structures of Innovation in Silicon Valley's High-Technology Industry," report no. 4, "Industrial Suppliers/Services in Silicon Valley" (Silicon Valley Research Group, University of California at Santa Cruz, January 1993, unpublished report).

15. David J. Teece, "Foreign Investment and Technological Development in Silicon Valley," *California Management Review* 34 (Winter 1992): 88–106; David C. Mowery and David J. Teece, "Japan's Growing Capabilities in Industrial Technology: Implications for U.S. Managers and Policymakers," *California Management Review* 36 (Winter 1993): 9–34.

16. Flavia Martinelli and Erica Schoenberger, "Oligopoly Is Alive and Well: Notes for a Broader Discussion of Flexible Accumulation," in *Industrial Change and Regional Development: The Transformation of New Industrial Spaces*, ed.

Georges Benko and Mick Dunford (London: Belhaven Press/Pinter, 1991), p. 121.

17. Mowery and Teece, "Japan's Growing Capabilities," p. 18.

18. Norman J. Glickman and Douglas P. Woodward, *The New Competitors* (New York: Basic Books, 1989).

19. Much of the following is drawn from AnnaLee Saxenian, "The Genesis of Silicon Valley," in *Silicon Landscapes*, ed. Peter Hall and Ann Markusen (Boston: Allen & Unwin, 1985), pp. 20–34; and "The Urban Contradictions of Silicon Valley," in *Sunbelt-Snowbelt: Urban Development and Regional Restructuring*, ed. Larry Sawers and William K. Tabb (New York: Oxford University Press, 1984), pp. 163–99.

20. The extent of occupational stratification within Silicon Valley's high-tech industry is explored further (with much accompanying statistical documentation) in Richard Gordon and Linda M. Kimball, "High Technology, Employment, and the Challenges to Education," in European Economic Commission, Conference on Job Creation, Innovation and New Technologies, Brussels, September 1985, unpublished MS available from the Center for the Study of Global Transformations, University of California at Santa Cruz.
    My own research on the electronics industry of Massachusetts during the 1980s revealed a similarly striking degree of stratification. See Kenneth Geiser and Bennett Harrison, "The High-Tech Industry Comes Down to Earth," *Boston Globe*, Sunday, June 23, 1985, pp. A1, A18–19.

21. Saxenian, "Urban Contradictions," pp. 175–77.

22. John C. Dvorak, "Inside Track," *PC Magazine*, December 22, 1992, p. 93.

23. Alice H. Amsden, "Beyond Shock Therapy: Why Eastern Europe's Recovery Starts in Washington," *American Prospect* (Spring 1993): 87.

24. Stephen J. Appold, "Firm Demography and Urban Growth," working paper 91-22, Heinz School of Public Policy and Management, Carnegie Mellon University, June 1991.

25. Ann Markusen and Joel Yudken, *Beyond the Cold War Economy* (New York: Basic Books, 1992); and Markusen, Peter Hall, Scott Campbell, and Sabina Deitrick, *The Rise of the Gunbelt* (New York: Oxford University Press, 1991). While this distinction between the specialties of northern and southern California emphasizes the differences, there were always close connections between the two regions. Thus, a whole complex of companies with their main West Coast facilities in southern California pioneered in the post–World War

II development of military-financed and military mission–related computing, including Northrup, Raytheon, Bendix, and the RAND Corporation. The California Institute of Technology in Los Angeles was much involved in these developments. One of the links to what would later become Silicon Valley was Hewlett-Packard, then a subcontractor to Northrup (Kenneth Flamm, *Creating the Computer* [Washington, D.C.: Brookings Institution, 1988], pp. 65–66).

26. Richard Gordon and Joel Krieger, "Anthropocentric Production Systems and U.S. Manufacturing Models in the Machine Tool, Semiconductor, and Automobile Industries," vol. 18, FOP 262, Research Paper Series, International Research Network on Culture and Production (CAPIRN), Forecasting and Assessment in Science and Technology (FAST) Project, Science Research and Development Programme, Commission of the European Communities, Brussels, December 1990, pp. 67–68. The role of the state is further explored, and applied specifically to the case of Silicon Valley, in Gordon, "State, Milieu, Network."

27. Markusen and Yudken, *Beyond the Cold War Economy*, p. 179. Markusen and her colleagues' principal analytic point is that it is erroneous to treat the growth of many industrial districts as substantially endogenous, or inner-directed. Elsewhere, they write: "Most work written on the new industrial districts of Orange County, Silicon Valley, and Route 128 downplays or omits altogether the significance of government, with its peculiar demands, as market. The highly lauded 'postindustrial' organization of manufacturing in these regions, with its flexible specialization and small-batch, custom-made products, is erroneously ascribed to the commercial sector, when in fact these attributes originated in their defense-based industries" (*Rise of the Gunbelt*, p. 248).

28. John Markoff, "IBM in Chip Deal with Toshiba and Siemens," *New York Times*, July 13, 1992, p. D1.

29. Teece, "Foreign Investment in Silicon Valley," p. 104.

30. Whatever one's preferred characterization of Silicon Valley, one thing is certain as we approach the end of the twentieth century. Silicon Valley as a region is in trouble. Between 1987 and 1992, the valley and the region around it lost over 40,000 manufacturing jobs. Some of this loss could be chalked up to increasing productivity, a good long-run sign. But much of it was directly connected to the cutbacks in federal military spending, thereby revealing just how important defense production and research continue to be for the valley. Moreover, Silicon Valley high-tech producers continue to outsource work to facilities located beyond California and even offshore, and some have been picking up and moving their operations out of Santa Clara county, altogether (John Markoff, "Silicon Valley Faces a Midlife Crisis," *New York Times*, September 28, 1992, p. C1; Robert Reinhold, "High Hurdles to Recovery for Struggling California," *New York Times*, December 21, 1992, p. A1).

CHAPTER 6: "FLEXIBILITY" AND THE EMERGENCE OF LARGE FIRM–LED PRODUCTION NETWORKS

1. Walter W. Powell, "Neither Market nor Hierarchy: Network Forms of Organization," in *Research in Organizational Behavior*, ed. Barry M. Straw and Larry L. Cummings (Greenwich, Conn.: JAI Press, 1990), pp. 295–336.

2. T. P. Hill, *Profits and the Rate of Return* (Paris: Organization for Economic Cooperation and Development, 1979).

3. Andrew Glyn, Andrew Hughes, Alain Lipietz, and Agit Singh, "The Rise and Fall of the Golden Age," in *The End of the Golden Age*, ed. Stephen Marglin and Juliet Schor (New York: Oxford University Press, 1989), pp. 39–125.

4. Barry Bluestone and Bennett Harrison, *The Deindustrialization of America* (New York: Basic Books, 1982); Samuel Bowles, David Gordon, and Thomas Weisskopf, *Beyond the Waste Land* (New York: Anchor, 1983); Glyn et al., "Rise and Fall"; Bennett Harrison and Barry Bluestone, *The Great U-Turn: Corporate Restructuring and the Polarizing of America* (New York: Basic Books, 1988); Alain Lipietz, "New Tendencies in the International Division of Labor: Regimes of Accumulation and Modes of Regulation," in *Production, Work, Territory*, ed. Allan Scott and Michael Storper (London: Allen & Unwin, 1986), pp. 16–40; Michael J. Piore and Charles Sabel, *The Second Industrial Divide* (New York: Basic Books, 1984); Anna Pollert, ed. *Farewell to Flexibility?* (New York: Oxford University Press, 1989); Karel Williams, John Williams, Colin Haslam, and Andrew Wardlow, "Facing Up to Manufacturing Failure," in *Reversing Industrial Decline?*, ed. Paul Hirst and Jonathan Zeitlin (Oxford: Berg, 1988), pp. 71–94.

5. Bluestone and Harrison, *Deindustrialization of America*.

6. On the idea of "time compression," see David Harvey, *The Condition of Postmodernity* (Oxford: Basil Blackwell, 1989); and Erica Schoenberger, "From Fordism to Flexible Accumulation: Technology, Competitive Strategies, and International Location," *Environment and Planning D: Society and Space* 6 (September 1988): 245–62. In his Harvard doctoral dissertation, Charles Sabel called this the phenomenon of the "accelerating product cycle" (Sabel, *Work and Politics* [Cambridge: Cambridge University Press, 1982]).

7. Philip Cooke, "Flexible Integration, Scope Economies, and Strategic Alliances: Social and Spatial Mediations," *Environment and Planning D: Society and Space* 6 (September 1988): 281–300.

8. Armen Alchian and H. Demsetz, "Production, Information Costs, and Economic Organization," *American Economic Review* 62 (December 1972):

777–95; Ronald H. Coase, "The Nature of the Firm," *Economica* 4 (November 1937): 386–405; Kirk Monteverde and David J. Teece, "Supplier Switching Costs and Vertical Integration in the Automobile Industry," *Bell Journal of Economics* 13 (Spring 1982): 206–13; David. T. Levy, "The Transactions Cost Approach to Vertical Integration: An Empirical Examination," *Review of Economics and Statistics* 67 (August 1985): 438–45; Richard E. Caves and Ralph M. Bradburd, "The Empirical Determinants of Vertical Integration," *Journal of Economic Behavior and Organizations* 9 (April 1988): 265–79; Oliver E. Williamson, *The Economic Institutions of Capitalism* (New York: Free Press, 1985); Williamson, *Markets and Hierarchies* (New York: Free Press, 1975).

9. Richard Walker, "The Geographical Organization of Production Systems," *Environment and Planning D: Society and Space* 6 (December 1988): 388.

10. For a review of the debate, see Bennett Harrison, "Industrial Districts: Old Wine in New Bottles?" *Regional Studies* 26 (October 1992): 469–83.

11. Powell, "Neither Market nor Hierarchy," p. 323.

12. Ibid., p. 319.

13. Of course, all of this emphasis on what is institutionally and structurally "new" about the post-1970s global competitive environment can be taken too far. There is a great temptation for scholars, in particular, to seek to identify great historical disjunctures, and to talk about the emergence of new "dominant tendencies." It is, therefore, worth being reminded that there are many *continuities* with the past, as well—a point made forcefully for many years by the geographers Andrew Sayer, in Britain, and Richard Walker, in the United States. See Sayer and Walker, *The New Social Economy: Reworking the Division of Labor* (Cambridge: Basil Blackwell, 1992).
   Thus, for example, while the big ideas in the current era may mainly be about accelerating product cycles, niche-seeking firms, and perpetual innovation, business strategy theorists such as Carnegie Mellon University's Jeffrey Williams remind us that there exist whole classes of successful firms that continue to make it by attaining dominant market shares over the manufacture of quite standardized, not especially innovative products (Williams, "How Sustainable Is Your Comparative Advantage?" *California Management Review* 34 [Spring 1992]: 29–51).

14. Ronald Dore, *Flexible Rigidities: Industrial Policy and Structural Adjustment in the Japanese Economy 1970–1980* (Stanford, Calif.: Stanford University Press, 1986); Bryn Jones, "Flexible Automation and Factory Politics: The United Kingdom in Comparative Perspective," in Hirst and Zeitlin, *Reversing Industrial Decline?*; Maryellen R. Kelley, "Programmable Automation and the Skill Question: A Re-Interpretation of the Cross-National Evidence," *Human*

Systems Management 6 (November 1986): 223–41; Kelley, "Alternative Forms of Work Organization under Programmable Automation," in *The Transformation of Work?* ed. Stephen Wood (London: Unwin-Hyman, 1989), pp. 235–46; Kelley, "Unionization and Job Design Under Programmable Automation," *Industrial Relations* 28 (Spring 1989): 174–87; Kelley, "New Process Technology, Job Design and Work Organization: A Contingency Model," *American Sociological Review* 55 (April 1990): 191–208; Kelley and Lan Xue, "Does Decentralization of Programming Responsibilities Increase Efficiency? An Empirical Test," in *Ergonomics of Advanced Manufacturing and Hybrid Automated Systems II*, ed. Waldemar Karwowski and Mansour Rahimi (New York: Elsevier, 1990), pp. 379–86; Horst Kern and Michael Schumann, *Das Ende der Arbeitsteilung? Rationalisierung in der industriellen Produktion* (Munich, Germany: Beck, 1984); Arndt Sorge and Wolfgang Streeck, "Industrial Relations and Technical Change: The Case for an Extended Perspective," in *New Technology and Industrial Relations*, ed. Richard Hyman and Wolfgang Streeck (Cambridge: Basil Blackwell, 1988), pp. 19–47.

15. Bluestone and Harrison, *Deindustrialization of America*; Harrison and Bluestone, *Great U-Turn*; Harrison and Bluestone, "Wage Polarisation in the U.S. and the 'Flexibility' Debate," *Cambridge Journal of Economics* 14 (September 1990): 351–73; Kochan, Katz, and McKersie, *American Industrial Relations*; Daniel J. B. Mitchell, "Shifting Norms in Wage Determination," *Brookings Papers on Economic Activity* 2 (1985): 575–608; Guy Standing, "Labour Flexibility: Towards a Research Agenda," working paper no. 3, World Employment Programme Research, International Labour Office, Geneva, 1986; Standing and Victor Tokman, eds., *Towards Social Adjustment: Labour Market Issues in Structural Adjustment* (Geneva: International Labour Office, 1991).

16. Eileen Appelbaum, "Restructuring Work: Temporary, Part Time, and At-home Employment," in *Computer Chips and Paper Clips: Technology and Women's Employment*, ed. Heidi Hartmann (Washington D.C.: National Academy Press, 1987); Appelbaum and Peter Albin, "Employment, Occupational Structure, and Educational Attainment in the United States, 1973, 1979, and 1987," report to the Organization for Economic Cooperation and Development, Commission on Services, Paris, 1988; Susan Christopherson, "Flexibility in the U.S. Service Economy and the Emerging Spatial Division of Labour," *Transactions of the Institute of British Geographers* 14 (1989): 131–43; Virginia L. duRivage, ed., *New Policies for the Part-Time and Contingent Workforce* (Armonk, N.Y.: Sharpe, for the Economic Policy Institute, 1992), containing papers by Appelbaum, Carre, Tilly, and the editor; Ronald G. Ehrenberg, Pamela Rosenberg, and Jeanne Li, "Part-Time Employment in the United States," in *Employment, Unemployment, and Labor Utilization*, ed. Robert A. Hart (Boston: Unwin Hyman, 1988), pp. 256–87. In the same volume, see also Katherine G. Abraham, "Flexible Staffing Arrangements and Employers' Short-Term Adjustment Strategies," pp. 288–313.

17. Paul Blyton, G. Ursell, A. Gorham, and S. Hill, *Human Resource Management in Canadian and U.K. Work Organizations: Strategies for Work Force Flexibility*, vol. 1, *Report to the Canadian High Commission* (London: Canadian High Commission, 1988); Marion Cross, "A Study of Contracting-out of Maintenance Services in UK Industry," unpublished MS, City University Business School, London, 1989; Bennett Harrison and Maryellen R. Kelley, "Outsourcing and the Search for 'Flexibility'," *Work, Employment and Society* 7 (June 1993): 213–35; Kelley and Harrison, "The Subcontracting Behavior of Single vs. Multi-Plant Enterprises in U.S. Manufacturing: Some Implications for Regional Economic Development," *World Development* 18 (September 1990): 1273–94; Katherine G. Abraham and Susan K. Taylor, "Firms' Use of Outside Contractors: Theory and Evidence" (Cambridge, Mass.: National Bureau of Economic Research, September 1993), working paper 4468.

18. Compare Barry Bluestone, Peter Jordan, and Mark Sullivan, *Aircraft Industry Dynamics* (Boston: Auburn House Press, 1981).

19. Bo Carlsson and Erol Taymaz, "Flexible Technology and Industrial Structure in the U.S.," *Small Business Economics*, forthcoming; Gary Loveman and Werner Sengenberger, "Introduction: Economic and Social Reorganization in the Small and Medium-Sized Enterprise Sector," in *The Re-Emergence of Small Enterprises: Industrial Restructuring in Industrialized Countries*, ed. Sengenberger, Loveman, and Michael J. Piore (Geneva: International Institute for Labour Studies, International Labour Office, 1990), pp. 1–61.

20. Paul Osterman, *Employment Futures* (New York: Oxford University Press, 1988); Jeffrey Pfeffer and James Baron, "Taking the Workers Back Out: Recent Trends in the Structuring of Employment," in   *Research in Organizational Behavior*, vol. 10, ed. Barry Staw and L. L. Cummings (Greenwich, Conn.: JAI Press, 1988), pp. 257–303.

21. For two excellent textbook-level overviews of the theory and evolution of global production networks, see John H. Dunning, *Multinational Enterprises and the Global Economy* (Reading, Mass.: Addison-Wesley, 1993), esp. chap. 16; and Peter Dicken, *Global Shift: The Internationalization of Economic Activity*, 2nd ed. (New York: Guilford Press, 1992), esp. chap. 7.

22. Joseph J. Badaracco, Jr., *The Knowledge Link* (Boston: Harvard Business School Press, 1991), p. 19. See also Badaracco, "Changing Forms of the Corporation," in *The U.S. Business Corporation: An Institution in Transition*, ed. John R. Meyer and James M. Gustafson (Cambridge, Mass.: Ballinger, 1988), pp. 67–91.

23. Badaracco, *The Knowledge Link*; Roberto P. Camagni, "Firm, Networks, and the Hidden Costs of Cooperative Behavior," Bocconi University, Milan and University of Padua, Italy, 1992, unpublished MS; Camagni, "Local 'Milieu,' Uncertainty and Innovation Networks: Towards a New Dynamic Theory of Eco-

nomic Space," in *Innovation Networks*, ed. Camagni (London: Belhaven Press, 1991), pp. 121–44; David Mowery, ed., *International Collaborative Ventures in U.S. Manufacturing* (Cambridge, Mass.: Ballinger, 1988); Powell, "Neither Markets nor Hierarchies"; Michael J. Storper and Bennett Harrison, "Flexibility, Hierarchy and Regional Development: The Changing Structure of Industrial Production Systems and Their Forms of Governance in the 1990s," *Research Policy* 20 (October 1991): 407–22; Storper, "Production Organization, Technological Learning and International Trade: The Role of Production Flexibility in Contemporary Development," Lewis Center for Regional Policy Studies, University of California at Los Angeles, January 1991, unpublished MS; Werner Sengenberger and Duncan Campbell, eds., *Is the Single Firm Vanishing? Inter-Enterprise Networks, Labour and Labour Institutions* (Geneva: International Institute for Labour Studies of the International Labour Office, 1992).

24. Powell, "Neither Markets nor Hierarchies," p. 305.

25. Ibid., p. 304.

26. This language, and these ideas, are most recently associated with business strategy theorists David J. Teece of the University of California at Berkeley and Eric von Hippel of M.I.T. See, for example, David J. Teece, "Foreign Investment and Technological Development in Silicon Valley," *California Management Review* 34 (Winter 1992): 101–2; Eric von Hippel, *The Sources of Innovation* (New York: Oxford University Press, 1988); von Hippel, "The Impact of 'Sticky Data' on Innovation and Problem Solving," working paper 3147-90-BPS, Sloan School of Management, M.I.T., April 1990. But the initial discovery of the crucial importance of tacit knowledge about technology embedded in the organizational "routines" of specific business firms, and of the difficulties in trying to transmit such knowledge through conventional interfirm licensing contracts, is generally attributed to Stanford's Kenneth Arrow and, subsequently, to Columbia University's Richard Nelson and his (then, Yale University) colleague, Sidney Winter. See Kenneth J. Arrow, "The Economic Implications of Learning by Doing," *Review of Economic Studies* 29 (1962): 155–73; Richard Nelson and Sidney Winter, *An Evolutionary Theory of Economic Change* (Cambridge, Mass.: Harvard University Press, 1982).

27. Powell, "Neither Markets nor Hierarchies," p. 303.

28. Camagni, "Firm Networks," p. 7.

29. Powell,"Neither Markets nor Hierarchies," p. 305.

30. Chris DeBresson and Fernand Amesse, "Networks of Innovators: A Review," *Research Policy* 20 (October 1991): 364.

31. Powell,"Neither Markets nor Hierarchies," p. 323.

32. Alice H. Amsden, *Asia's Next Giant: South Korea and Late Industrialization* (New York: Oxford University Press, 1989).

33. Bluestone et al., *Aircraft Industry Dynamics*; David A. Hounschell, *From the American System to Mass Production, 1800–1932* (Baltimore: Johns Hopkins University Press, 1984).

34. The British geographers John Shutt and Richard Whittington bring to this particular discussion a concern for how relationships between the suppliers and their former parent evolve *after* the latter's disintegration. They distinguish between three ideal types of relation between the (typically larger) customer firm and the (generally smaller) "spin-off" (Shutt and Whittington, "Large Firm Strategies and the Rise of Small Units: The Illusion of Small Firm Job Generation," working paper no. 15, School of Geography, University of Manchester, North West Industry Research Unit, December 1984).

Productive decentralization refers to the disintegration of large operations into a series of smaller branches and subsidiaries, perhaps with different company names above the front door but with legal ownership retained by the original "parent." By contrast, under a regime of *detachment*, the large firms cease to directly own their suppliers but continue to actually or potentially dominate them by their ability to allocate finance and licenses or through franchising arrangements. Finally, under true *vertical disintegration*, the large firms may still retain a degree of control over ostensibly independent subcontractors through the exercise of monopsony power (that is, being the sole or principle customer for the supplier), or by controlling the supplier's access to third-party credit or political favors.

35. Compare Suzanne Berger, "The Traditional Sector in France and Italy," in *Dualism and Discontinuity in Industrial Societies*, ed. Berger and Michael Piore (New York: Cambridge University Press, 1980), pp. 88–131; Bertrand Chaillou, "Definition et typologie de la sous-traitance," *Revue Economique* 28 (April 1977): 262–85; Andrew Friedman, *Industry and Labour* (London: Macmillan, 1977); Dimitri Germidis, ed., *International Subcontracting: A New Form of Investment* (Paris: Development Centre of the OECD, 1980); Harrison and Kelley, "Outsourcing"; Kelley and Harrison, "Subcontracting Behavior"; Michael J. Piore, "Dualism as a Response to Flux and Uncertainty," in Berger and Piore, *Dualism and Discontinuity*, pp. 23–54; Alejandro Portes and Saskia Sassen-Koob, "Making It Underground: Comparative Material on the Informal Sector in Western Market Economies," *American Journal of Sociology* (July 1987): 30–61; Jill Rubery and Frank Wilkinson, "Outwork and Segmented Labour Markets," in *The Dynamics of Labour Market Segmentation*, ed. Wilkinson (New York: Academic Press, 1981); Allen Scott, "Location and Linkage Systems: A Survey and Reassessment," *Annals of Regional Science* 17 (March 1983): 1–39; M. Sharpston, "International Subcontracting," *Oxford Economic Papers* 27 (March 1975): 94–135; Michael Taylor and Nigel Thrift, "Industrial Linkage and the Segmented Economy," *Environment and Planning, Series A* 14 (April 1982): 1601–32; and Susumu Watanabe,

"Subcontracting, Industrialization and Employment Creation," *International Labour Review* 104 (July–August 1971): 51–76.

36. Walker, "Geographical Organization," pp. 389–90.

37. Badaracco, *Knowledge Link*; Philip Cooke, "Flexible Integration, Scope Economies, and Strategic Alliances: Social and Spatial Mediations," *Environment and Planning D: Society and Space* 6 (September 1988): 281–320; Mowery, *Collaborative Ventures*.

38. Kindleberger, "The 'New' Multinationalization of Business," *ASEAN Economic Bulletin* 5 (Summer 1988): 113–24.

39. Dicken, *Global Shift*, p. 213.

40. Mowery, "Introduction," in Mowery, *Collaborative Ventures*.

41. Camagni, "Firm Networks," p. 7.

42. Bluestone and Harrison, *Deindustrialization of America*.

43. Badaracco, *Knowledge Link*.

44. Ibid., p. 13.

45. James G. March and Herbert A. Simon, *Organizations* (New York: Wiley, 1958).

46. Stephen H. Hymer, *The International Operations of National Firms: A Study of Direct Foreign Investment* (Cambridge, Mass.: M.I.T. Press, 1976; published posthumously); and Hymer, "The Multinational Corporation and the Law of Uneven Development," in *Economics and World Order*, ed. Jagdish N. Bhagwati (New York: Macmillan, 1972), pp. 113–40.

47. Quoted by Dicken, *Global Shift*, pp. 223–24.

48. Edna Bonacich, Lucie Cheng, Norma Chinchilla, Norma Hamilton, and Paul Ong, *The Globalization of the Garment Industry in the Pacific Rim* (Philadelphia: Temple University Press, 1993); Gary Gereffi and Miguel Korzeniewicz, eds., *Commodity Chains and Global Capitalism* (Westport, Conn.: Greenwood Press, 1993).

49. The following discussion is greatly elaborated in Storper and Harrison, "Flexibility, Hierarchy."

50. For a primer (and a history of thought) on these matters, see Bela Gold,

"Changing Perspectives on Size, Scale and Returns: An Interpretative Essay," *Journal of Economic Literature* 19 (March 1981): 5–33.

51. Gary Herrigel, "Large Firms, Small Firms, and the Governance of Flexible Specialization: Baden Wuerttemberg and the Socialization of Risk," in *Country Competitiveness: Technology and the Organizing of Work*, ed. Bruce Kogut (New York: Oxford University Press, 1992).

52. See Storper and Harrison, "Flexibility, Hierarchy," for a lengthy discussion of the competitive conditions—from the number of firms in the system to the degree of asset specificity—that give rise to particular core-ring configurations in the real world.

53. Susan Christopherson and Michael Storper, "The Effects of Flexible Specialization on Industrial Politics and the Labor Market: The Motion Picture Industry," *Industrial and Labor Relations Review* 42 (April 1989): 331–47; Michael J. Enright, "Organization and Coordination in Geographically Concentrated Industries," in *Coordination and Information: Historical Perspectives on the Organization of Enterprise*, ed. Daniel Raff and Naomi Lamoreaux (Chicago: University of Chicago Press, forthcoming).

54. Michael L. Dertouzos, Richard K. Lester, and Robert M. Solow, *Made in America: Regaining the Competitive Edge* (Cambridge, Mass.: M.I.T. Press, 1989), pp. 270–77.

55. Herrigel, "Large Firms, Small Firms"; Charles Sabel, Gary Herrigel, Richard Kazis, and Richard Deeg, "How to Keep Mature Industries Innovative," *Technology Review* 90 (April 1987): 27–35; Sabel, Horst Kern, and Herrigel, "Collaborative Manufacturing" (Department of Political Science, Massachusetts Institute of Technology, March 31, 1989, unpublished MS).

56. Christopherson and Storper, "Effects of Flexible Specialization"; Enright, "Organization and Coordination."

57. Herrigel, "Large Firms, Small Firms"; Sabel, Kern, and Herrigel, "Collaborative Manufacturing"; Philip Cooke and Kevin Morgan, "Growth Regions under Duress: Renewal Strategies in Baden-Württemberg and Emila-Romagna," in *Holding Down the Global: Possibilities for Local Economic Policy*, ed. Ash Amin and Nigel Thrift (New York: Oxford University Press, forthcoming).

58. Bluestone, Jordan, and Sullivan, *Aircraft Industry Dynamics*; Mowery, *Collaborative Ventures*.

59. Martin Kenney and Richard Florida, *Beyond Mass Production* (New York: Oxford University Press, 1992).

60. Eike Schamp, "Towards a Spatial Reorganisation of the German Car Industry? The Implications of New Production Concepts," in *Industrial Change and Regional Development: The Transformation of New Industrial Spaces*, ed. Georges Benko and Mick Dunford (London: Belhaven Press/Pinter, 1991), pp. 159–70.

61. Christopherson, "U.S. Service Economy."

62. Raymond E. Miles and Charles C. Snow, "Causes of Failure in Network Organizations," *California Management Review* 34 (Summer 1992): 53–72.

63. In principle, other tendencies are possible. For example, Storper is presently studying the aircraft manufacturing complex in Toulouse, France, which up until now has been strongly dependent on its relationship to Airbus Industrie. He thinks that this sector is in the process of diversifying its range of activities and the number of customers it supplies, which could eventually move the "Toulouse system" in the direction of less rather than more hierarchy.

64. Amin and Thrift, *Holding Down the Global*; Flavia Martinelli and Erica Schoenberger, "Oligopoly Is Alive and Well: Notes for a Broader Discussion of Flexible Accumulation," in Benko and Dunford, *New Industrial Spaces*, chap. 6; Saskia Sassen, *Global Cities: London, Tokyo, New York* (Cambridge: Basil Blackwell, 1992); Walker, "Geographical Organization."

65. Michael Porter, *The Competitive Advantage of Nations* (New York: Free Press, 1990).

66. That is why, together with my Carnegie Mellon University colleague Maryellen Kelley, I am presently engaged in a multiyear National Science Foundation–sponsored effort to actually measure the extent and significance of agglomerative externalities in one key sector of the American economy: the metalworking industries that produce everything from aircraft engines and automobiles to scientific instruments and coffee grinders. Our initial hypotheses and research plan are contained in "Production Networks and Locational Context," working paper 92-13, Heinz School of Public Policy and Management, Carnegie Mellon University, February 1992. Initial findings are presented in "Spatially Distributed and Proximate Inter-Organizational Networks, Agglomeration, and Technological Performance in U.S. Manufacturing," working paper 93-20, Heinz School of Public Policy and Management, Carnegie Mellon University, April 1993.

CHAPTER 7: LARGE FIRM–CENTERED NETWORKED PRODUCTION SYSTEMS IN JAPAN AND EUROPE

1. Michael L. Gerlach, *Alliance Capitalism* (New York: Oxford University Press, 1992); James R. Lincoln, Michael L. Gerlach, and Peggy Takahashi,

"*Keiretsu* Networks in the Japanese Economy: A Dyad Analysis of Intercorporate Ties," *American Sociological Review* 57 (October 1992): 561–85.

2. Marie Anchordoguy, "A Brief History of Japan's Keiretsu," *Harvard Business Review* 68 (July–August 1990): 58, a special report contained within Charles H. Ferguson, "Computers and the Coming of the U.S. Keiretsu," *Harvard Business Review* 68 (July–August 1990): 55–70.

3. Eric Van Kooij, "Industrial Networks in Japan: Technology Transfer to SMEs [Small and Medium Sized Enterprises]," *Entrepreneurship and Regional Development* 2 (Fall 1990): 279–301.

4. Gerlach, *Alliance Capitalism*.

5. Jane Perlez, "Toyota and Honda Create Global Production System," *New York Times*, March 26, 1993, p. A1.

6. In addition to Lincoln, Gerlach, and Takahashi, "*Keiretsu* Networks," and Gerlach, *Alliance Capitalism*, a partial bibliography would include Masahiko Aoki, ed., *Anatomy of the Japanese Firm* (Rotterdam: North Holland, 1984); Aoki, "The Japanese Firm in Transition," in *The Political Economy of Japan*, ed. K. Yamamura and Y. Yasuba (Stanford: Stanford University Press, 1987), pp. 263–88; Banri Asanuma, "The Contractual Framework for Parts Supply in the Japanese Automotive Industry," *Japanese Economic Studies* 13 (Summer 1985): 54–78; Asanuma, "Manufacturer-Supplier Relationships in Japan and the Concept of Relation-Specific Skill," *Journal of the Japanese and International Economies* 3 (March 1989): 1–30; Michael Cusamano, *The Japanese Automobile Industry: Technology and Management at Nissan and Toyota* (Cambridge, Mass.: Harvard University Press, 1985); Ronald Dore, *Flexible Rigidities: Industrial Policy and Structural Adjustment in the Japanese Economy 1970–1980* (London: Athlone Press, 1986); David Friedman, *The Misunderstood Miracle* (Ithaca, N.Y.: Cornell University Press, 1988); Masahiko Ikeda, "International Comparison of Production Systems between Europe and Japan," Chuo University, Department of Economics, Tokyo, unpublished MS, January 1986; Ikeda, "Production Networks of Big Firms and Smaller Subcontractors," Chuo University, Department of Economics, Tokyo, unpublished MS, May 1987; Ken-ichi Imai, "Evolution of Japan's Corporate and Industrial Networks," in *Industrial Dynamics*, ed. Bo Carlsson (Boston: Kluwer, 1989); Martin Kenney and Richard Florida, *Beyond Mass Production* (New York: Oxford University Press, 1992); John McMillan, "Managing Suppliers: Incentive Systems in Japanese and U.S. Industry," *California Management Review* 32 (Summer 1990): 38–55; Toshihiro Nishiguchi, "Strategic Dualism: An Alternative in Industrial Societies" (Ph.D. diss., Oxford University, 1990); Y. Sato, "The Subcontracting Production (Shitauke) System in Japan," *Keio Business Review* 21 (January 1983): 1–25; Richard Schoenberger, *Japanese Manufacturing Techniques* (New York: Free Press, 1982); Eric van Kooij, "Japanese Subcontracting at a Crossroads," *Small Business Economics* 3 (June 1991): 145–54.

7. McMillan, "Managing Suppliers," p. 51.

8. Van Kooij, "Industrial Networks in Japan," p. 292. Three out of five firms in this network were officially classified by MITI's Small and Medium Enterprise Agency as being "small or medium sized" enterprises, based on total capital stock. But then two out of five participants were in fact "large" enterprises.

9. James P. Womack, Daniel T. Jones, and Daniel Roos, *The Machine That Changed the World* (New York: Rawson/Macmillan, 1990), p. 168.

10. Van Kooij, "Industrial Networks in Japan," p. 286.

11. Lincoln, Gerlach, and Takahashi, "*Keiretsu* Networks," p. 569.

12. I have in mind Shutt and Whittington's implicit criticism, discussed in chapter 6, of those who simply assume that spin-offs in the process of vertical disintegration are independent of their former parent. The Japanese experience shows that, as in so many evolutionary developments in the real world, "it depends." Recall from chapter 2 that even Sengenberger and Loveman had uncovered empirical evidence from Germany that the "independence" of spin-offs might not so easily be read off from the fact of their legal status alone.

13. Lincoln, Gerlach, and Takahashi, "*Keiretsu* Networks," p. 564.

14. Anchordoguy, "Japan's Keiretsu," p. 59.

15. Lincoln, Gerlach, and Takahashi, "*Keiretsu* Networks," p. 566.

16. Ibid., p. 581.

17. McMillan, "Managing Suppliers," p. 41.

18. Based on research conducted by Mari Sako and by David Whittaker, cited in U.S. Congress, Office of Technology Assessment, *Making Things Better: Competing in Manufacturing*, report OTA-ITE-443 (Washington, D.C.: U.S. Government Printing Office, February 1990), p. 135.

19. Peter Dicken, *Global Shift: The Internationalization of Economic Activity*, 2nd ed. (New York: Guilford Press, 1992); Richard Florida and Martin Kenney, "Transplanted Organizations: The Transfer of Japanese Industrial Organization to the U.S.," *American Sociological Review* 56 (June 1991): 381–98; Norman J. Glickman and Douglas P. Woodward, *The New Competitors: How Foreign Investors Are Changing the American Economy* (New York: Basic Books, 1989); Kenney and Florida, "How Japanese Industry Is Rebuilding the Rust Belt," *Technology Review* 94 (February/March 1991): 25–33; Perlez, "Toyota and Honda."

20. Lincoln, Gerlach, and Takahashi, "*Keiretsu* Networks," p. 582.

21. On the role of the state in Japanese economic development, see Marie Anchordoguy, "Mastering the Market: Japanese Government Targeting of the Computer Industry," *International Organization* 42 (Summer 1988): 509–43; Anchordoguy, *Computers Inc: Japan's Challenge to IBM* (Cambridge, Mass.: Harvard University Press, 1989); Chalmers Johnson, *MITI and the Japanese Economic Miracle: The Growth of Industrial Policy, 1925–1975* (Stanford, Calif.: Stanford University Press, 1982); Johnson, "MITI, MPT, and the Telecom Wars: How Japan Makes Policy for High Technology," in *Politics and Productivity: How Japan's Development Strategy Works*, ed. Johnson, Laura D'Andrea Tyson, and John Zysman (New York: Harper Business, 1990), pp. 177–240.

22. Ash Amin, "Big Firms versus the Regions in the Single European Market," in *Cities and Regions in the New Europe*, ed. Mick Dunford and George Kafkalal (London: Belhaven, 1992).

23. Flavia Martinelli and Erica Schoenberger, "Oligopoly Is Alive and Well: Notes for a Broader Discussion of Flexible Accumulation," in *Industrial Change and Regional Development: The Transformation of New Industrial Spaces*, ed. Georges Benko and Mick Dunford (London: Belhaven Press/Pinter, 1991), chap. 6, p. 127.

24. Laboratorio di Politica Industriale, *Acquisizioni Fusioni Concorrenza* (Bologna: NOMISMA, June 1989), p. 74.

25. "Feeding Frenzy on the Continent," *Business Week*, May 18, 1992, p. 65.

26. As the editors of *Business Week* put it: "European managers don't have time to expand by internal growth, the old-fashioned way. The No. 1 strategy today is growth-by-acquisition" ("Europe's Giants Are Hungrier Than Ever," *Business Week*, July 17, 1989, p. 144).

27. "A Helping Hand for Europe's High-Tech Heavies," *Business Week*, July 13, 1992, p. 43.

28. "Siemens and Matsushita Plan Component Venture," *International Herald Tribune*, June 16, 1989, p. 15.

29. "A Waltz of Giants Sends Shock Waves Worldwide," *Business Week*, March 19, 1990, p. 59.

30. "An American Eagle Talks Turkey with the Taiwanese," *Business Week*, December 2, 1991, p. 55.

31. Luc Soete, "National Support Policies for Strategic Industries: The International Implications," working paper 91-006, Maastricht Economic Research Institute on Innovation and Technology (MERIT), Maastricht (Netherlands), January 1991, p. 18.

32. The following section draws on Michael Calingaert, "Government-Business Relations in the European Community," *California Management Review* 35 (Winter 1993): 118–33; Lynn Krieger Mytelka, "States, Strategic Alliances, and International Oligopolies: The European ESPRIT Programme," in *Strategic Partnerships, States, Firms and International Competition*, ed. Mytelka (London: Pinter, 1991), pp. 182–210; and Todd Watkins, "A Technological Communications Costs Model of R&D Consortia as Public Policy," *Research Policy* 20 (March 1991): 87–107.

33. Mytelka, "Strategic Partnerships," table 9.4.

34. Ibid., p. 190.

35. Amin, "Big Firms versus the Regions."

36. "How Europe Swings the Big Science Tab," *Business Week*, March 22, 1993, pp. 62–64.

37. "European High Tech Is Sinking Mighty Low," *Business Week*, September 17, 1990, p. 52; "Can Europe Survive the Chip Wars?" *Business Week*, December 21, 1992, pp. 56–57.

38. "Zoom! Airbus Comes on Strong," *Business Week*, April 22, 1991, p. 49.

39. "Now for the Really Big One," *Economist*, January 9, 1993, pp. 57–58.

CHAPTER 8: INTERFIRM PRODUCTION NETWORKS
IN THE UNITED STATES

1. Ray Marshall and Marc Tucker, *Thinking for a Living* (New York: Basic Books, 1992).

2. "Learning from Japan," *Business Week*, January 22, 1992, pp. 52–60.

3. This section is based on Bennett Harrison and Maryellen R. Kelley, "The New Industrial Culture," *American Prospect* (Winter 1991): 54–61.

4. John McMillan, "Managing Suppliers: Incentive Systems in Japanese and U.S. Industry," *California Management Review* 32 (Summer 1990): 49.

324

5. Susan Helper, "How Much Has Really Changed between U.S. Automakers and Their Suppliers?" *Sloan Management Review* 32 (Summer 1991): 15–28.

6. Quoted in Harrison and Kelley, "New Industrial Culture," p. 58.

7. Quoted in ibid., p. 59.

8. Maryellen R. Kelley and Todd A. Watkins, *The Defense Industrial Network: A Legacy of the Cold War*, Heinz School of Public Policy and Management, Carnegie Mellon University, August 1992, unpublished MS.

9. Chris Farrell, "Where Will Merger Mania Strike Next?" *Business Week*, December 18, 1989, p. 32. The incidence and job-creating significance of foreign acquisitions of domestic U.S. companies is the main subject of Norman J. Glickman and Douglas P. Woodward, *The New Competitors: How Foreign Investors Are Changing the American Economy* (New York: Basic Books, 1989).

10. The following is based on Barry Bluestone, Peter Jourdan, and Mark Sullivan, *Aircraft Industry Dynamics* (Boston: Auburn House, 1981); Artemis March, "The U.S. Commercial Aircraft Industry and Its Foreign Competitors,"vol. 1, *Working Papers*, M.I.T. Commission on Industrial Productivity, 1989; and David Mowery, "Joint Ventures in the U.S. Commercial Aircraft Industry," in *International Collaborative Ventures in U.S. Manufacturing*, ed. Mowery (Cambridge, Mass.: Ballinger, 1988), pp. 71–110.

11. The French government owns 90 percent of SNECMA's equity. Just to make the organization chart even more complicated, the other 10 percent is owned by Pratt & Whitney, GE's main U.S. competitor in the jet engine industry!

12. Bluestone, Jourdan, and Sullivan, *Aircraft Industry Dynamics*, pp. 67–68.

13. William C. Taffel, "Advantageous Liasons," *Technology Review* 96 (May/June 1993): 28–36.

14. This story has been recounted many times by now. Compare Barry Bluestone and Bennett Harrison, *The Deindustrialization of America: Plant Closings, Community Abandonment and the Dismantling of Basic Industry* (New York: Basic Books, 1982); John Hoerr, *And the Wolf Finally Came: The Decline of the American Steel Industry* (Pittsburgh: University of Pittsburgh Press, 1988); William Serrin, *Homestead: The Glory and Tragedy of an American Steel Town* (New York: New York Times Books, 1992).

15. Martin Kenney and Richard Florida, *Beyond Mass Production* (New York: Oxford University Press, 1992); "Japanese Steel Programs with U.S. Steelmakers," news release, Japan Steel Information Center, New York City, April 1991.

16. Except where otherwise indicated, the following case study is based on Steven Klepper, "Collaboration in Robotics," in Mowery, *Collaborative Ventures*, pp. 233–66.

17. Joseph L. Badaracco, *The Knowledge Link* (Boston: Harvard Business School Press, 1991), p. 113.

18. Artemis March, "The U.S. Machine Tool Industry and Its Foreign Competitors," vol. 2, *Working Papers*, M.I.T. Commission on Industrial Productivity, Cambridge, Mass., 1989, p. 36.

19. Badaracco, *Knowledge Link*, p. 113.

20. Ibid., p. 114.

21. Klepper, "Collaboration in Robotics," p. 254.

22. International Trade Administration, U.S. Department of Commerce, *U.S. Industrial Outlook '92: Business Forecasts for 350 Industries* (Washington, D.C.: U.S. Government Printing Office, 1992), pp. 18-5–18-6.

23. By contrast, Ford's response was, for a time, to push its strategy of building a globally sourced, highly modularized "world car." When this strategy failed, it and Chrysler, too, turned increasingly to alliances with foreign auto manufacturers.

The recent history of the U.S. auto industry's efforts to cope, first with rising fuel costs and later with the American public's perception that foreign (especially Japanese) cars were of higher quality, has been much analyzed elsewhere. Surely no sector of the American economy has been more closely studied. See, for example, Alan Altshuler, Martin Anderson, Daniel Jones, and James Womack, *The Future of the Automobile* (Cambridge, Mass.: M.I.T. Press, 1984); Barry Bluestone and Irving Bluestone, *Negotiating the Future* (New York: Basic Books, 1992); Bluestone and Harrison, *Deindustrialization of America*; Harry Katz, *Changing Gears* (Cambridge, Mass.: M.I.T. Press, 1985); James P. Womack, Daniel T. Jones, and Daniel Roos, *The Machine That Changed the World* (New York: Rawson/Macmillan, 1990); and James Womack, "The U.S. Automobile Industry in an Era of International Competition: Performance and Prospects," vol. 1, *Working Papers*, M.I.T. Commission on Industrial Productivity, Cambridge, Mass., 1989.

24. Badaracco, *Knowledge Link*, p. 61.

25. Clair Brown and Michael Reich, "When Does Union-Management Cooperation Work? A Look at NUMMI and GM-Van Nuys," *California Management Review* 31 (Summer 1989): 26–44.

At least in principle, the purpose of team forms of work organization is to cross-train workers in a variety of skills, so that they may operate a wider variety of machines, detect flaws in each others' work, and spell one another on the line. In practice, some critics contend that, while GM certainly does *rotate* workers at NUMMI among jobs requiring different tasks, those tasks have been greatly diminished in breadth, complexity, and challenge. In other words, job *rotation* need not imply job *enlargement*. See Mike Parker, *Inside the Circle: A Union Guide to QWL* (Boston: South End Press, 1985).

26. Badaracco, *Knowledge Link*, p. 62.

27. Ibid., p. 65.

28. Ibid., p. 122. For a detailed case study of Saturn, see Bluestone and Bluestone, *Negotiating the Future.*

29. "Crisis at GM: Turmoil at the Top Reflects the Depth of Its Troubles," *Business Week*, November 9, 1992, pp. 84–86.

30. Womack, Jones, and Roos, *Machine That Changed*, p. 167.

31. U.S. Congress, Office of Technology Assessment, *Making Things Better: Competing in Manufacturing*, report OTA-ITE-443 (Washington, D.C.: U.S. Government Printing Office, February 1990), p. 135.

32. Helper, "U.S. Automakers and Their Suppliers," pp. 24–25. Similar evidence has been found in the British auto industry, where partial implementation of JIT and an emphasis on parts quality at the expense of anticipation of production problems is reported to be creating costs that are effectively being pushed onto the suppliers. See Al Rainnie, "Just-in-Time, Sub-Contracting, and the Small Firm," *Work, Employment, and Society* 5 (September 1991): 353–76.
Even in the much-vaunted German auto production system, there are signs of trouble. "In particular, the powerful engineering departments in Mercedes, VW and Porsche . . . seem to be constantly altering the design specifications which the purchasing department has given to suppliers, with the result that engineering costs tend to be much higher than they need be otherwise." Moreover, "some OEMs want all cost savings [achieved by their suppliers] passed on to the OEM, an attitude whioch has led some suppliers to conceal savings desite the 'high-trust' relationship" (Kevin Morgan, "Do Networks Result in Innovations? The Automotive and Machine Tool Industries in Europe," in *Significant Others: Exploring the Potentials of Manufacturing Networks, Background Papers*, ed. Brian Bosworth and Stuart Rosenfeld [Chapel Hill: Regional Technology Strategies, 1993], p. 10). This recent evidence calls for at least a qualification of impressions formed several years earlier about collaborative customer-supplier relations in the auto industry of Baden-Württembourg by Charles Sabel, Horst Kern, and Gary Herrigel, "Collaborative Manufacturing"

(Department of Political Science, M.I.T., March 31, 1989, unpublished MS).

Of course, by the 1990s the German auto industry was also facing worldwide overcapacity and global recession, which only serves to remind us that all of these innovative organizational practices must always be assessed in the context of current macroeconomic conditions, that is, the ability to sell the product.

33. "GM Tightens the Screws," *Business Week*, June 22, 1992, pp. 30–31.

34. Maryellen R. Kelley and Harvey Brooks, "External Learning Opportunities and the Diffusion of Process Innovations to Small Firms: The Case of Programmable Automation," *Technological Forecasting and Social Change* 39 (April 1991): 103–25; Kelley and Brooks, "From Breakthrough to Follow-Through," *Issues in Science and Technology* [National Academy of Sciences] (Spring 1989): 42–47.

35. Office of Technology Assessment, *Making Things Better*.

36. Michael Porter, *Capital Choices: Changing the Way America Invests in Industry* (Washington, D.C.: Council on Competitiveness, June 1992). Also see my review and critique of this report: Bennett Harrison, "Where Private Investment Fails," *American Prospect* (September 1992): 906–14.

37. James Poterba and Lawrence H. Summers, "Time Horizons of American Firms: New Evidence from a Survey of CEOs" (Council on Competitiveness and the Harvard Business School, October 1991, unpublished MS).

38. Porter, *Capital Choices*, p. 26.

39. David J. Ravenscraft and F. M. Scherer, *Mergers, Sell-Offs, and Economic Efficiency* (Washington, D.C.: Brookings Institution, 1987).

40. Here, Porter is building on Robert Kaplan's, Alfred Chandler's, William Abernathy's, and Robert Hayes's original critique of the importation by American corporations in the 1960s and 1970s of financial reporting techniques devised by independent British public accountants.

41. Gunnar Myrdal, *Rich Lands and Poor* (New York: Harper & Row, 1957).

42. E. Duttweiler, ed., *Factbook 1991* (New York: New York Stock Exchange, 1991); cited by Porter, *Capital Choices*, p. 26.

CHAPTER 9: THE DARK SIDE OF FLEXIBLE PRODUCTION

1. Robert Kuttner, "The Declining Middle," *Atlantic*, July 1983, pp. 60–69; Barry Bluestone and Bennett Harrison, *The Deindustrialization of America* (New York: Basic Books, 1982).

2. For one journalist's account of the history of the debate, see James Lardner, "The Declining Middle," *New Yorker*, May 3, 1993, pp. 108–14.

3. Lester C. Thurow, "A Surge in Inequality," *Scientific American*, September 1987, pp. 30–37.

4. Barry Bluestone and Bennett Harrison, *The Great American Job Machine: The Proliferation of Low Wage Employment in the U.S. Economy*, report prepared for the Joint Economic Committee, November 1986.

5. Bluestone and Harrison, "The Grim Truth about the Job 'Miracle'," *New York Times*, February 1, 1987, p. F1. The Reagan administration reacted three weeks later, in the form of a rejoinder by the then-commissioner of the U.S. Bureau of Labor Statistics, Janet L. Norwood: "The Job Machine Has Not Broken Down," *New York Times*, February 22, 1987, p. F3.

6. Warren T. Brooks, "Low-Pay Jobs: The Big Lie," *Wall Street Journal*, March 25, 1987, p. 1.

7. One especially influential paper that presaged a growing agreement on the basic facts was by Gary Burtless, "Earnings Inequality over the Business and Demographic Cycles," in *A Future of Lousy Jobs?* ed. Burtless (Washington D.C.: Brookings Institution, 1990), pp. 77–122.

8. Barry Bluestone, "The Great U-Turn Revisited: Economic Restructuring, Jobs, and the Redistribution of Earnings," in *Jobs, Earnings, and Employment Growth Policies in the United States*, ed. John D. Kasarda (Boston: Kluwer, 1990), pp. 7–27; Bluestone and Bennett Harrison, "The Growth of Low-Wage Employment: 1963–86," *American Economic Review: Proceedings* 78 (May 1988): 124–28; Lucy Gorham and Bennett Harrison, *Working below the Poverty Line* (Washington, D.C.: Aspen Institute, 1990); Harrison and Bluestone, "Wage Polarisation in the U.S. and the 'Flexibility' Debate," *Cambridge Journal of Economics* 14 (September 1990): 351–73; Harrison and Bluestone, *The Great U-Turn: Corporate Restructuring and the Polarizing of America* (New York: Basic Books, 1988); Harrison and Gorham, "Growing Inequality in Black Wages in the 1980s and the Emergence of an African-American Middle Class," *Journal of Policy Analysis and Management* 11 (February 1992): 235–53, reprinted in *The City in Black and White*, ed. George Galster and Edward Hill (New Brunswick: Center for Urban Policy Research, Rutgers University, 1992), pp. 56–71; Chris Tilly, Bluestone, and Harrison, "What Is Making American Wages More Unequal?" in *Proceedings of the Thirty-ninth Annual Meeting of the Industrial Relations Research Association* (Madison, Wis.: IRRA, May 1987), pp. 338–48.

9. U.S. Bureau of the Census, "Workers with Low Earnings: 1964 to 1990," *Current Population Reports, Consumer Income*, series P-60, no. 178 (Washington, D.C.: U.S. Government Printing Office, March 1992), pp. 5, 8.

10. Lawrence Mishel and Jared Bernstein, *The State of Working America, 1992* (Armonk, N.Y.: Sharpe, 1993), p. 146.

11. Frank Levy and Richard Murnane, "U.S. Earnings Levels and Earnings Inequality: A Review of Recent Trends and Proposed Explanations," *Journal of Economic Literature* 30 (September 1992): 1333–81.

12. Lynn A. Karoly, "The Trend in Inequality Among Families, Individuals and Workers in the United States: A Twenty-Five Year Prospective" (Santa Monica, Calif.: RAND Corporation, 1990), reprinted in *Uneven Tides: Rising Inequality in America*, ed. Sheldon Danziger and Peter Gottschalk (New York: Russell Sage, 1993), pp. 19–97.

13. Levy and Murnane, "U.S. Earnings Levels," p. 1333.

14. Ibid., p. 1371.

15. Evidence is accumulating that earnings are polarizing in Europe, too. See Francis Green, Andrew Henley, and Euclid Tsakalotos, "Income Inequality in Corporatist and Liberal Economies: A Comparison of Trends within OECD Countries," *Studies in Economics*, no. 92-13 (Canterbury, U.K.: University of Kent, November 1992); T. Elfring and R. Kloosterman, "The Dutch 'Job Machine': The Fast Growth of Low-Wage Jobs in Services, 1979–1986" (Erasmus University, Rotterdam School of Management, Rotterdam, the Netherlands, 1989); Katherine McFate, "Poverty, Inequality, and the Crisis of Social Policy: Summary of Findings" (Joint Center for Political and Economic Studies, Washington D.C., September 1991); Bob Rowthorn, "Centralisation, Employment and Wage Dispersion," *Economic Journal* 102 (May 1992): 506–23; Tim Smeeding, Michael O'Higgins, and Lee Rainwater, eds., *Poverty, Inequality, and Income Distribution in Comparative Perspective: the Luxembourg Income Study* (London: Harvester Wheatsheaf, 1990); and Michael White, *Against Unemployment* (London: Policy Studies Institute, 1991).

    Even in Sweden, whose labor market institutions have throughout the twentieth century been more explicitly committed to promoting equity that those of any other country, wage differentials among bargaining units (work places), which had been narrowing since the 1940s, stopped declining after 1977–78. See Guy Standing, *Unemployment and Labour Market Flexibility: Sweden* (Geneva: International Labour Office, 1988).

    The newest and most systematic analyses of international trends in earnings inequality are to be found in Richard B. Freeman, ed., *Working under Different Rules* (New York: Russell Sage, 1994).

16. Mishel and Bernstein, *Working America*, p. 167. See also the discussion of education and earnings in Harrison and Gorham, "Growing Inequality in Black Wages."

17. Lawrence Katz and Kevin Murphy, "Changes in Relative Wages, 1963–1987: Supply and Demand Factors," *Quarterly Journal of Economics* 107 (February 1992): 35–78.

18. Levy and Murnane, "U.S. Earnings Levels," p. 1372.

19. Richard B. Freeman, "How Much Has De-unionization Contributed to the Rise in Male Earnings Inequality?" in Danziger and Gottschalk, *Uneven Tides*, p. 159.

Together with Katz and George Borjas, his colleagues at the NBER, Freeman also conjectures that the rapid growth of imports during the 1980s, part of the mounting trade deficit, may have contributed to the problem of rising inequality. Essentially, the mix of goods and services imported into this country, especially from such developing nations as India and Mexico, disproportionately embody low- and midrange skilled labor, even as the United States exports such high-tech products as aircraft and computers. It is as though the domestic supply of unskilled and semiskilled labor had suddenly grown by a substantial fraction, with the predictable depressing effect on the wage claims of U.S. workers lacking higher-level skills. But again, Freeman would be the last to claim that this explanation sufficiently accounts for the severity of the surge in inequality (Borjas, Freeman, and Katz, "On the Labor Market Effects of Immigration and Trade," in *Immigration and the Work Force: Economic Consequences for the United States and Source Areas*, ed. Freeman and Borjas [Chicago: University of Chicago Press, 1992], pp. 213–44).

20. In what follows, I am drastically telescoping what is in fact a vast literature on the subject of dual labor markets, to which I and many other economists and sociologists of my generation made some contribution.

Without question, the single most influential theorist of dual labor markets was and is the M.I.T. economist Michael J. Piore—first, in his early collaboration with Peter Doeringer, then in several papers written alone or with M.I.T. colleagues. Particularly important in retrospect are Peter B. Doeringer and Michael J. Piore, *Internal Labor Markets and Manpower Analysis* (Lexington, Mass.: Heath, 1971; reprinted with a new introduction, Armonk, N.Y.: Sharpe, 1985); and Piore, "The Technological Foundations of Dualism and Discontinuity," in *Dualism and Discontinuity in Industrial Societies*, ed. Suzanne Berger and Michael J. Piore (New York: Cambridge University Press, 1980), pp. 55–87.

The interested reader might also want to look at the following papers and books, and at the extensive bibliographies therein, but be advised that this is an extremely abbreviated list: Robert Averitt, *The Dual Economy: the Dynamics of American Industrial Structure* (New York: Norton, 1968); Barry Bluestone and Mary Huff Stevenson, "Industrial Transformation and the Evolution of Dual Labor Markets," in *The Dynamics of Labor Market Segmentation*, ed. Frank Wilkinson (New York: Academic Press, 1981); Richard Edwards, *Contested Terrain* (New York: Basic Books, 1979); David M. Gordon, *Theories of Poverty and Underemployment* (Lexington, Mass.: Lexington Books, 1973); Gordon, Richard

Edwards, and Michael Reich, *Segmented Work, Divided Workers* (New York: Oxford University Press, 1982); Bennett Harrison, *Education, Training, and the Urban Ghetto* (Baltimore: Johns Hopkins University Press, 1972), chap. 5; Harrison and Andrew Sum, "The Theory of 'Dual' or Segmented Labor Markets," *Journal of Economic Issues* 13 (September 1979): 687–706; and Paul Osterman, ed., *Internal Labor Markets* (Cambridge, Mass.: M.I.T. Press, 1984).

This body of work eschewed what its contributors saw as the false precision of formal mathematics in favor of a more institutionally rich, if thereby often "fuzzy," characterization of labor market structures. Much later, during the 1980s, several neoclassical labor economists attempted to account for the stubborn persistence of labor market duality with more conventional theoretical and statistical tools. Compare Jeremy Bulow and Lawrence Summers, "A Theory of Dual Labor Markets with Applications to Industrial Policy, Discrimination, and Keynesian Unemployment," *Journal of Labor Economics* 4 (July 1986): 376–414; William T. Dickens and Kevin Lang, "A Test of Dual Labor Market Theory," *American Economic Review* 75 (September 1985): 792–805; and Dickens and Lang, "Labor Market Segmentation Theory: Reconsidering the Evidence" (Department of Economics, University of California at Berkeley, August 1991, unpublished MS).

21. Chris Tilly, "Dualism in Part-Time Employment," *Industrial Relations* 31 (Spring 1992): 331.

22. Ibid., p. 346.

23. John Goldthorpe, "The End of Convergence: Corporatist and Dualist Tendencies in Modern Western Societies," in *Order and Conflict in Contemporary Capitalism*, ed. Goldthorpe (Oxford: Clarendon Press, 1984), pp. 315–43.

24. John Atkinson, "Recent Changes in the Internal Labour Market Structure in the U.K.," in *Technology and Work*, ed. Wout Buitelaar (Aldershot, U.K.: Avebury, 1988), pp. 133–49.

Atkinson's work had already created something of a storm among British labor scholars and activists by the time I arrived at Oxford University in the spring of 1989 to begin a term as a Fulbright fellow. His rather impressionistic writing alienated mainstream scholars, and his normative views alarmed and antagonized the academic Left. Looking back on the brouhaha, now, it is clear that he was on to something.

25. Marino Regini, "Industrial Relations in the Phase of 'Flexibility,'" (Department of Sociology, University of Milano, Spring 1987, unpublished MS).

26. Fiorenza Belussi, "Benetton Italy: Beyond Fordism and Flexible Specialization to the Evolution of the Network Firm Model," in *Information Technology and Women's Employment: The Case of the European Clothing Industry*, ed.

S. Mitter (Berlin: Springer Verlag, 1989). These ideas appear yet again in Flavia Martinelli and Erica Schoenberger, "Oligopoly Is Alive and Well: Notes for a Broader Discussion of Flexible Accumulation," in *Industrial Change and Regional Development: The Transformation of New Industrial Spaces,* ed. Georges Benko and Mick Dunford (London: Belhaven Press/Pinter, 1991), pp. 117–33; and in Knuth Dohse, Ulrich Jurgens, and Thomas Malsch, "From 'Fordism' to 'Toyotism'? The Social Organization of the Labor Process in the Japanese Automobile Industry," *Politics and Society* 14 (1985): 115–46.

27. Jeffrey Pfeffer and James Baron, "Taking the Workers Back Out: Recent Trends in the Structuring of Employment," in *Research in Organizational Behavior,* vol. 10, ed. Barry Staw and L. L. Cummings (Greenwich, Conn.: JAI Press), 1988, p. 36.

28. The tendential devolution of ILMs in the U.S. has been carefully documented by M.I.T.'s Paul Osterman and Columbia University's Thierry Noyelle. See Paul Osterman, *Employment Futures* (New York: Oxford University Press, 1988); and Thierry Noyelle, *Beyond Industrial Dualism* (Boulder, Colo.: Westview Press, 1987).

29. Noyelle, *Beyond Industrial Dualism,* pp. 16–17.

30. Maury B. Gittleman and David R. Howell, "Job Quality, Labor Market Segmentation, and Earnings Inequality: Effects of Economic Restructuring in the 1980s by Race and Gender" (Graduate School of Management, New School for Social Research, New York City, July 1992, unpublished MS).

31. Gordon, Edwards, and Reich, *Segmented Work, Divided Workers.*

32. Gittleman and Howell, p. i.

33. Richard Belous, *The Contingent Economy: The Growth of the Temporary, Part-Time and Subcontracted Workforce* (MacLean, Va.: National Planning Association, 1989).

34. Katherine G. Abraham, "Flexible Staffing Arrangements and Employers' Short-Term Adjustment Strategies," in *Employment, Unemployment, and Labor Utilization,* ed. Robert A. Hart (Boston: Unwin Hyman, 1988), pp. 288–311; Eileen Appelbaum, "Restructuring Work: Temporary, Part Time, and At-home Employment," in *Computer Chips and Paper Clips: Technology and Women's Employment,* ed. Heidi Hartmann (Washington D.C., National Academy Press, 1987), pp. 268–310; Appelbaum and Peter Albin, "Employment, Occupational Structure, and Educational Attainment in the United States, 1973, 1979, and 1987," report to the Organization for Economic Cooperation and Development, Commission on Services, Paris, 1988; Susan Christopherson, "Flexibility in the U.S. Service Economy and the Emerging Spatial Division of Labour,"

Transactions of the Institute of British Geographers 14 (1989): 131–43; Ronald
G. Ehrenberg, Pamela Rosenberg, and Jeanne Li, "Part-Time Employment in
the United States," in Hart, *Employment, Unemployment*, pp. 256–87.

An important recent resource document is Virginia L. duRivage, ed., *New
Policies for the Part-Time and Contingent Workforce* (Armonk, N.Y.: Sharpe, for
the Economic Policy Institute, 1992), containing papers by Appelbaum,
Françoise Carre, Tilly, and the editor. See also Katherine G. Abraham and
Susan K. Taylor, "Firms' Use of Outside Contractors: Theory and Evidence"
(Cambridge, Mass.: National Bureau of Economic Research, September 1993),
working paper 4468.

35. Polly Callaghan and Heidi Hartmann, *Contingent Work: A Chart Book on
Part-Time and Temporary Employment* (Washington, D.C.: Economic Policy
Institute, for the Institute for Women's Policy Research, 1991), p. 5.

36. Ehrenberg, Rosenberg, and Li, "Part-Time Employment," p. 274.

37. Mishel and Bernstein, *Working America*, p. 155.

38. Anne Polivka and Thomas Nardone, "On the Definition of Contingent
Work," *Monthly Labor Review* 112 (December 1989): 9–16.

39. Gina Kolata, "More Children Are Employed, Often Perilously," *New York
Times*, June 21, 1992, p. 1.

40. Kathleen Christensen, "The Two-Tiered Workforce in U.S. Corporations,"
in *Turbulence in the American Workplace*, ed. Peter B. Doeringer with Kathleen
Christensen, Patricia M. Flynn, Douglas T. Hall, Harry C. Katz, Jeffrey H.
Keefe, Christopher J. Ruhm, Andrew M. Sum, and Michael Useem (New York:
Oxford University Press, 1991).

41. Of 2,775 corporate human resources or personnel executives in nine
industries to whom the survey instrument was mailed, a total of 521
returned usable questionnaires, for a response rate of about 19 percent. No
attempt was made to ascertain the presence or extent of response bias—
whether those who *did* reply fairly represent the behavior of those who did
not (ibid., p. 154).

42. The following section is drawn from Michael T. Donaghu and Richard
Barff, "NIKE Just Did It: International Subcontracting and Flexibility in Athletic Footwear Production," *Regional Studies* 24 (December 1990): 537–52.

43. Andrew Pollack, "Japan's Companies Moving Production to Sites Overseas," *New York Times*, Sunday, August 29, 1993, p. 1.

44. Donaghu and Barff, "Nike Just Did It," p. 545.

45. Joseph L. Badaracco, Jr., "Changing Forms of the Corporation," in *U.S. Business Corporation: An Institution in Transition*, John R. Meyer and James M. Gustafsom (Cambridge, Mass.: Ballinger, 1988), p. 84; also Badaracco, *The Knowledge Link* (Boston: Harvard Business School Press, 1991). On the limits to such labor-management cooperation programs, see Maryellen R. Kelley and Bennett Harrison, "Unions, Technology, and Labor-Management Cooperation," in *Unions and Economic Competitiveness*, ed. Lawrence Mishel and Paula B. Voos (Armonk, N.Y.: Sharpe, 1992), pp. 247–86. For an impassioned advocacy of such cooperation, see Barry Bluestone and Irving Bluestone, *Negotiating the Future: A Labor Perspective on American Business* (New York: Basic Books, 1992).

46. Walter F. Powell, "Neither Markets nor Hierarchies: Network Forms of Organization," in *Research in Organizational Behavior* 12 (1990): 321.

47. Alain Lipietz, "New Tendencies in the International Division of Labor: Regimes of Accumulation and Modes of Regulation," in *Production, Work, Territory*, Allen Scott and Michael Storper (London: Allen & Unwin, 1986), pp. 16–40.

A propos the global campaign by capital for more flexible wage contracts, Lipietz, one of the architects of French regulation theory, writes: "The[ir] idea is to optimize the microeconomic capacity of the firm to adapt to volatility in demand, and to ensure a higher share of profit within value added. But this microeconomic strategy falls into a fallacy of composition . . . at the national level; with lower wages and less rigidity in aggregate demand, problems are likely to appear on the demand side (let alone the social unrest), thus leading to a return of business cycles, and to a further ex-post fall in profitability" (Danielle Leborgne and Alain Lipietz, "New Technologies, New Modes of Regulation: Some Spatial Implications," *Environment and Planning D: Society and Space* 6 [September 1988]: 270).

48. Harrison and Bluestone, *Great U-Turn*, chap. 6.

49. S. Deakin and Frank Wilkinson, "Labour Law, Social Security, and Economic Inequality" (London: Institute of Employment Rights, February 1989), p. 44.

50. Ibid. International comparisons of wage and productivity growth in the service sector provide one piece of evidence in favor of the Deakin and Wilkinson hypothesis. Thurow reports that, whereas "private service workers in the U.S. are paid only 67 percent as much as those in manufacturing, in Japan they are paid 93 percent as much and in Germany 85 percent as much as manufacturing workers. . . . With lower wages in services, American firms had less need to use more capital than their foreign counterparts who were forced to pay wages more nearly equal to those in manufacturing" (Lester C. Thurow, "Toward a High-Wage, High Productivity Service Sector," working paper, Washington, D.C., Economic Policy Institute, 1989).

The consequence of this behavior shows up in differential rates of productivity growth in the service economies of the various countries. Thus, for example, labor productivity growth between 1972 and 1983 averaged –0.5 percent per year in U.S. producer services and at best 0.3 percent per year in U.S. consumer services. In Germany, these growth rates were 2.5 percent and 1.5 percent, respectively. See Louise Waldstein, "Service Sector Wages, Productivity and Job Creation in the U.S. and Other Countries," working paper, Washington, D.C., Economic Policy Institute, 1989, table 13.

51. Richard B. Freeman and Lawrence F. Katz, "Rising Wage Inequality: The United States vs. Other Advanced Countries," in Freeman, ed., *Working Under Different Rules.*

52. While this "high road–low road" metaphor was made popular by President Bill Clinton during the 1992 presidential campaign in the United States, and while many writers now employ it regularly, it is the technology policy theorist Chris Freeman, at the Science Policy Research Unit, University of Sussex, who first suggested this imagery some years ago.

53. Richard B. Freeman and Joel Rogers, "Who Speaks for Us? Employee Representation in a Non-Union Labor Market," in Bruce E. Kaufman and Morris M. Kleiner, eds., *Employee Representation: Alternative and Future Directions* (Madison, Wis.: Industrial Relations Research Association, 1993), pp. 13–79.

54. Samuel Bowles, David Gordon, and Thomas Weisskopf, *Beyond the Waste Land* (New York: Anchor, 1983).

55. Sylvia Nasar, "U.S. Industry's Sluggish Pace: Revision Finds Less Growth of Capacity," *New York Times*, May 15, 1993, p. 17.

56. Quoted in Gene Koretz, "U.S. Industry May Be Mean, But It Also Looks Too Lean," *Business Week*, June 14, 1993, p. 22.

CHAPTER 10: ECONOMIC DEVELOPMENT POLICY IN A WORLD OF LEAN AND MEAN PRODUCTION

1. Epigraph taken from Gary Gereffi and Gary G. Hamilton, "Modes of Incorporation in an Industrial World: The Social Economy of Global Capitalism" (paper presented to the Annual Meeting of the American Sociological Association, Washington, D.C., August 1990), pp. 1–2.

2. This much is now conventional wisdom, thanks largely to the definitive research of the Harvard Business School's Alfred D. Chandler, Jr. See *The Visible Hand: The Managerial Revolution in American Business* (Cambridge, Mass.: Belknap Press, Harvard University Press, 1977) and *Scale and Scope: The*

Dynamics of Industrial Capitalism (Cambridge, Mass.: Belknap Press, Harvard University Press, 1990).

3. Michael Storper and Richard Walker, *The Capitalist Imperative: Territory, Technology, and Industrial Growth* (Oxford: Basil Blackwell, 1989).

4. For discussions of the fundamental underlying problem—the structural imbalance between spatially fixed governments and geographically mobile corporations—see Raymond Vernon, *Sovereignty at Bay* (New York: Basic Books, 1971), and Barry Bluestone and Bennett Harrison, *The Deindustrialization of America* (New York: Basic Books, 1982).

5. This paradox is a central theme in Gordon L. Clark, *Unions and Communities under Siege* (New York: Cambridge University Press, 1989).

6. Robert B. Reich, "Who Is Us?" *Harvard Business Review* 68 (January–February 1990): 53–64; and Reich, *The Work of Nations: Preparing Ourselves for 21st Century Capitalism* (New York: Knopf, 1991).

7. Reich also rehearses the by now well-known statistics on the growth of direct foreign investment (DFI) by the companies of other countries in U.S. assets. Thus, by 1989, foreign-owned firms controlled substantial fractions (varying from about a fifth to more than half) of the American electronics, chemical, automobile, and entertainment industries, and lesser but growing shares of the American banking, steel, machine tool, robotics, and telecommunications industries. Moreover, he rightly says, to get a clearer picture of the extent of the internationalization of economic activity, we must add to these numbers on DFI the growing extent of cross-border technology licensing and strategic alliances among the big firms, and the growth of the number of individual contractors—computer programmers, financial and engineering consultants, and the like—who live in London, Tokyo, Stuttgart, Paris, and New York but sell their services to companies all over the world. See Reich, *Work of Nations.*

John Dunning is an economist at the U.K.'s University of Reading and Rutgers University in New Jersey. Together with Harvard's Vernon, he has dominated the study of the evolution of the multinational corporation for two generations. In a recent paper, Dunning reinforces Reich's description of the extent of the global reach of the big companies. He tells us that, according to United Nations estimates:

> global direct investment in the period 1980–1988 rose [one and a half] times faster than trade; and there are now up to 35,000 companies which between them own or control 150,000 foreign affiliates. The number of strategic alliances runs into tens of thousands; and the number of subcontracting agreements into hundreds of thousands.... MNEs [multinational enterprises] account for about three-quarters of world trade [and]

an increasing proportion of intermediate goods and services are produced and traded within the same MNEs. . . . Adding the actual value added of foreign based MNEs in a country to that of the foreign output of its home based MNEs, and expressing the result as a percentage of private GNP, the resulting percentages exceed 50% in the case of such countries as the UK, Netherlands, Belgium, Switzerland, Canada, Singapore and Hong Kong; more than 30% in the case of Germany, France, Australia and Italy; and more than 20% in the case of Japan, the US, Taiwan, Brazil and Nigeria. No less important is the fact that these ratios have been rising over the last ten years (Dunning, "The Global Economy and the National Governance of Economies: A Plea for a Fundamental Re-Think" [Graduate School of Management, Rutgers University, Newark, N.J., November 1992, unpublished MS], pp. 7–10).

8. Tyson, "They Are Not Us: Why American Ownership Still Matters," *American Prospect* (Winter 1991): 37–49; Reich, "Who Do We Think They Are?" *American Prospect* (Winter 1991): 49–53.

9. Compare Paul R. Krugman, *Geography and Trade* (Cambridge, Mass.: M.I.T. Press, 1991); Krugman, ed., *Strategic Trade Policy and the New International Economics* (Cambridge, Mass.: M.I.T. Press, 1986); Paul M. Roemer, "Capital Accumulation in the Theory of Long-Run Growth," in *Modern Business Cycle Theory*, ed. Robert J. Barro (Cambridge, Mass.: Harvard University Press, 1989), pp. 51–127; Roemer, "Growth Based on Increasing Returns Due to Specialization," *American Economic Review* 77 (May 1987): 56–62; Laura D'Andrea Tyson, *Who's Bashing Whom?* (Washington, D.C.: Institute for International Economics, 1992); and Chalmers Johnson, Laura D'Andrea Tyson, and John Zysman, eds., *Politics and Productivity: The Real Story of How Japan Works* (Cambridge, Mass.: Ballinger, 1989). For a (somewhat) more popular treatment of the logic underlying certain aspects of the new trade theory, see W. Brian Arthur, "Positive Feedbacks in the Economy," *Scientific American*, February 1990, pp. 92–99.

10. For an early statement of the argument, see Michael E. Porter, "How Competitive Forces Shape Strategy," *Harvard Business Review* 57 (March–April 1979): 137–45. Following the publication of what have become two standard textbooks in management education, *Competitive Strategy* (New York: Free Press, 1980) and *Competitive Advantage* (New York: Free Press, 1985), Porter's magnum opus appeared in 1990. This hefty volume, drawing on case studies on ten countries and more than one hundred industries, is *The Competitive Advantage of Nations* (New York: Free Press, 1990).

11. The following draws on *The Competitive Advantage of Nations*, especially chap. 3. I am touching on only that small subset of the elements in Porter's model that are especially relevant to the "who is 'us'?" and "home

base" questions. The complete construct goes far beyond what I am describing here.

12. Here, Porter is drawing heavily on the insights of Eric von Hippel, *Sources of Innovation* (New York: Oxford University Press, 1988). On the theory of user-led innovation, see also Maryellen R. Kelley and Todd Watkins, "The Defense Industrial Network: A Legacy of the Cold War" (Heinz School of Public Policy and Management, Carnegie Mellon University, August 1992, unpublished MS).

13. Porter, *Competitive Advantage of Nations*, p. 19.

14. Ibid., p. 606.

15. William Lazonick, "Industry Clusters versus Global Webs: Organizational Capabilities in the U.S. Economy" (Department of Economics, Barnard College, Columbia University, revised unpublished MS of March 15, 1992), p. 6.

16. Ibid., pp. 32–33. I share with Lazonick some surprise that Reich should now choose to ignore the obvious association between the startling growth of the high end of the American income distribution during the 1980s and the culmination of decades of unproductive financial speculation and asset manipulation. After all, it was Reich, himself, who, in an earlier book, first coined the apt expression "paper entrepreneurialism" to distinguish productive from unproductive activity in the economy. Robert B. Reich, *The Next American Frontier* (New York: Times Books, 1983).

17. The classic statement is Raymond Vernon, *Sovereignty at Bay* (New York: Basic Books, 1971).

18. Raymond Vernon, "Government Control over Its Multinational Corporations: The U.S. Case" (Cambridge, Mass.: Center for Business and Government, John F. Kennedy School of Government, Harvard University, 1985), p. 8.

19. Raymond Vernon, "Sovereignty at Bay: Ten Years After," in *Multinational Corporations: The Political Economy of Foreign Direct Investment*, ed. Theodore H. Moran (Lexington, Mass.: Lexington Books, 1985), p. 258. Moran himself describes nation-states as "acknowledging a growing impotence as they match each other in the intensification of threats against multinationals that do not obey them" (ibid., p. 274).

20. A good general reference is John H. Dunning, *Multinational Enterprises and the Global Economy* (Reading, Mass.: Addison-Wesley, 1993), chap. 21. On the particular example of efforts to legislate international rules on advanced notification to workers of factory closures, see Bennett Harrison, "The International Movement for Prenotification of Plant Closures," *Industrial Relations* 23 (Fall 1984): 387–409.

21. David J. Teece, "Competition, Cooperation, and Innovation: Organizational Arrangements for Regimes of Rapid Technological Progress," *Journal of Economic Behavior and Organization* 18 (May 1992): 23.

22. Thomas M. Jorde and David J. Teece, "Innovation and Cooperation: Implications for Competition and Antitrust," *Journal of Economic Perspectives* 4 (Summer 1990): 75–96.

23. Charles H. Ferguson and Charles R. Morris, *Computer Wars: How the West Can Win in a Post-IBM World* (New York: Times Books, 1992), p. 241.

24. Kelley and Watkins, "Defense Industrial Network."

25. John E. Ullman, "A National Program to Convert Military Enterprises to Civilian Production Is Crucial to Our Industrial Future," *Technology Review* 94 (August–September 1991): 57–63.

26. Kelley and Watkins, "Defense Industrial Network"; John Alic, Lewis Branscomb, Harvey Brooks, Ashton Carter, and Gerald Epstein, *Beyond Spinoff* (Boston: Harvard Business School Press, 1992).

27. Jay Stowsky and Burgess Laird, "Conversion to Competitiveness: Making the Most of the National Labs," *American Prospect* (Fall 1992): 91–98.

28. Leon Trilling, "Attitude Adjustment," *Technology Review* 96 (May–June 1993): 32–33.

29. Maryellen R. Kelley and Harvey Brooks, "From Breakthrough to Follow-through," *Issues in Science and Technology* [journal of the National Academy of Sciences] (Spring 1989): 42–47; Kelley and Brooks, "External Learning Opportunities and the Diffusion of Process Innovations to Small Firms," *Technological Forecasting and Social Change* 39 (March–April 1991): 103–25; Philip Shapira, "Helping Small Manufacturers Modernize," *Issues in Science and Technology* (Fall 1990): 49–54; Shapira, *Modernizing Manufacturing* (Washington, D.C.: Economic Policy Institute, 1990); Philip Shapira, J. David Roessner, and Richard Barke, *Federal-State Collaboration in Industrial Modernization* (Atlanta: School of Public Policy, Georgia Institute of Technology, July 1992); U.S. Congress, Office of Technology Assessment, *Making Things Better: Competing in Manufacturing*, publication OTA-ITE-443 (Washington, D.C.: U.S. Government Printing Office, February 1990).

30. Ray Marshall and Marc Tucker, *Thinking for a Living: Education and the Wealth of Nations* (New York: Basic Books, 1992), p. 69.

31. Maryellen R. Kelley and Ashish Arora, *Towards a National System of Industrial Modernization Services: Program Design and Evaluation Issues* (Heinz

School of Public Policy and Management, January 1994, unpublished MS); Leo Reddy, "Industrial Strength Aid for Small Business," *Technology Review* 96 (July 1993): 54–59; and Shapira, Roessner, and Barke, *Federal-State Collaboration*. The network of NIST Manufacturing Technology Centers also has its own newsletter, published for it by an organization called the Modernization Forum, itself a spin-off from the Michigan Industrial Technology Institute (ITI), one of the nation's leading industry-government partnerships in the field.

32. OTA, *Making Things Better*, p. 59.

33. "Southern Industrial Competitiveness Project Launched," *Southern Growth: Journal of the Southern Growth Policies Board* 17 (Winter 1991): 8. Others who have assumed positions of leadership in this burgeoning movement to deliberately create and nurture industrial districts in this country are M.I.T.'s Charles Sabel, the Georgia Tech planning professor Philip Shapira, Stuart Rosenfeld, Brian Bosworth, C. Richard Hatch, Jacques Koppel, and Ann Heald, formerly an executive with the German Marshall Fund of the United States and now an administrator at the University of Maryland.

34. Stuart A. Rosenfeld, *Competitive Manufacturing: New Strategies for Regional Development* (New Brunswick, N.J.: Center for Urban Policy Research, Rutgers University, 1992); Rosenfeld, with Philip Shapira and J. Trent Williams, *Smart Firms in Small Towns* (Washington, D.C.: Aspen Institute, 1992); Brian Bosworth and Stuart Rosenfeld, *Significant Others: Exploring the Potential of Manufacturing Networks* (Chapel Hill: Regional Technology Strategies, 1993).

35. June Holley and Roger Wilkens, "A Market Driven Approach to Flexible Manufacturing Networks" (Athens, Ohio: Appalachian Center for Economic Networks, November 1991), p. viii.

36. Examples of efforts to develop such linkages within the metalworking, plastics, textiles, and furniture sectors of Massachusetts and Pennsylvania are to be found in Charles Sabel, "Studied Trust: Building New Forms of Cooperation in a Volatile Economy," in *Industrial Districts and Local Economic Regeneration*, ed. Frank Pyke and Werner Sengenberger (Geneva: International Institute for Labour Studies, International Labour Office, 1992), pp. 215–50. See also Bosworth and Rosenfeld, *Significant Others*; and Gregg Lichtenstein, *A Catalogue of U.S. Manufacturing Networks*, NIST GCR 92–616 (Washington, D.C.: National Institute for Standards and Technology, U.S. Department of Commerce, September 1992).

37. Compare the newsletter of the Federation for Industrial Retention and Renewal, published out of Chicago. At last count, FIRR—an organization that evolved out of community resistance to plant closures in the early 1980s—had thirty-three affiliates around the country.

38. C. Richard Hatch, "The Power of Manufacturing Networks," *Transatlantic Perspectives* [magazine of the German Marshall Fund] (Winter 1991): 3–6.

39. For example, Patrizio Bianchi and Nicola Bellini, "Public Policies for Local Networks of Innovators," *Research Policy* 20 (October 1991): 487–97.

40. Fiorenza Belussi and Massimo Festa, "The Veneto Model of Networked Companies: From Post-Fordism to Toyotism" (IRES Veneto, Mestre, 1990, unpublished MS); Bianchi and Bellini, "Public Policies"; Sabastiano Brusco, "The Idea of the Industrial District: Its Genesis," in *Industrial Districts and Inter-Firm Co-operation in Italy,* ed. Frank Pyke, Giaccomo Becattini and Werner Sengenberger (Geneva: International Institute for Labour Studies, International Labour Office, 1990), pp. 10–19; F. Pasquini and M. G. Giordani, "The Incubator and Regional Development Policies: A Critical Review of Recent Italian Experience" (Laboratorio di Politica Industriale, NOMISMA, Bologna, April 1989, unpublished MS).

41. David Osborne, *Laboratories of Democracy* (Boston: Harvard Business School Press, 1988).

42. Quoted by Irwin Feller, "American State Governments as Models for National Science Policy," *Journal of Policy Analysis and Management* 11 (Spring 1992): 303.

43. Ibid. But see Kelley and Arora, *Towards a National System,* for new information on coordination between federal and local ITP activities.

44. Michael I. Luger and Harvey A. Goldstein, *Technology in the Garden: Research Parks and Regional Economic Development* (Chapel Hill: University of North Carolina Press, 1991); and Doreen Massey, Paul Quintas, and David Wield, *High Tech Fantasies: Science Parks in Society, Science and Space* (London: Routledge, 1992).

45. Michael L. Gerlach, *Alliance Capitalism: The Social Organization of Japanese Business* (Berkeley: University of California Press, 1992).

46. Virginia L. duRivage, ed., *New Policies for the Part-Time and Contingent Workforce* (Armonk, N.Y.: Sharpe, for the Economic Policy Institute, 1992).

47. Richard B. Freeman and Joel Rogers, "Who Speaks for Us? Employee Representation in a Non-Union Labor Market," in Bruce E. Kaufman and Morris M. Kleiner, eds., *Employee Representation: Alternative and Future Directions* (Madison, Wis.: Industrial Relations Research Association, 1993), pp. 13–79.

48. Joseph B. White and Bradley A. Stertz, "Crisis Is Galvanizing Detroit's Big Three," *Wall Street Journal,* May 2, 1991, p. B1.

CHAPTER 11: POSTSCRIPT

1.   On this score, I urge readers to feast upon a remarkable new book by Manuel Castells, published in the interim: *The Rise of the Networked Society* (Cambridge and London: Blackwell, 1996). This is the first in a three-volume series called "The Information Age: Economy, Society, and Culture." Later in this postscript, I will return to the idea that if globalization means anything, it is that the whole amounts to more than the sum of its parts, and that what we are experiencing is surely *not* a straightforward continuation—a working out of some fundamental underlying logic (such as world systems theory)—of the expansion of eighteenth- and nineteenth-century European and American industrial capitalism in the twentieth and into the twenty-first centuries.

2.   "Small Is Beautiful! Big Is Best!: An On-Line Debate," *Inc.*, Special Issue on the State of Small Business, 1995; "Symposium on Harrison's 'Lean and Mean,'" *Small Business Economics* 7, no. 5 (October 1995).

3.   Bennett Harrison, Marcus Weiss, and Jon Gant, *Building Bridges: Community Development Corporations and the World of Employment Training* (New York: Ford Foundation, 1995); Bennett Harrison and Marcus Weiss, *Workforce Development Networks* (Thousand Oaks, CA: Sage, 1998).

4.   And, in fact, one is in the works. See my forthcoming monograph (again coauthored with Barry Bluestone), *Sabotaging Prosperity* (New York: Twentieth Century Fund and Houghton Mifflin).

5.   E.g., David Hirschberg, "Small Business Blarney: What Does It Take to Kill a Bad Number?," posted on *Slate*, the on-line magazine from Microsoft, on October 18, 1996.

6.   "It is estimated that, over the next decade, the top ten discount chains will control 90 percent, and the top ten specialty retailers 40 percent, of their respective markets"; statistics cited in Thomas R. Bailey and Annette D. Bernhardt, "In Search of the High Road in a Low-Wage Industry," *Politics and Society* 25, no. 2 (1997): 179 201.

7.   New York City Office of the Comptroller, *Economic Notes*, November 1996.

8.   "New York City's Hidden Asset: Small Business," Special Advertising Section, *Business Week*, May 12, 1997.

9.   David P. Angel, *Restructuring for Innovation: The Remaking of the U.S. Semiconductor Industry* (New York: The Guilford Press, 1994), pp. 5–6.

10.   James C. Cooper and Kathleen Madigan, "No Slowdown until the Labor Market Cools Off," *Business Week*, April 21, 1997, p. 31.

11. Cited in Peter Coy, "The Best Kind of Affirmative Action," *Business Week*, May 19, 1997, p. 35.

12. For recent in-depth explorations of these questions, see Eileen Appelbaum and Rosemary Batt, *The New American Workplace* (Ithaca, NY: ILR Press, 1994); Bailey and Bernhardt, "In Search of the High Road," op. cit.; Peter Cappelli, "Rethinking Employment," *British Journal of Industrial Relations*, 33, no. 4 (December 1995): 563–602; Peter Cappelli, Laurie Bassi, Harry Katz, David Knoke, Paul Osterman, and Michael Useem, *Change at Work* (New York: Oxford University Press, 1997); Paul Osterman, "How Common Is Workplace Transformation and How Can We Explain Who Does It?," *Industrial and Labor Relations Review*, January 1994; and Paul Osterman, "Skills, Training, and Work Organization in American Establishments," *Industrial Relations*, April 1995.

13. The following draws especially heavily on the recent writing of Peter Cappelli. There, and elsewhere, readers can also learn about other new developments, from the growth of teams to the breakdown of narrow job descriptions and the enlargement of many existing jobs, with those who *do* have work being required/enabled to perform a greater variety of tasks, often using productivity-enhanced new technology. The changes that are occurring in the organization of work in America are not all bad—provided, of course, that one can break into the system.

14. Cited in Cappelli, "Rethinking Employment," p. 15. Such permanent layoffs ("displacements") are currently actually higher for managers than for other occupations, after taking other characteristics into account.

15. Ibid., p. 17.

16. For such evidence, see Dave Marcotte, "Evidence of a Fall in the Wage Premium of Job Security," Center for Governmental Studies, Northern Illinois University, 1994, unpublished MS; Barry Bluestone and Stephen Rose, "Overworked *and* Underemployed," *American Prospect*, March–April 1997; and Stephen Rose, "The Decline of Employment Stability in the 1980s," National Commission on Employment Policy, Washington, DC, 1995.

17. Cited in *Business Week*, January 27, 1997, p. 20.

18. Moshe Buchinsky and Jennifer Hunt, "Wage Mobility in the United States," Working Paper 5455, National Bureau of Economic Research, Cambridge, MA, February 1996, p. 30.

19. Peter Gottschalk and Robert Moffitt, "The Growth of Earnings Instability in the U.S. Labor Market," *Brookings Papers on Economic Activity*, no. 2, 1994. Some of this increasing volatility may be due to the growing incidence of performance-based pay and bonuses, and, as such, is built into the new work systems by design.

By contrast, the growing structural (permanent-trend) component could turn out to be a reflection of declining job tenure, although (as Cappelli explains) "tenure" can be a confusing indicator of job security since it is affected by changes in both quits (mostly voluntary) and layoffs (usually involuntary). Unfortunately, the Census Bureau stopped collecting data distinguishing between quits and layoffs back in 1981, at the beginning of the Reagan administration.

20.   Annette Bernhardt, Martina Morris, Mark Handcock, and Marc Scott, "Job Instability and Wage Inequality: Preliminary Results from Two NLS Cohorts," Institute on Education and the Economy, Teachers College, Columbia University, Feb. 11, 1997, unpublished MS, p. 15.

21.   Cappelli, op. cit., p.12.

22.   Bennett Harrison, Maryellen Kelley, and Jon Gant, "Innovative Firm Behavior and Local Milieu: Exploring the Intersection of Agglomeration, Industrial Organization, and Technological Change," *Economic Geography*, July 1996; Maryellen Kelley and Susan Helper, "Firm Size and Capabilities, Regional Agglomeration, and the Adoption of New Technology," *Economics of Innovation and New Technology*, forthcoming; Maryellen Kelley and Susan Helper, "The Organizational Context and the Use of Computer Applications in Industry: The Influence of Regional Agglomerations and Institutions," paper presented to the Annual Meetings of the American Sociological Association, Toronto, August 1997.

23.   D. H. Whittaker, *Small Firms in the Japanese Economy* (New York: Cambridge University Press, 1997), p. 106. In her 1996 M.I.T. Ph.D. dissertation, Lynn McCormick makes a similar observation about the two-edged nature of customer–supplier relations in the metalworking cluster of Chicago.

24.   Whittaker, op. cit., p. 59.

25.   Michael Storper, "Territories, Flows, and Hierarchies in the Global Economy," in *Spaces of Globalization*, ed. Kevin R. Cox (New York: The Guilford Press, 1997), p. 35.

26.   Peter Dicken, *Global Shift*, 2d ed. (New York: The Guilford Press, 1992), p. 17.

27.   For a collection of papers that frame their critiques explicitly around debates over the theorization of geographical relations, see Cox, *Spaces of Globalization*, op. cit., esp. the editor's rather dispeptic introduction and the chapter entitled "Globalization and the Politics of Distribution: A Critical Assessment."

28.   Henry Figueroa, "Does Globalization Matter?," *In These Times*, March 31, 1997, p. 17.

29. Dicken, op. cit., p. 19.

30. David M. Gordon, "The Global Economy: New Edifice or Crumbling Foundations?," *New Left Review* 168 (1988).

31. Thea Lee, "Trade and Inequality," in *Restoring Broadly Shared Prosperity*, ed. Ray Marshall (Washington, DC: Economic Policy Institute/Austin, TX: Lyndon B. Johnson School of Public Affairs, University of Texas, 1997), p. 40. The Cornell study was subsequenly published privately as Kate Bronfenbrenner, "The Effects of Plant Closing or Threat of Plant Closing on the Right of Workers to Organize," New York State School of Industrial and Labor Relations, Cornell University, Sept. 30, 1996. These were no mere anecdotes; Bronfenbrenner found that fully 62 percent of all U.S. firms surveyed used the threat of relocation as a bargaining chip.

32. David E. Broder, "I Told You So," *New York Times*, July 13, 1997, p. E17.

33. Dean Baker, Gerald Epstein, and Robert Pollin, eds., *Globalization and Progressive Economic Policy: What Are the Real Constraints? What Are the Real Possibilities?*, forthcoming.

34. Dicken, op. cit., p. 53.

35. Stephen Baker, "The Bridges Steel Is Building," *Business Week*, June 2, 1997, p. 39.

36. Saskia Sassen, *Losing Control? Sovereignty in an Age of Globalization* (New York: Columbia University Press, 1996), pp. 39–40.

37. Storper, "Territories, Flows and Hierarchies," op. cit., p. 26.

38. Susan S. Fainstein, "Global Cities and Local Communities: The Cases of New York and London," *Competition and Change* 1, no. 2 (1995): 140.

39. Erik Swyngedouw, "Neither Global nor Local: 'Glocalization' and the Politics of Scale," in Cox, *Spaces of Globalization*, op. cit., and "From Global Order to 'Glocal Disorder'," in *The Global Economy in Transition*, ed. W. Daniels and William F. Lever (Harlow, UK: Addison Wesley Longman, forthcoming). The idea of a simultaneous movement toward globalization *and* localization also appears in Todd Swanstrom with David Wright, "Rising U.S. Income Inequality: A Social Capital Approach," Rockefeller Institute of Government, State University of New York at Albany, April 26, 1996, unpublished MS.

40. Charles P. Sohner, "Passing the Buck," *In These Times*, January 20, 1997.

41. Chris Tilly, "Workfare's Impact on the New York City Labor Market: Lower

Wages and Worker Displacement," unpublished MS, Russell Sage Foundation, March 21, 1996, p. 2.

42. On public works related job creation, see Felix Rohatyn and Robert R. Kiley, "Welfare-to-Work Means Public Works," *Home Economics* (monthly newsletter of the New York City Partnership Policy Center), February 1997. On the creation of local "community service employment jobs" that can simultaneously provide useful neighborhood services and absorb at least some of those on workfare, and on gaining EITC coverage for those on workfare, see Mark Alan Hughes, "Critical Issues Facing Philadelphia's Neighborhoods: Welfare Reform," Public/Private Ventures, 1997, unpublished MS. On the prospects for reviving targeted public service job creation at the national level, see Cliff Johnson, "The Role of Public Job Creation in Income Security Policy," Center on Budget and Policy Priorities, Washington, DC, 1997, unpublished MS. And on the danger of those being forced off welfare displacing union labor in the local public sector, see Annette Fuentes, "Slaves of New York," *In These Times*, December 23, 1996.

43. E.g., Erin Flynn and Robert Forrant, "Facilitating Firm-level Change: The Role of Intermediary Organizations in the Manufacturing Modernization Process," Jobs for the Future, Boston, February 1995, unpublished MS.

44. *Research Policy* 25 (1996). See especially Maryellen R. Kelley and Ashish Arora, "The Role of Institution-Building in U.S. Industrial Modernization Programs."

45. Bennett Harrison, Amy K. Glasmeier, and Karen R. Polenske, "National, Regional, and Local Economic Development Policy: New Thinking about Old Ideas," working paper, Taubman Center for State and Local Government, Kennedy School of Government, Harvard University, revised January 1996; Bennett Harrison and Amy K. Glasmeier, "Why Business Alone Won't Redevelop the Inner City," *Economic Development Quarterly*, February 1997; Edward W. Hill, Bennett Harrison, and Marcus S. Weiss, "Rethinking National Economic Development Policy," Levin College of Urban Affairs, Cleveland State University, May 1997, unpublished MS.

46. Harrison, Weiss, and Gant, *Building Bridges*, op. cit.

47. Edwin Melendez and Bennett Harrison, "Matching the Disadvantaged to Job Opportunities: Structural Explanations for the Past Successes of the Center for Employment Training," *Economic Development Quarterly*, November 1997.

48. Harrison and Weiss, *Workforce Development Networks*, op. cit.

49. Rochelle L. Stanfield, "Just Connect," *National Journal*, May 31, 1997, pp. 1082–1084.
50. For example, *Labor Research Review's* issue of spring/summer 1995 (no. 23) is

devoted to a collection of excellent papers on "Confronting Global Power." A prominent new specialized journal with an all-star editorial board is *Working USA*. And thanks to a grant from the MacArthur Foundation, journalist David Moberg has been writing a series of penetrating articles on issues confronting the new labor movement for the magazine *In These Times*.

51.   David Moberg, "Work Ethics," *In These Times*, March 31, 1997.

52.   I cannot resist the note that my work with Bluestone, which Cox and others on the left somehow seem to think ignores or negates the efficacy of local politics and overhypes the global, in fact grew out of the political work many of us were doing in the early 1980s, with unions and community groups, around deindustrialization. Indeed, the original sponsors of that research were the United Auto Workers, the United Steelworkers, the United Rubber Workers, and the Service Employees International Union! The UAW published a "stewards" version of *The Deindustrialization of America* that could fit into the hip pocket of an organizer's jeans. At last count, some 30,000 of these had been distributed to union activists around the country.

53.   David Ashton and Francis Green, *Education, Training, and the Global Economy* (Cheltenham, UK: Brookfield, 1996), p. 6.

54.   Will Hutton, "Relaunching Western Economies: The Case for Regulating Financial Markets," *Foreign Affairs*, November–December 1996.

55.   Bluestone and Harrison, *Sabotaging Prosperity*, op. cit.

56.   But it must be more complicated than that, since a number of business associations and organs—especially *Business Week* magazine—have for some time been calling for faster aggregate domestic growth. See also Jerry J. Jasinowski, "The Case for Higher Growth: Technology, Disinflation, and New Economic Policies," Manufacturing Institute of the National Association of Manufacturers, Washington, DC, July 1996.

57.   Robert Pollin, "Can Domestically Expansionary Policy Succeed in a Globally Integrated Environment? An Examination of Alternatives," unpublished MS, Department of Economics, University of California at Riverside, June 1996, pp. 38–39.

INDEX

Abraham, Katherine, 201, 204
Acquisitions. *See* Mergers and acquisitions
Acs, Zoltan, 55, 56, 57, 59, 60, 68, 75, 78,
  296*n*24
Admiral, 118
Advanced Manufacturing Research Facility,
  237
Advanced Micro Devices (AMD), 110, 121,
  232
Advanced Research Projects Agency (ARPA),
  167, 179, 234, 237, 308*n*13
Advanced Technology Centers, New Jersey, 243
Advanced Technology Program, 273
AEG, 167
Airbus Industrie, 131, 165–66, 169–71, 174
Aircraft manufacturing industry, 39, 73, 250;
  Airbus project and, 165–66, 169–71, 174;
  CFM International alliance in, 175–78
Air Force, 236
Air France, 164, 175–76
Akebono, 179
Akhasic, 114
Alliances: big firms and, 8, 9; "core" firms in,
  140; cross-border, 175–79; European firms
  and, 5, 10, 165–67; global demand related
  to, 30–31; globalization of knowledge in,
  139; government regulation of, 27, 34,
  231–33; *keiretsu* and, 156; knowledge links
  in, 140–41; mergers and acquisitions and,
  141; motives for, 166; production networks

and, 27, 135–41; revitalization of big firm
  sector and, 10; Silicon Valley firms and,
  121; television manufacturing industry
  and, 6
American Airlines, 265
American Economic Association, 191
American Machinist (magazine), 60, 286*n*55
American Management Association, 258
Amin, Ash, 105, 263
Amsden, Alice, 117
Anchordoguy, Marie, 152–53
Angel, David, 256
Antitrust movement, 230–31
Apollo, 114
Apple, 4, 174
Appold, Stephen, 117
Arcangell, Fablo, 63
Armco, 177
Arora, Ashish, 67
Arrow, Kenneth, 136, 140
Ashton, David, 276
Asia: collaboration with workers and restructur-
  ing experiments in, 10–11; industrial dis-
  tricts and, 15; Silicon Valley outsourcing to,
  107, 116
Aspen Institute, 191
AT&T, 4, 6, 164, 258
Atkinson, John, 197–98
Audretsch, David, 55, 56, 57, 59, 60, 68, 78,
  296*n*24

small firm development model and, 16–17, 53; third shift in business organization and, 221–22
Policy Studies Institute, 62
Politechnico, 63
Political parties: small firm development model and, 16–17; social democratic, 10, 16
Pollin, Robert, 268, 277, 278
Porsche, 145, 146
Porter, Michael, 28–29, 148, 159, 184, 224–27
Portes, Alejandro, 102
Portugal, 76
Post-Fordists, 262
Post–World War II economic system, 257
Poterba, James, 184
Powell, Walter W., 65, 67, 68, 125, 129, 132, 133–34, 141, 210
Prab, 178
PRATEL, 100
Pratese, 104
Prato textile industry case study, 80, 95–102, 143, 262
Pratt and Whitney (P&W), 169, 175
Problem-solving committees, 11, 174, 183
Product cycle theory, 73, 112–13
Product development: biotechnology firms and, 66; commercialization of innovative ideas and, 56; Silicon Valley firms and, 112; small firms and, 56–58, 295n13. See also Innovation
Production networks, 145–46; access to information in, 132–33, 315n26; Airbus Industrie example of, 169–71; Benetton textile industry case study of, 89–95, 304n19; big firms and, 8, 9, 12, 134; biotechnology firms and, 66–68; computer systems for, 10; concentration without centralization in, 171; contradictory combinations of aspects of, 210–11; "core" firms in, 140; corporate restructuring and, 173–74; craft-type industries and, 134; decentralization of production systems and, 9; developmental tendencies of, 146–47; development within firms of, 133–34; as dominant organizing principle, 148; eligibility decisions regarding participation in, 27; European intercorporate, 162–71; finance shared in, 183; flexibility and, 131–33, 220–21; geographic clustering in, 33, 134; global demand related to, 30–31; globalization of knowledge in, 139; as "global webs," 9; government regulation of, 34, 229–31; income distribution issues and, 29–30; in Japan, 150–62; interfirm relationships in, 248; lean production strate-

gy in, 9; local and regional economic development policies and, 31–34; loyalty to a specific national government and, 27–28, 223; managers and, 183–87; polarization of wages and, 195–98, 330n20; public policy to support industrial environment for, 29; revitalization of big firm sector and, 12; Sasib Group case study of, 86; Silicon Valley and, 106–7, 120; small firms and, 22, 134; social costs of, 133; strategic alliances in, 27, 135–41, 210; subcontracting (outsourcing) in, 135; types of, 133–41; in U.S., 150, 172–88; vertical disintegration in, 135, 316n34
Production systems: automobile industry and global, 6; basic definitions in, 142–43; collaboration with workers and restructuring experiments in, 10–11; competition and fragmentation of, 99–102; compression of time in, 127; computer-based technologies in, 10, 21; concentration without centralization in, 9; decentralization of, 8–9, 76; dualism in Nike case study of, 206–9; economies of scale and size of units in, 13–14; flexibility in, 127–49; governance structures in, 144–45; industrial districts and localized suppliers in, 23; input-output systems in, 143; manufacturing-to-services shift and changes in size of, 39, 40, 41; offshore operations in, 28–29, 71; Prato region case study of textile industry and, 97–98; small firms used by big companies as part of, 18; strategic alliances for, 10; territorially based, 141–47. See also Core competencies; Lean production strategy; Outsourcing
Production unit, definition of, 142
Productivity: Italian industrial districts and, 104–5; labor market dualism and, 212, 334n50; offshore operations and, 28–29
Profitability, corporate, 125–26, 213, 223, 227
Programmable automation (PA), 59; Benetton textile industry case study of, 91–92; revitalization of big firm sector and, 10; size of firm and use of, 21, 55, 61–62, 286n55; statistical models on use of, 59–60, 295n18; use of, in Europe and Japan, 62–63, 296n24
Programming of software, 68–72; factories for development of, 69–71; maintenance in, 71–72; work organization in, 68–69, 72
Public opinion: debate over income inequalities in, 191; small firm development model in, 16
Public policy. See Policy makers
Putting out system, in Italy, 90, 92–94, 97–98